T0299736

Theoretical and Computational Fluid Mechanics

Theoretical and Computational Fluid Mechanics: Existence, Blow-up, and Discrete Exterior Calculus Algorithms centralizes the main and current topics in theoretical and applied fluid dynamics at the intersection of a mathematical and non-mathematical environment. This book is accessible to anyone with a basic level of understanding of fluid dynamics and yet still engaging for those of a deeper understanding.

This book is aimed at theorists and applied mathematicians from a wide range of scientific fields, including the social, health and physical sciences. It provides a step-by-step guide to the construction of solutions of both elementary and open problems of viscous and non-viscous models and for the applications of such models for the functional analysis and real analysis of data.

Features

- Offers a self-contained treatment that does not require a previous background in fluid dynamics

- Suitable as a reference text for graduate students, researchers and professionals and could easily be used as a teaching resource

- Provides various examples using Maple, and to a lesser extent MATLAB® programming languages.

Numerical Analysis and Scientific Computing Series

Series Editors:
Frederic Magoules and Choi-Hong Lai

About the Series

This series, comprising of a diverse collection of textbooks, references, and handbooks, brings together a wide range of topics across numerical analysis and scientific computing. The books contained in this series will appeal to an academic audience, both in mathematics and computer science, and naturally find applications in engineering and the physical sciences.

For more information about this series please visit:
https://www.crcpress.com/Chapman–HallCRC-Numerical-Analysis-and-Scientific-Computing-Series/book-series/CHNUANSCCOM

Theoretical and Computational Fluid Mechanics

Existence, Blow-up, and Discrete Exterior Calculus Algorithms

First Edition

Terry E. Moschandreou
University of Western Ontario, Canada

Keith Afas
University of Western Ontario, Canada

Khoa Nguyen
University of Western Ontario, Canada

CRC Press
Taylor & Francis Group
Boca Raton London New York

CRC Press is an imprint of the
Taylor & Francis Group, an **informa** business

A CHAPMAN & HALL BOOK

First edition published 2024
by CRC Press
2385 NW Executive Center Drive, Suite 320, Boca Raton FL 33431

and by CRC Press
4 Park Square, Milton Park, Abingdon, Oxon, OX14 4RN

CRC Press is an imprint of Taylor & Francis Group, LLC

© 2024 Terry Moschandreou, Keith Afas, Khoa Nguyen

Library of Congress Cataloging-in-Publication Data
Names: Moschandreou, Terry E., author.
Title: Advances in theoretical and computational fluid mechanics : existence, blow-up, and discrete exterior calculus algorithms / Terry Moschandreou, University of Western Ontario, Canada, Keith Afas, University of Western Ontario, Canada, Khoa Nguyen, University of Western Ontario, Canada.
Description: First edition. | Boca Raton : C&H/CRC Press, 2024. | Series: Numerical analysis and scientific computing series | Includes bibliographical references and index.
Identifiers: LCCN 2023046128 (print) | LCCN 2023046129 (ebook) | ISBN 9781032589251 (hbk) | ISBN 9781032589480 (pbk) | ISBN 9781003452256 (ebk)
Subjects: LCSH: Fluid mechanics.
Classification: LCC QA901 .M68 2024 (print) | LCC QA901 (ebook) | DDC 532.001–dc23/eng/20231102
LC record available at https://lccn.loc.gov/2023046128
LC ebook record available at https://lccn.loc.gov/2023046129

ISBN: 9781032589251 (hbk)
ISBN: 9781032589480 (pbk)
ISBN: 9781003452256 (ebk)

DOI: 10.1201/9781003452256

Typeset in Minion
by codeMantra

Terry E. Moschandreou

This book is dedicated to my parents, with much love and also to the memory of my uncle John Skoupas who taught mathematics in Greece.

Keith C. Afas

With much love, this book is dedicated...

...To Jesse Afas and Marlene Villanueva, who have inspired me with love and wisdom, to chase my dreams both with fervent determination derived from wonder, and prudence derived from foresight.

...To Behnam Afas, who dedicated his life as a scholar, so that the lives of others will never be forgotten; who always recognized and believed in the gift of research in myself; and who will always be an inspiration, by a full life led, of unapologetically embodying one's research.

Khoa Nguyen

To the memory of my parents.

Contents

TERRY E. MOSCHANDREOU

Preface

In the origins of hydrodynamics and acoustics, largely mathematical tools have been developed over the span of many decades. In them during the eighteenth and nineteenth centuries originated, and for them was developed, much theory of partial-differential equations and kinematics of continuous media. The mathematical treatment for fluid dynamics in general can lead to introspective analysis and methods for which deeper truths can be derived and maintained. Both inviscid and viscous fluid dynamics are treated in this light. It is the merging of theory and application that is of greatest benefit for a researcher of a continuum and statistical approaches in fluid dynamics. The aim of this book is to provide a basic and self-contained introduction to the ideas underpinning Fluid Dynamics, in both a mathematical and numerical setting. Central to this is that teaching and research should be extended to presently unsolved and open problems in fluid dynamics. Hence, readers of our proposed book would include Ph.D. students and researchers from quantitatively oriented fields of the biomedical sciences, physical sciences, as well as statistics and mathematics and would include scientists and research workers either from academic or from industrial settings. This book will be aimed for practitioners and theorists alike. For the practitioner, often interested in model building and analysis, we provide the foundational ideas. Here we proffer some applications special to fluids, first of a general kind and then for the classical fluids named after Euler, Navier and Stokes. Mathematically exact work is laid out with subsequent applications that are important in engineering, physics and the biomedical sciences. Any fluid dynamics model must be well motivated, and we will discuss what the objectives for such models are. Model development is important and should be based on so-called evolutionary models in nonlinear dynamics. So then a large reservoir of theory emerges here and in this setting especially in the study of Phase portraits, bifurcations and Chaos. As already mentioned, the field of applied and theoretical fluid dynamics is currently a strong area of research. Many theorists and practitioners are investigating and using the tools developed over the last few decades in a wide range of scientific fields. However, while there are now a vast number of books that are solely devoted to the topic of Fluid dynamics, and which detail the current state of research, all of these books are targeted for either mathematicians working in an abstract functional analytic setting and inaccessible to many scientists who do not have the time to devote to a complete understanding of the language of mathematics because they are on the other side of the mathematics' spectrum and are quite literate in full numerical theory and techniques. It is therefore important to develop a book that merges the two sides of study for the field of fluid dynamics one that theoretically answers and unlocks unsolved open problems to a significant extent and provides insight into the models that are important in fluid dynamics with a clear down to earth and physical understanding. Indeed, there are more user-friendly books on the general topic of mathematical fluid analysis, which devote only a few chapters on the most general nonlinear models of fluid dynamics such as for the Euler and Navier–Stokes equations. These books mostly cover traditional models while excluding some important modern models which have not yet been solved and fail to look into the nature and solution of the most general governing equations of fluid dynamics. The objective of this book is to centralize the main and current topics in theoretical and applied fluid dynamics at the intersection or merging of a mathematical and non-mathematical environment, so that it will be readable to anyone with a basic level of understanding of fluid dynamics and those of a deeper understanding. This book will be aimed for theorists, as well as for applied mathematicians, from a wide range of scientific fields, including the social, health and physical sciences. This book will provide a step-by-step guide to the construction of solutions of both elementary and open problems of viscous and non-viscous models, and for the applications of such models for the functional analysis and real analysis of data as for example in turbulence. This book will provide a self-contained coverage and will not require a previous background in fluid dynamics. The reader should have an understanding of elementary real analysis and technical understanding of computational tools used in the area of study. These are mainly Maple and MATLAB. This book will illustrate among the many some important nonlinear models with a step-by-step approach using specific languages of Maple and to a lesser extent MATLAB. We provide software code for readers who wish to practice and produce graphics of important flow phenomena in general fluid dynamics. The code will be written in a way so that it can be extended to handle new nonlinear models that one may construct on their own. Software code that can be used to perform data analysis for both deterministic models and some turbulent ones using any one of the nonlinear models covered in this book is provided. This book can be used to teach Master's and/or Ph.D. level graduate students and can be used by researchers, from any field of the social, health, and physical sciences. The basic contents of this book can be taught to students. Researchers can use it as a handbook for nonlinear partial differential equations in fluids and models thereof.

Terry E. Moschandreou
London, Canada

MATLAB® is a registered trademark of The MathWorks, Inc. For product information, please contact:

The MathWorks, Inc.
3 Apple Hill Drive
Natick, MA 01760-2098 USA
Tel: 508-647-7000
Fax: 508-647-7001
E-mail: info@mathworks.com
Web: http://www.mathworks.com

Authors

Dr. Terry E. Moschandreou, Ph.D.

Dr. Moschandreou has taught mathematics at the University of Western Ontario in the School of Mathematical and Statistical Sciences where he worked for several years. He earned his PhD in Applied Mathematics from the University of Western Ontario in 1996. The greater part of his professional life has been spent at the University of Western Ontario(teaching 8 years mathematics and fluid dynamics courses) and Fanshawe College in London (where he has taught physics courses), Ontario, Canada. Dr. Moschandreou is also currently working for Thames Valley District School Board in London, Ontario, Canada, where he teaches elementary and high school students Mathematics and Science. For a short period, he worked at the National Technical University of Athens, Greece. Dr. Moschandreou is the author of several research articles in blood flow and oxygen transport in the microcirculation, general fluid dynamics, and theory of differential equations. Also, he has contributed to the field of finite element modelling of the upper airways in sleep apnea as well as surgical brain deformation modeling. More recently, he has been working with the partial differential equations of multiphase flow and level set methods as used in fluid dynamics. Finally, Dr. Moschandreou proposes to have a solution to the Millennium Prize Problem (main findings submitted in the period 2018–2022) for the breakdown of solutions to the Periodic Navier–Stokes Equations on the 3-Torus.

Mr. Keith C. Afas, M.E.Sc.

Mr. Afas is a graduate researcher at the University of Western Ontario (UWO), who has been engaged in applied mathematics and mathematical modelling research since 2014, leading to multiple first-author and co-author publications, posters, conference presentations, and book chapters in a variety of topics. He has received his Master's degree in Biomedical Engineering from the University of Western

Ontario in 2022, and has received Bachelor degrees in Medical Biophysics and Medical Cell Biology from the University of Western Ontario in 2021 and 2019, respectively. Currently, Mr. Afas is pursuing a Biomedical Engineering PhD at the University of Western Ontario. Mr. Afas' primary research interests are multiscale mathematical modeling of the circulatory system, and differential geometric models explaining the genesis of cell shape. Mr. Afas' past work includes the analysis of partial differential equations governing intraluminal arteriolar O2 and bulk-tissue skeletal muscle O2 transport. Mr. Afas' past work also includes creating a differential geometry framework to develop tensor-invariant models for lipid membrane deformation. Mr. Afas' current work involves applying soft condensed matter field theory to determined erythrocyte (red blood cell) membrane geometry to theoretically characterize the release of adenosine triphosphate from erythrocytes undergoing shear, with the goal of uncovering mechanisms behind microcirculatory blood flow regulation.

Dr. Khoa Nguyen, B.Eng.,MSc.PhD

Dr. Khoa Nguyen has taught applied mathematics at Western University at London, Ontario, Canada since 2001 to present. Most of his students are engineering and science students. His interests are varying from physics to engineering, mathematics and applications of mathematics to these fields. He has had two publications on Physical Review D with his collaborators and a book on numerical methods in C and MATLAB with his colleague. He has a wife and two sons. They live in London, Ontario, Canada.

## CHAPTER 1

Introduction to Fluid Dynamics

Terry E. Moschandreou and Khoa Nguyen

TVDSB and Western University

1.1 INTRODUCTION

With external forces described for the equations of fluid dynamics that will be considered, one very important force is friction. There exist stresses in neighboring layers of fluid which are referred to as shear stress. It is important to consider this type of force for the general equations of viscous fluid dynamics and in general for viscous flow. Stresses (including normal stresses) in general are the forces that exist inside the fluid. If there is a boundary, the stress can be classified according to the direction of the force whether it is normal to or parallel to the boundary. The boundary can either be an actual physical boundary or be an imaginary surface. In Chapter 8, where some reduction methods for the Periodic Navier-Stokes equations using symbolic computation are examined, the 3-Torus is considered which is a boundary-less 3-manifold. Thus, shear forces may exist for surfaces between neighboring sheets or layers of fluid.

Suppose we have a fluid with pressure p acting on a surface with unit normal $\mathbf{n}$, pointing *toward* the fluid domain,

Definition 1.1 *Normal stress: The normal stress is*

$$\tau_p = -p\mathbf{n}. \tag{1.1}$$

DOI: 10.1201/9781003452256-1

The pressure applied to any part of a fluid will be transmitted equally in all directions through the fluid as Pascal's Law states. As a result, the net effect will be zero. It is when there exists a pressure *gradient*, then this is the real force that drives fluid flow. Considering a flow in a tube where the pressure P_2 is higher at the left end in comparison to the right P_1, Then this gives a *body force* $-\nabla p$ that drives the fluid from left to right.

Considering shear stress in the horizontal direction and two infinite plates with fluid in between. The bottom plate is kept at rest, and the top plate moves with velocity U.

The shear stress is the force per unit area of the top plate and is the horizontal force required to apply to keep the top plate in motion. It is the force required to overcome the friction or viscous counter-force and to have the plate move at the prescribed positive velocity.

Definition 1.2 *Tangential stress: The tangential stress τ_s is the force (per unit area) required to move the top plate at speed U.*

This is also the force we need to exert on the bottom plate to keep it still. Or the horizontal force at any point in fluid in order to maintain the velocity gradient.

For a Newtonian fluid, this stress is proportional to the velocity gradient,

Definition 1.3 *For a Newtonian fluid, we have that*

$$\tau_s \propto \frac{U}{h}.$$

Definition 1.4 *Dynamic viscosity: The dynamic viscosity μ of the fluid is the constant of proportionality in*

$$\tau_s = \mu \frac{U}{h}. \tag{1.2}$$

For a general flow, let $u(\mathbf{x})$ be the velocity of the fluid at position $\mathbf{x}$. Then, the velocity gradient is

$$\frac{\partial u(\mathbf{x})}{\partial \mathbf{n}}.$$

The tangential stress is the shear stress of a fluid which can be defined as a unit area amount of force acting on the fluid parallel to a very small element of the surface.

$$\tau_s = \mu \frac{\partial u(\mathbf{x})}{\partial \mathbf{n}}, \tag{1.3}$$

and is in the direction of the tangential component of velocity. Again, the normal vector $\mathbf{n}$ points *into* the fluid.

1.1.1 Steady Parallel Viscous Flow

We consider two types of important flows,

Definition 1.5 *Steady flow: The flow of a fluid is steady if its velocity, pressure and all the numerical values relating to its substance (e.g. density and viscosity) are independent of time at every point in the flow field. A steady flow is in principle only possible in a condition of equilibrium (steady state).*

Definition 1.6 *Parallel flow: A parallel flow is a flow where the fluid only flows in one dimension (say the x direction) and only depends on the direction perpendicular to a plane (say the x − z plane). So the velocity can be written as*

$$\mathbf{u} = (u(y), 0, 0).$$

Here the velocity does not depend on the x direction.

Next we derive Euler's equation which is an equation that describes non-viscous flow. Here the viscosity of the fluid is not taken into consideration.

Newton's second law of motion tells that the sum of the forces acting on a volume of fluid V is equal to the rate of change of its momentum. Since $\frac{D(\mathbf{u})}{Dt}$ is the acceleration of the fluid particles, or fluid elements, within V, one has

$$\int_V \rho \frac{D\mathbf{u}}{Dt} dV = F_S + \int_V \rho \mathbf{g} dV \tag{1.4}$$

where $p(x) > 0$ is the fluid pressure and

$$F_S = \int_S -p\mathbf{n} dS$$

The term F_S in Eq. (1.4) is the force exerted on the fluid into the volume V by the fluid on the other side of the surface S.

Applying the divergence theorem to Eq. (1.4) we have that,

$$\int_V \rho \frac{D\mathbf{u}}{Dt} dV = \int_V (-\nabla p + \rho \mathbf{g}) \, dV \qquad (1.5)$$

Now both integrands must be identical since the volume V is arbitrary.

So we obtain Euler's equation,

$$\rho \frac{D\mathbf{u}}{Dt} = \rho \left[\frac{\partial \mathbf{u}}{\partial t} + (\mathbf{u} \cdot \nabla) \mathbf{u} \right] = -\nabla p + \rho \mathbf{g} \qquad (1.6)$$

Here we have neglected viscous effects as mentioned.

In general, Newtonian fluids satisfy one of the following boundary conditions, The no-slip condition says that at any solid surface the relative velocity of the Newtonian fluid with respect to the surface is zero.

$$u = 0. \qquad (1.5)$$

alternatively, a tangential stress τ is imposed on the fluid. In this case,

$$-\mu \frac{\partial u}{\partial n} = \tau.$$

For fluid-solid boundaries, the fluid sticks to the solid boundary. The stress condition occurs when we have a fluid-fluid boundary where the tangential stress is matched between the two fluids. We will now look at two important examples.

Definition 1.7 *Couette flow: Couette flow is the flow of a viscous fluid in the space between two plates, one of which is moving tangentially relative to the other. The relative motion of the plates creates a shear stress on the fluid and produces flow. We assume that this is a steady flow, and there is no pressure gradient. Thus, we get*

$$\frac{\partial^2 u}{\partial y^2} = 0.$$

Further we have that the no-slip condition says $u = 0$ on $y = 0$; $u = U$ on $y = h$. The solution is thus

$$u = \frac{Uy}{h}.$$

Poiseuille flow is pressure-induced flow (channel Flow) in a long duct. We have a high pressure P_2 on the left, and a low pressure $P_1 < P_2$ on the right. Including gravity, we have the following equations,

$$-\frac{\partial p}{\partial x} + \mu \frac{\partial^2 u}{\partial y^2} = 0 \tag{1.7}$$

$$-\frac{\partial p}{\partial y} - g\rho = 0 \tag{1.8}$$

The boundary conditions are $u = 0$ at $y = 0, h$. The second equation in Eq. (1.8) implies

$$p = -g\rho y + f(x)$$

for some function f. Substituting this into the first equation in Eq. (1.7) gives

$$\mu \frac{\partial^2 u}{\partial y^2} = f'(x).$$

The left is a function of y only, while the right depends only on x. This method of solution reappears throughout the text. So both must be constant, say C. Using the boundary conditions, we get

$$\mu \frac{\partial^2 u}{\partial y^2} = f'(x) = C = \frac{P_2 - P_1}{L},$$

where L is the length of the tube. Then, we find

$$u = \frac{C}{2\mu} y(h - y).$$

The velocity is greatest at the middle of the tube, where $y = \frac{h}{2}$.

1.1.2 Properties of a Flow

The volume flux is the rate of volume flow across a unit area and has dimensions of L/T or $(L^3/(TL^2))$.

Definition 1.8 *Volume flux: The volume flux is the volume of fluid traversing a cross-section per unit time. This is given by*

$$q = \int_0^h u(y)\, \partial y \tag{1.9}$$

per unit transverse width.

The volumetric flux for Couette and Poiseuille flows is now considered,

Definition 1.9 *For Couette flow, we have*

$$q = \int_0^h \frac{Uy}{h} \, \partial y = \frac{Uh}{2}. \tag{1.10}$$

For Poiseuille flow, we have

$$q = \int_0^h \frac{C}{2\mu} y(h-y) \, \partial y = \frac{Ch^3}{12\mu}. \tag{1.11}$$

Vorticity is a pseudovector field that describes the local spinning motion of a continuum near some point. A complete definition of vorticity in terms of Liutex vector, shear and strain will be given in Chapter 8; a new generation of vorticity methods and definitions will also be considered.

Definition 1.10 *Vorticity: The* vorticity *is defined by*

$$\boldsymbol{\omega} = \nabla \times \mathbf{u}.$$

In our case, since we have

$$\mathbf{u} = (u(y,t), 0, 0),$$

we have

$$\boldsymbol{\omega} = \left(0, 0, -\frac{\partial u}{\partial y}\right). \tag{1.12}$$

For the case of the Couette flow, the vorticity is $\boldsymbol{\omega} = \left(0, 0, -\frac{U}{h}\right)$. Here the vorticity is uniform. For the case of Poiseuille flow, we have

$$\boldsymbol{\omega} = \left(0, 0, \frac{C}{\mu}\left(y - \frac{h}{2}\right)\right). \tag{1.13}$$

Recall that the tangential stress τ_s is the tangential force per unit area exerted by the fluid on the surface, given by

$$\tau_s = \mu \frac{\partial u}{\partial \mathbf{n}},$$

where $\mathbf{n}$ is pointing into the fluid.

For Couette flow, we have

$$\tau_s = \begin{cases} \mu\frac{U}{h} & y = 0 \\ -\mu\frac{U}{h} & y = h \end{cases}. \tag{1.14}$$

For Poiseuille flow, we have

$$\tau_s = \begin{cases} \frac{Ch}{2} & y = 0 \\ \frac{Ch}{2} & y = h \end{cases} \tag{1.15}$$

1.1.3 Exact Equation Leads to the Diffusion Equation

Considering the following where we have unsteady flow. Consider fluid initially at rest in $y > 0$, resting on a flat surface. At time $t = 0$, the boundary $y = 0$ starts to move at constant speed U. There is no force and no pressure gradient.

Substituting into the x momentum equation, we get

$$\frac{\partial u}{\partial t} = \nu\frac{\partial^2 u}{\partial y^2}, \tag{1.16}$$

where $\nu = \frac{\mu}{\rho}$. This is the diffusion equation, with the diffusivity ν.

Definition 1.11 *Kinematic viscosity: The* kinematic viscosity *is*

$$\nu = \frac{\mu}{\rho}. \tag{1.17}$$

Reynold's transport theorem
Consider the volume integral

$$J = \int\int\int_{V(t)} \alpha(\vec{x}, t)\, dV$$

where $\alpha(\vec{x}, t)$ is a property (such as mass density or momentum) of a fluid at position $\vec{x}$ at time t. We want to compute

$$\frac{dJ}{dt} = \frac{d}{dt}\int\int\int_{V(t)} \alpha(\vec{x}, t)\, dV$$

We note that the change ΔJ in Δt comes from two factors. The first one is from the change of α with respect to time. That is,

$$\Delta J = \Delta t \int\int\int_{V(t)} \frac{\partial \alpha}{\partial t} \, dV$$

The second one is from the movement of $V(t)$ to a new position of $V(t + \Delta t)$. Let $S(t)$ be the boundary of $V(t)$ at time t. The change of volume $V(t + \Delta t) - V(t)$ is equal to

$$\Delta t \int\int_{S(t)} \alpha(\vec{x}, t) v_n \, dS$$

where v_n is the velocity of the fluid in the direction of $\hat{n}$, the outward unit normal vector to $S(t)$. Hence, $v_n = \vec{v} \cdot \hat{n}$. Then, the above integral becomes

$$\Delta t \int\int_{S(t)} \alpha(\vec{x}, t) \vec{v} \cdot \hat{n} \, dS$$

Adding these contributions, we have

$$\Delta J = \Delta t \left(\int\int\int_{V(t)} \frac{\partial \alpha(\vec{x}, t)}{\partial t} \, dV + \int\int_{S(t)} \alpha(\vec{x}, t) \vec{v} \cdot \hat{n} \, dS \right)$$

Hence,

$$\frac{dJ}{dt} = \lim_{\Delta t \to 0} \frac{\Delta J}{\Delta t} = \int\int\int_{V(t)} \frac{\partial \alpha}{\partial t} \, dV + \int\int_{S(t)} \alpha(\vec{x}, t) \vec{v} \cdot \hat{n} \, dS$$

which is Reynold's transport theorem.

If $\alpha(\vec{x}, t) = \rho(\vec{x}, t)$, the mass density of a fluid at position $\vec{x}$ at time t, then

$$\int\int\int_{V(t)} \rho(\vec{x}, t) \, dV$$

is the mass of a body of fluid of volume $V(t)$. If $V(t)$ is a fluid volume which consist of the same mass in motion, then the mass is conserved. That is,

$$\frac{d}{dt} \int\int\int_{V(t)} \rho(\vec{x}, t) \, dV = 0$$

Hence, by Reynold's transport theorem,

$$\int\int\int_{V(t)} \frac{\partial\rho}{\partial t}\,dV + \int\int_{S(t)} \rho(\vec{x},t)\vec{v}\cdot\hat{n}\,dS = 0$$

Using Divergence theorem, the second integral is converted to a volume integral

$$\int\int\int_{V(t)} \frac{\partial\rho}{\partial t}\,dV + \int\int\int_{V(t)} \nabla\cdot(\rho\vec{v})\,dV = 0$$

$$\int\int\int_{V(t)} \left(\frac{\partial\rho}{\partial t} + \nabla\cdot(\rho\vec{v})\right)\,dV = 0$$

Since $V(t)$ is chosen arbitrarily, we must have

$$\frac{\partial\rho}{\partial t} + \nabla\cdot(\rho\vec{v}) = 0$$

which is the continuity equation.

The boundary layer concept

In 1904, Prandtl introduced the concept of boundary layer. He visualized a rigid body in a flow as a combination of two regions. The first one is inside a tiny layer surrounding the body where viscosity is a dominant factor. The other one is outside the first layer where flow can be treated as a potential flow. A boundary layer is a thin fluid layer adjacent to the surface of a body in which strong viscous effects exist.

Boundary layer thickness parameters

We define the distance $y = \delta$ at which $u = 0.99U$ where y is the distance of the fluid from the body, u is the velocity of the fluid at distance y and U is the far away velocity of the fluid.

Since fluid inside the boundary layer flows much slower than the free stream flow, the volume flow in this region is less than that of the free stream. The effect is the same as the stream line at the wall is shifted by a distance δ^*. δ^* is referred as the displacement boundary layer thickness.

That is,

$$\int_0^\delta (U - u)\,dy = U\delta^* \tag{1.18}$$

or

$$\delta^* = \int_0^\delta \left(1 - \frac{u}{U}\right) dy \qquad (1.19)$$

A third type of boundary thickness is called momentum thickness, denoted as θ. The thickness of a layer, at zero velocity, has the same momentum loss as that of the actual boundary layer. Hence, the momentum loss with thickness θ is

$$\rho U^2 \theta$$

On the other hand, this momentum loss due to the boundary layer is

$$\int_0^\delta \rho(U - u)u\, dy \qquad (1.20)$$

Equating the above two, we have

$$\rho U^2 \theta = \int_0^\delta \rho(U - u)u\, dy \qquad (1.21)$$

Therefore,

$$\theta = \int_0^\delta \frac{u}{U}\left(1 - \frac{u}{U}\right) dy$$

We note that

$$\theta < \delta^* < \delta$$

Question:
Find δ^* and θ for a flow described as follows:

$$u = U(1 - 10^{-\frac{2y}{\delta}})$$

Hint: $\int 10^{ax}\, dx = \frac{10^{ax}}{a \ln 10}$

Answer: $\delta^* = 0.215\, \delta$, $\theta = 0.1064\, \delta$

The boundary layer equations
In this section, we derive the boundary layer equations from the Navier-Stokes equations and the continuity equation. They are,

$$u\frac{\partial u}{\partial x} + v\frac{\partial u}{\partial y} = -\frac{1}{\rho}\frac{\partial p}{\partial x} + \nu\left(\frac{\partial^2 u}{\partial x^2} + \frac{\partial^2 u}{\partial y^2}\right)$$

$$u\frac{\partial v}{\partial x} + v\frac{\partial v}{\partial y} = -\frac{1}{\rho}\frac{\partial p}{\partial y} + \nu\left(\frac{\partial^2 v}{\partial x^2} + \frac{\partial^2 v}{\partial y^2}\right)$$

$$\frac{\partial u}{\partial x} + \frac{\partial v}{\partial y} = 0$$

where ν is the kinematic viscosity of the fluid and it is defined by

$$\nu = \frac{\mu}{\rho}$$

where μ is the dynamic viscosity of the fluid.

Near the edge of the plate, both x and δ are small and have the same order of magnitude. However, as x becomes large δ is small compared with x due to the boundary layer being thin from our assumption. That is,

$$\frac{\delta}{x} \ll 1$$

Since $u \sim U$ immediately outside the boundary layer and the horizontal velocity u increases from 0 to U over the very small transverse distance δ, $\partial u/\partial y$ is large and the order of magnitude is $\partial u/\partial y \sim U/\delta$ by comparison with $\partial u/\partial x$ which is of order U/x. That is,

$$\frac{\partial u}{\partial y} \gg \frac{\partial u}{\partial x}$$

From the continuity equation, both $\partial u/\partial x$ and $\partial v/\partial y$ have the same order of magnitude, U/x. Hence,

$$\frac{\partial u}{\partial y} \gg \frac{\partial v}{\partial y} \tag{1.22}$$

Since $v = 0$ on the plate and $\partial v/\partial y$ is small it follows that v must be very small throughout the boundary, or $u \gg v$. Next, we want to find the order of magnitude of v. Since the order of magnitude of $\partial v/\partial y$ is U/x and that of $\partial/\partial y$ is $1/\delta$, the order of magnitude of v is $U\delta/x$. Thus,

$$
\begin{aligned}
u &\sim U \\
v &\sim \frac{U\delta}{x} \\
\frac{\partial}{\partial x} &\sim \frac{1}{x} \\
\frac{\partial}{\partial y} &\sim \frac{1}{\delta}
\end{aligned}
$$

Using the order of magnitude, the Navier-Stokes equations can be described as

$$U\frac{U}{x} + \frac{U\delta}{x}\frac{U}{\delta} = -\frac{1}{\rho}\frac{\partial p}{\partial x} + \nu\frac{U}{x^2} + \nu\frac{U}{\delta^2}$$

$$U\frac{U\delta}{x^2} + \frac{U\delta}{x}\frac{U}{x} = -\frac{1}{\rho}\frac{\partial p}{\partial y} + \nu\frac{U\delta}{x^3} + \nu\frac{U\delta}{x\delta^2}$$

Simplifying,

$$\frac{U^2}{x} + \frac{U^2}{x} = -\frac{1}{\rho}\frac{\partial p}{\partial x} + \nu\frac{U}{x^2} + \nu\frac{U}{\delta^2}$$

$$\frac{\delta}{x}\left(\frac{U^2}{x} + \frac{U^2}{x}\right) = -\frac{1}{\rho}\frac{\partial p}{\partial y} + \frac{\delta}{x}\left(\nu\frac{U}{x^2} + \nu\frac{U}{\delta^2}\right)$$

Considering the first equation, the inertia terms on the left have the same order of magnitude. On the right side, the first viscous term, $\partial^2 u/\partial x^2$ is much smaller than the second viscous term, $\partial^2 u/\partial y^2$. Hence, we can drop the first one out of the equation. The first equation now becomes

$$u\frac{\partial u}{\partial x} + v\frac{\partial u}{\partial y} = -\frac{1}{\rho}\frac{\partial p}{\partial x} + \nu\frac{\partial^2 u}{\partial y^2} \tag{1.23}$$

Considering the second equation, the inertia terms on the left are multiplied by δ/x which results

$$u\frac{\partial v}{\partial x} + v\frac{\partial v}{\partial y} \approx 0 \tag{1.24}$$

while on the right, two viscous terms are also multiplied by the same factor δ/x which gives

$$\frac{\partial^2 v}{\partial x^2} + \frac{\partial^2 v}{\partial y^2} \approx 0 \tag{1.25}$$

Hence, the second Navier-Stokes equation becomes

$$0 = -\frac{1}{\rho}\frac{\partial p}{\partial y}$$

This indicates p is independent of y. Hence, p is a function of x only. That makes the first modified Navier-Stokes equation become

$$u\frac{\partial u}{\partial x} + v\frac{\partial u}{\partial y} = -\frac{1}{\rho}\frac{dp}{dx} + \nu\frac{\partial^2 u}{\partial y^2} \tag{1.26}$$

Since p is independent of y, p inside and outside the boundary at a given x are the same. Furthermore, we assume outside the boundary layer the flow is inviscid and $u = U = constant$. As a result, Bernoulli equation can be satisfied. That is,

$$p + \frac{1}{2}\rho U^2 = constant$$

Hence, $p = constant$ everywhere. This gives

$$\frac{dp}{dx} = 0$$

This makes sense because if p is a function of x then the flow cannot have uniform velocity. The presence of a pressure gradient causes the velocity of the flow to either increase or decrease. Thus, the modified Navier-Stokes equation becomes

$$u\frac{\partial u}{\partial x} + v\frac{\partial u}{\partial y} = \nu\frac{\partial^2 u}{\partial y^2} \qquad (1.27)$$

In summary, the boundary equations for a flat plate are

$$u\frac{\partial u}{\partial x} + v\frac{\partial u}{\partial y} = \nu\frac{\partial^2 u}{\partial y^2}$$
$$\frac{\partial u}{\partial x} + \frac{\partial v}{\partial y} = 0$$

subject to the boundary conditions

$$u(x,0) = 0$$
$$v(x,0) = 0$$
$$u(x,y) = U \text{ as } y \to \infty$$

The first two boundary conditions are called no-slip conditions.

The Blasius solution
To combine two equations in one, we introduce the stream function Ψ by letting

$$u = \frac{\partial \Psi}{\partial y} \text{ and } v = -\frac{\partial \Psi}{\partial x}$$

We note that the continuity equation is satisfied immediately. The modified Navier-Stokes equation becomes

$$\frac{\partial \Psi}{\partial y}\frac{\partial^2 \Psi}{\partial x \partial y} - \frac{\partial \Psi}{\partial x}\frac{\partial^2 \Psi}{\partial y^2} = \nu \frac{\partial^3 \Psi}{\partial y^3} \tag{1.28}$$

which is a third-order non-linear partial differential equation.

To solve the above PDE, we introduce a new variable

$$\eta = \frac{Ay}{x^n}$$

where A and n are parameters to be determined.

Let

$$\frac{u}{U} = f(\eta)$$

then

$$u = \frac{\partial \Psi}{\partial y} = U f(\eta)$$

Therefore,

$$
\begin{aligned}
\Psi(x,y) &= \int U f(\eta)\, dy = U \int f(\eta)\frac{dy}{d\eta}d\eta = U \int f(\eta)(\frac{x^n}{A})\, d\eta \\
&= \frac{Ux^n}{A}\underbrace{\int f(\eta)d\eta}_{F(\eta)} = \frac{Ux^n}{A}F(\eta)
\end{aligned}
$$

Therefore,

$$u = \frac{\partial \Psi}{\partial y} = \frac{Ux^n}{A}F'(\eta)\frac{d\eta}{dy} = \frac{Ux^n}{A}F'(\eta)\frac{A}{x^n} = UF'(\eta)$$

Similarly,

$$\frac{\partial u}{\partial x} = UF''(\eta)\frac{d\eta}{dx} = UF''(\eta)(-nAx^{-n-1}y) = -\frac{nAUy}{x^{n+1}}F''(\eta)$$

and

$$\frac{\partial u}{\partial y} = UF''(\eta)\frac{d\eta}{dy} = UF''(\eta)(\frac{A}{x^n}) = \frac{UA}{x^n}F''(\eta)$$

Using $\Psi(x, y)$, we have

$$
\begin{aligned}
v = -\frac{\partial \Psi}{\partial x} &= -\frac{U}{A}[nx^{n-1}F(\eta) + x^n F'(\eta)\frac{d\eta}{dx}] \\
&= -\frac{U}{A}[nx^{n-1}F(\eta) + x^n F'(\eta)(-nAx^{-n-1}y)] \\
&= \frac{U}{A}[-nx^{n-1}F(\eta) + \frac{nAy}{x}F'(\eta)]
\end{aligned}
$$

and

$$
\begin{aligned}
\frac{\partial^2 u}{\partial y^2} &= \frac{\partial}{\partial y}(\frac{UA}{x^n}F''(\eta)) = \frac{UA}{x^n}F'''(\eta)\frac{d\eta}{dy} \\
&= \frac{UA}{x^n}F'''(\eta)(\frac{A}{x^n}) \\
&= \frac{UA^2}{x^{2n}}F'''(\eta)
\end{aligned}
$$

Substituting these into the modified Navier-Stokes equation, we obtain

$$
UF'(\eta)(-\frac{nAUy}{x^{n+1}}F''(\eta)) + \frac{U}{A}[-nx^{n-1}F(\eta) + \frac{nAy}{x}F'(\eta)](\frac{UA}{x^n}F''(\eta)) = \nu(\frac{UA^2}{x^{2n}}F'''(\eta))
$$

Expanding,

$$
-\frac{nAU^2y}{x^{n+1}}F'(\eta)F''(\eta) - \frac{nU^2}{x}F(\eta)F''(\eta) + \frac{nAU^2y}{x^{n+1}}F'(\eta)F''(\eta) = \nu\frac{UA^2}{x^{2n}}F'''(\eta)
$$

We note that the first term and the third term on the left cancel each other. Hence,

$$
-\frac{nU^2}{x}F(\eta)F''(\eta) = \nu\frac{UA^2}{x^{2n}}F'''(\eta)
$$

Transposing the term on the left to the right side, we obtain

$$
\frac{nU^2}{x}F(\eta)F''(\eta) + \nu\frac{UA^2}{x^{2n}}F'''(\eta) = 0 \tag{1.29}
$$

Dividing both sides by $\nu U A^2/x^{2n}$, we obtain

$$
F'''(\eta) + \frac{nUx^{2n-1}}{\nu A^2}F(\eta)F''(\eta) = 0
$$

We want the aforementioned expression to be independent of x, so we let

$$
2n - 1 = 0
$$

or

$$n = \frac{1}{2}$$

Then the aforementioned expression becomes

$$F'''(\eta) + \frac{1}{2}\frac{U}{\nu A^2}F(\eta)F''(\eta) = 0$$

Choosing A such that $U/\nu A^2 = 1$, we obtain

$$A = \sqrt{U/\nu}$$

Hence, the Blasius problem is reduced to solving the non-linear ordinary differential equation

$$F'''(\eta) + \frac{1}{2}F(\eta)F''(\eta) = 0 \qquad (1.30)$$

subject to the boundary conditions

$$F(0) = 0, \; F'(0) = 0, \; and \; F(\infty) = 1 \qquad (1.31)$$

Once, $F(\eta)$ is solved, we have

$$\Psi(x,y) = \sqrt{\nu U x}F(\eta)$$

where

$$\eta = \sqrt{\frac{U}{\nu x}}y$$

and

$$
\begin{aligned}
u &= UF'(\eta) \\
v &= \sqrt{\nu U}[-\frac{1}{2\sqrt{x}}F(\eta) + \frac{U}{2x}F'(\eta)]
\end{aligned}
$$

At $y = 0$, $u(x,0) = 0$ and $v(x,0) = 0$ which gives $\eta = 0$ and

$$u(x,0) = 0 \Rightarrow F'(0) = 0$$

and

$$v(x,0) = 0 \Rightarrow \sqrt{\nu U}[-\frac{1}{2\sqrt{x}}F(0) + \frac{U}{2x}F'(0)] = 0$$

Since $F'(0) = 0$, we must have

$$F(0) = 0$$

As $y \to \infty$, $\eta \to \infty$ and $u \to U$. Hence, as $\eta \to \infty$,

$$UF'(\infty) = U$$

Hence,

$$F'(\infty) = 1$$

In summary, we want to solve the following boundary problem

$$F'''(\eta) + \frac{1}{2}F(\eta)F''(\eta) = 0 \qquad (1.32)$$

subject to the boundary conditions

$$F(0) = 0, \ F'(0) = 0, \ F'(\infty) = 1 \qquad (1.33)$$

We solve the above boundary value problem by numerical methods. We will use Runge-Kutta method of order fourth and shooting method (the secant method). First, we rewrite this non-linear third-order ode as a set of first-order odes as follows:

$$\begin{aligned} G(\eta) &= F'(\eta) \\ H(\eta) &= G'(\eta) \end{aligned}$$

then the ode gives

$$H'(\eta) = -\frac{1}{2}F(\eta)H(\eta)$$

subject to

$$F(0) = 1, \ G(0) = 0, \ G(\infty) = 1$$

To initiate RK4sys (the Runge-Kutta method of order 4) we need $F(0)$, $G(0)$ and $H(0)$. Unfortunately, we do not have $H(0)$ at hand. So we must assign some value to $H(0)$. Unless we are lucky, RK4sys will give us the correct solution. However, this is not the case. Hence, we should assign some values to $H(0)$. Here comes the shooting method. With the first two attempts for $H(0)$, the secant method will give us a better estimate for $H(0)$ until we obtain the correct solution. Here is the table of values of the solution. From the table, we note that $u = 0.99U$ when

$\eta = 5.0$. As $\eta > 5$, $G(\eta)$ is close to 1. Also from the table, the infinite distance is about $\eta = 7$ or $\eta = 8$. It is at this location that $H(0)$ is to be determined so that the condition $G(\infty) = 1$ is satisfied. Our MATLAB codes use the secant method to improve the accuracy of $H(0)$ by using the estimates of $G(8)$ from the first two attempts for $H(0)$. From the table, we find the correct value of $H(0)$ which is

$$H(0) = 0.33160277$$

The velocity profile of the solution can be obtained readily by running the MATLAB code in the Appendix on the Blasius solution. Recall

$$\frac{u}{U} = F'(\eta)$$

where

$$\eta = \sqrt{\frac{U}{\nu x}}y$$

From the Blasius solution (see the Figure 1.1)

```
             Table of values of the Blasius solution
      ========================================================
          η          F              G              H
      ========================================================
        0.00     0.00000000     0.00000000     0.33160277
        0.50     0.04143604     0.16565836     0.33045952
        1.00     0.16534569     0.32933166     0.32257712
        1.50     0.36963643     0.48613790     0.30220456
        2.00     0.64915299     0.62894749     0.26646540
        2.50     0.99499938     0.75032594     0.21723738
        3.00     1.39501252     0.84505124     0.16129357
        3.50     1.83540156     0.91203475     0.10778317
        4.00     2.30294982     0.95452867     0.06428133
        4.50     2.78684930     0.97855030     0.03403173
        5.00     3.27951253     0.99060043     0.01594465
        5.50     3.77634408     0.99595228     0.00660084
        6.00     4.27493209     0.99805437     0.00241285
        6.50     4.77417528     0.99878420     0.00077855
        7.00     5.27363474     0.99900813     0.00022173
        7.50     5.77315736     0.99906884     0.00005574
        8.00     6.27269629     0.99908339     0.00001237
```

Figure 1.1 Table to determine correct value of $H(0)$.

$$5 = \sqrt{\frac{U}{\nu x}}\delta$$

Hence,

$$\delta = \frac{5x}{\sqrt{Re}}$$

where Re is the Reynolds number which is defined as

$$Re = \frac{Ux}{\nu}$$

We note that as the Reynolds number becomes large, δ is very small. It follows that when the Reynolds number is large,

$$\frac{\delta}{x} \ll 1$$

which agrees with the assumption that the boundary layer is thin.
Similarly,

$$\delta^* = \int_0^\delta (1 - \frac{u}{U})\, dy = \int_0^5 (1 - F'(\eta))\frac{dy}{d\eta}\, d\eta = \frac{1}{\sqrt{\frac{U}{\nu x}}}\int_0^5 (1 - F'(\eta))\, d\eta$$

$$= \frac{1}{\sqrt{\frac{U}{\nu x}}}(\eta - F(\eta)) = \frac{1}{\sqrt{\frac{U}{\nu x}}}(5 - F(5)) = \frac{1}{\sqrt{\frac{U}{\nu x}}}(5 - 3.27951253)$$

$$= \frac{1.7205x}{\sqrt{Re}}$$

where the value of $F(5)$ is obtained from the Blasius solution.
Also,

$$\theta = \int_0^\delta \frac{u}{U}(1 - \frac{u}{U})\, dy = \int_0^5 F'(\eta)(1 - F'(\eta))\frac{dy}{d\eta}\, d\eta = \frac{1}{\sqrt{\frac{U}{\nu x}}}\int_0^5 F'(\eta)(1 - F'(\eta))\, d\eta$$

Using the table of values of the Blasius solution and numerical method (such as Trapezoidal rule or Simpson's rule where the MATLAB codes are listed in the appendix), we obtain

$$\int_0^5 F'(\eta)(1 - F'(\eta))\, d\eta = 0.6616$$

Hence,

$$\theta = \frac{0.6616x}{\sqrt{Re}}$$

Finally, the curious reader may wonder how the form of η has been chosen as

$$\eta = \frac{Ay}{x^n}$$

where A and n are parameters to be determined. In fact, the above variable is called a similarity variable. Its form is derived from the techniques of Dimension Analysis and Group Theory. For further detail, please see the references [13] and [14].

Karman's integral equation
From the Navier-Stokes equations and the continuity equation

$$u\frac{\partial u}{\partial x} + v\frac{\partial u}{\partial y} = -\frac{1}{\rho}\frac{\partial p}{\partial x} + \nu(\frac{\partial^2 u}{\partial x^2} + \frac{\partial^2 u}{\partial y^2})$$

$$u\frac{\partial v}{\partial x} + v\frac{\partial v}{\partial y} = -\frac{1}{\rho}\frac{\partial p}{\partial y} + \nu(\frac{\partial^2 v}{\partial x^2} + \frac{\partial^2 v}{\partial y^2})$$

$$\frac{\partial u}{\partial x} + \frac{\partial v}{\partial y} = 0$$

We rewrite the left-hand side of the first equation as

$$u\frac{\partial u}{\partial x} + v\frac{\partial u}{\partial y} = u\frac{\partial u}{\partial x} + \frac{\partial(uv)}{\partial y} - u\frac{\partial v}{\partial y} = u\frac{\partial u}{\partial x} + \frac{\partial(uv)}{\partial y} - u(-\frac{\partial u}{\partial x})$$

$$= 2u\frac{\partial u}{\partial x} + \frac{\partial(uv)}{\partial y}$$

where the last step is obtained by using the third equation (the continuity equation). Therefore, the first equation becomes

$$\frac{\partial u^2}{\partial x} + \frac{\partial(uv)}{\partial y} = -\frac{1}{\rho}\frac{dp}{dx} + \nu\frac{\partial^2 u}{\partial y^2}$$

Integrating both sides with respect to y from 0 to δ, we obtain

$$\int_0^\delta \frac{\partial u^2}{\partial x} \, dy + (uv)|_0^\delta = -\frac{1}{\rho}\int_0^\delta \frac{dp}{dx} \, dy + \nu\frac{\partial u}{\partial y}\Big|_0^\delta$$

Using $u(x,\delta) = U$, $u(x,0) = 0$ and $\partial u/\partial y(x,y = \delta) = 0$, the aforementioned equation becomes

$$\int_0^\delta \frac{\partial u^2}{\partial x} \, dy + U\int_0^\delta \frac{\partial v}{\partial y} \, dy = -\frac{\delta}{\rho}\frac{dp}{dx} - \nu\frac{\partial u}{\partial y}\Big|_{y=0}$$

Using the continuity equation, the second integral on the left is converted to,

$$\int_0^\delta \frac{\partial u^2}{\partial x} \, dy - U \int_0^\delta \frac{\partial u}{\partial x} \, dy = -\frac{\delta}{\rho}\frac{dp}{dx} - \nu \left. \frac{\partial u}{\partial y} \right|_{y=0}$$

or

$$\frac{d}{dx}\int_0^\delta u^2 \, dy - U\frac{d}{dx}\int_0^\delta u \, dy = -\frac{\delta}{\rho}\frac{dp}{dx} - \nu \left. \frac{\partial u}{\partial y} \right|_{y=0}$$

Applying Bernoulli equation

$$p + \frac{1}{2}\rho U^2(x) = constant$$

Differentiating both sides with respect to x,

$$\frac{dp}{dx} + \frac{1}{2}\rho(2U\frac{dU}{dx}) = 0$$

or

$$-\frac{1}{\rho}\frac{dp}{dx} = U\frac{dU}{dx}$$

Substituting this into the above integral equation, we obtain

$$\frac{d}{dx}\int_0^\delta u^2 \, dy - U\frac{d}{dx}\int_0^\delta u \, dy = U\delta\frac{dU}{dx} - \nu \left. \frac{\partial u}{\partial y} \right|_{y=0}$$

Recall

$$\tau = \mu\frac{\partial u}{\partial y}$$

On the flat plate ($y = 0$), this gives

$$\tau_0 = \mu \left. \frac{\partial u}{\partial y} \right|_{y=0}$$

then the integral equation becomes

$$\frac{d}{dx}\int_0^\delta u^2 \, dy - U\frac{d}{dx}\int_0^\delta u \, dy = U\delta\frac{dU}{dx} - \frac{\tau_0}{\rho}$$

where τ_0 is the shear stress on the plate. This equation is called the Karman integral equation. This equation is valid for either laminar or turbulent flows.

Separation of the boundary layer

Consider the boundary layer equation

$$u\frac{\partial u}{\partial x} + v\frac{\partial u}{\partial y} = -\frac{1}{\rho}\frac{dp}{dx} + \nu\frac{\partial^2 u}{\partial y^2}$$

At the wall, $u = 0$ and $v = 0$ (no slip conditions), the aforementioned equation becomes

$$0 = -\frac{1}{\rho}\frac{dp}{dx} + \nu\frac{\partial^2 u}{\partial y^2}$$

or

$$\frac{\partial^2 u}{\partial y^2} = \frac{1}{\mu}\frac{dp}{dx}$$

where we use

$$\mu = \rho\nu$$

In general, $u = U(x)$. By Bernoulli equation

$$\frac{dp}{dx} = \rho U\frac{dU}{dx}$$

Thus, the curvature of the velocity profile, $\partial^2 u/\partial y^2$, depends on the gradient of pressure. In the case of a positive pressure gradient causes a positive velocity profile curvature. It means the inflection point of the velocity profile is above the surface. That leads to a reversed (adverse) flow near the surface. Such a velocity profile can cause separation as shown in the figure below (Figure 1.2).

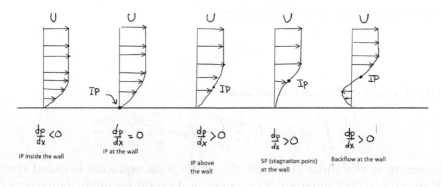

Figure 1.2 Flow in boundary layer and detachment.

In the case of a flat plate, $dp/dx = 0$, the inflection point is right at the wall ($y = 0$). Hence, there is no tendency for separation to occur unless the Reynolds number is high ($> 10^4$, [1]).

If $dp/dx < 0$, we have negative curvature. It means the inflection point is inside the wall. Thus, a negative pressure gradient prevents the existence of an inflection point above the surface. Hence, it inhibits the separation. In this case, we say the negative pressure gradient is favorable.

Thus, separation only occurs in the region of positive pressure gradient.

1.2 KINEMATICS IN FLUIDS

Kinematics of fluid flow deals with the motion of fluid particles. Here we consider the geometry of motion of fluid particles including the velocity and acceleration of the particles. The motion of a fluid is such that the fluid takes the shape of the volume it is enclosed by. For a fluid body, the fluid particles have motions independent of each other. A fluid particle may have a motion different from those particles surrounding it. We consider motion about a local and global region in our flow.

1.2.1 Material Time Derivative

We are interested in measuring changes in quantities such as velocity, pressure, or temperature. Taking a time derivative of a quantity f,

$$\frac{\partial f}{\partial t}.$$

does give an instantaneous rate of change. The derivative of a function of a real variable measures the sensitivity to change of the function value (output value) with respect to a change in its argument (input value). Derivatives are a fundamental tool of calculus. We consider first the Eulerian approach which deals with time derivatives of the quantity we are measuring and the Lagrangian approach which deals with individual particles and calculates the path of each particle separately with respect to each other. The Eulerian approach deals with a concentration of particles (continuum hypothesis assumed) and calculates the overall diffusion and convection of a number of particles. (1) Eulerian method has less computational cost in comparison with

Lagrangian one and (2) instead of positions of particles, Eulerian method works with concentration of particles which is appropriate for engineering applications. However, the question is that for what conditions are the results of the two approaches equivalent and when are the two methods different?

Consider the Lagrangian approach to fluid dynamics. We consider a trajectory $\mathbf{x}(t)$ through the field $\mathbf{f}$, and we want to know how the field varies as we move along the trajectory.

Along the trajectory $\mathbf{x}(t)$, the chain rule gives

$$\frac{\partial f(\mathbf{x}(t), t)}{\partial t} = \frac{\partial f}{\partial x}\frac{\partial x}{\partial t} + \frac{\partial f}{\partial y}\frac{\partial y}{\partial t} + \frac{\partial f}{\partial z}\frac{\partial z}{\partial t} + \frac{\partial f}{\partial t}$$
$$= \nabla f \cdot \dot{\mathbf{x}} + \frac{\partial f}{\partial t}.$$

Definition 1.12 *Material derivative: If* $\mathbf{x}(t) \in \mathbb{R}^3$ *is the Lagrangian path followed by a fluid particle, then* $\dot{\mathbf{x}}(t) = \mathbf{u}$ *by definition. Here we write*

$$\frac{df}{dt} = \frac{Df}{Dt}.$$

This is the material derivative.

In other words, we have

$$\frac{Df}{Dt} = \mathbf{u} \cdot \nabla f + \frac{\partial f}{\partial t}.$$

On the left of this equation, we have the Lagrangian derivative, which is the change in f as we follow the path. On the right of this equation, the first term is the advective derivative, which measures the change due to a change in position. The second term is $\frac{\partial f}{\partial t}$, the Eulerian time derivative, which is the change at time $t \in \mathbb{R}^+$.

1.2.2 Conservation of Mass

The first equation we consider is the conservation of mass.

We fix an arbitrary region of space $\mathcal{D}$ with boundary $\partial\mathcal{D}$ and outward normal $\mathbf{n}$. Suppose we have some flow through this volume,

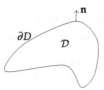

It is physically and mathematically true that the change in the mass inside volume $\mathcal{D}$ is equal to the total flow of fluid through the boundary (Flux is the rate of flow of a property such as velocity in this case). We can write this as

$$\frac{d}{dt}\int_{\mathcal{D}} \rho \, \partial V = -\int_{\partial \mathcal{D}} \rho \mathbf{u} \cdot \mathbf{n} \, \partial S. \qquad (1.34)$$

The negative sign is required because the outward flow of mass through the volume boundaries is leaving the volume($\mathbf{n}$ outward); therefore, the change in mass over time is negative, resulting in less mass. Since we picked the outward normal, the integral measures the outward flow of fluid.

Since the domain is fixed, we can interchange the derivative and the integral on the left; on the right, we can use the divergence theorem to get

$$\int_{\mathcal{D}} \left(\frac{\partial \rho}{\partial t} + \nabla \cdot (\rho \mathbf{u}) \right) \partial V = 0. \qquad (1.35)$$

Since $\mathcal{D}$ is arbitrary, we have

$$\frac{\partial \rho}{\partial t} + \nabla \cdot (\rho \mathbf{u}) = 0 \qquad (1.36)$$

everywhere in space.

In the mass conservation equation, we expand $\nabla \cdot (\rho \mathbf{u})$ to get

$$\frac{\partial \rho}{\partial t} + \mathbf{u} \cdot \nabla \rho + \rho \nabla \cdot \mathbf{u} = 0. \qquad (1.37)$$

We see that the first term is just the material derivative of ρ. Hence,

$$\frac{D\rho}{Dt} + \rho \nabla \cdot \mathbf{u} = 0. \qquad (1.38)$$

If there are no changes in density, then the material derivative $\frac{D\rho}{Dt}$ must vanish, so we define an incompressible fluid as follows:

Definition 1.13 *Incompressible fluid: A fluid is incompressible if the density of the fluid particle does not change. From this,*

$$\frac{D\rho}{Dt} = 0,$$

hence,

$$\nabla \cdot \mathbf{u} = 0. \tag{1.39}$$

This is known as the continuity equation.

For parallel flow, $\mathbf{u} = (u, 0, 0)$. So if the flow is incompressible, we must have $\frac{\partial u}{\partial x} = \nabla \cdot \mathbf{u} = 0$. So we can consider u of the form $u = u(y, z, t)$.

1.2.3 Kinematic Boundary Conditions

Fluids are not allowed to pass through boundaries that are either rigid or move with the fluid as in moving boundary value problems. For boundaries, we suppose they have a velocity $\mathbf{U}$ which change spatially. The fluid has a relative velocity to the boundary,

$$\mathbf{u}' = \mathbf{u} - \mathbf{U}. \tag{1.40}$$

The impenetrability of fluids at walls and boundaries in general is given by the condition,

$$\mathbf{u}' \cdot \mathbf{n} = 0.$$

A rigid boundary as in a wall, $\mathbf{U} = \mathbf{0}$. So the boundary condition is

$$\mathbf{u} \cdot \mathbf{n} = 0. \tag{1.41}$$

An example of a free material boundary is the surface of a water wave or interface between two immiscible fluids. We can define the surface,

$$z = F(x, y, t). \tag{1.42}$$

We can define the surface as a contour of

$$G(x, y, z, t) = z - F(x, y, t). \tag{1.43}$$

Then the surface is defined by $G = 0$. The normal is parallel to

$$\mathbf{n} \parallel \nabla F = (-F_x, -F_y, 1).$$

We also have,
$$\mathbf{U} = (0, 0, F_t).$$

Letting the velocity of the fluid be
$$\mathbf{u} = (u, v, w).$$

Then the boundary condition is
$$\mathbf{u} \cdot \mathbf{n} = \mathbf{U} \cdot \mathbf{n}.$$

and
$$-uF_x - vF_y + w = F_t. \tag{1.44}$$

so
$$w = uF_x + vF_y + F_t = \mathbf{u} \cdot \nabla F + \frac{\partial F}{\partial t} = \frac{DF}{Dt}. \tag{1.45}$$

Also,
$$\frac{DG}{Dt} = 0.$$

1.2.4 Streamfunction in the Incompressible Case

We suppose that the fluid is incompressible,
$$\nabla \cdot \mathbf{u} = 0.$$

It follows that there is a vector potential $\mathbf{A}$ such that the following definition arises,

Definition 1.14 *A vector potential is an* $\mathbf{A}$ *such that*
$$\mathbf{u} = \nabla \times \mathbf{A}. \tag{1.46}$$

For two-dimensional flow,
$$\mathbf{u} = (u(x, y, t), v(x, y, t), 0),$$

and $\mathbf{A}$ is of the form
$$\mathbf{A} = (0, 0, \psi(x, y, t)),$$

Take the curl and we get
$$\mathbf{u} = \left(\frac{\partial \psi}{\partial y}, -\frac{\partial \psi}{\partial x}, 0 \right). \tag{1.47}$$

Definition 1.15 *The ψ such that $\mathbf{A} = (0, 0, \psi)$ is the streamfunction.*

To see an important property of the stream function, consider contours $\psi = c$. These have normal $\mathbf{n} = \nabla\psi = (\psi_x, \psi_y, 0)$. We look at some properties of the streamfunction. Let's look at the contours $\psi = c$. These have normal

$$\mathbf{n} = \nabla\psi = (\psi_x, \psi_y, 0) \,.$$

It follows that,

$$\mathbf{u} \cdot \mathbf{n} = \frac{\partial\psi}{\partial x}\frac{\partial\psi}{\partial y} - \frac{\partial\psi}{\partial y}\frac{\partial\psi}{\partial x} = 0. \tag{1.48}$$

So the flow is perpendicular to the normal or tangent to the contours of ψ.

Definition 1.16 *Streamlines: The* streamlines *are the contours of the streamfunction ψ.*

Example: Find the stream function $\psi(x, y)$ for the uniform flow with velocity field,

$$\mathbf{u} = U\cos(\beta)\mathbf{i} + U\sin(\beta)\mathbf{j}$$

where U and β are constants.

Show that the contour lines $\psi = $ constant are the streamlines of this flow.

Solution: For the uniform flow, the streamfunction equations give,

$$u = \frac{\partial\psi}{\partial y} = U\cos(\beta)$$

$$v = -\frac{\partial\psi}{\partial x} = U\sin(\beta)$$

Integrating the second equation we have that,

$$\psi(x, y) = -Ux\sin(\beta) + f(y)$$

where f is an arbitrary function. Differentiating this equation with respect to y, we have,

$$\frac{\partial\psi}{\partial y} = 0 + f'(y) = U\cos(\beta)$$

from the first equation. Integrating $f'(y) = U\cos(\beta)$ gives,

$$f(y) = Uy\cos(\beta) + A.$$

where A is an arbitrary constant. Hence, the stream function is,

$$\psi(x, y) = -Ux\sin(\beta) + Uy\cos(\beta) + A$$

The lines $\psi = $ constant are defined by the equation,

$$-Ux\sin(\beta) + Uy\cos(\beta) + A = \text{constant} = B, \quad \text{a constant}$$

We write $(B - A)/(U\cos(\beta)) = C$(a constant)
to obtain,

$$y - x\tan(\beta) = C$$

These equations are parallel lines making an angle β with the x- axis.

1.3   DYNAMICS

1.3.1   Navier-Stokes Equations

The Navier-Stokes equations are partial differential equations that govern the motion of incompressible fluids. These equations are the basic equations of fluid mechanics. The Navier–Stokes equations are useful because they describe the physics of many phenomena of scientific interest. They may be used in meteorology, industrial fluid dynamics, ocean current flows and aerodynamics. The Navier–Stokes equations, in their full and simplified forms, help us with the design of aircraft and automobiles, the study of blood flow in the human circulation and many other areas. Coupled with Maxwell's equations, they can be used to model and study magnetohydrodynamics. Newton's second law gives,

Definition 1.17 *Navier-Stokes equation:*

$$\rho\frac{D\mathbf{u}}{Dt} = -\nabla P + \mu\nabla^2\mathbf{u} + \mathbf{f}. \tag{1.49}$$

The left-hand side of the equation is mass times acceleration, and the right is the sum of the individual forces, the pressure gradient, viscosity, and the body forces per unit volume, respectively.

In general, these are notoriously difficult to solve mathematically due to non-linearity that appear in the material derivative, where we have the term $\mathbf{u} \cdot \nabla \mathbf{u}$. The body force can be gravity or friction forces that are external forces. The friction forces will also be examined in Chapter 3 for the periodic Navier-Stokes equations. The acceleration of a fluid particle is the Lagrangian material derivative of the velocity. The derivation of the viscous term depends on, three stresses on each side of the cube, there is a normal direction and two tangential directions on each face. We then have the stress tensor. Note that $\nabla^2 \mathbf{u}$ can be written as

$$\nabla^2 \mathbf{u} = \nabla(\nabla \cdot \mathbf{u}) - \nabla \times (\nabla \times \mathbf{u}).$$

In an incompressible fluid, we have

$$\nabla^2 \mathbf{u} = -\nabla \times (\nabla \times \mathbf{u}) = -\nabla \times \boldsymbol{\omega},$$

where $\boldsymbol{\omega} = \nabla \times \mathbf{u}$ is the vorticity. In Cartesian coordinates, for $\mathbf{u} = (u_x, u_y, u_z)$, we have

$$\nabla^2 \mathbf{u} = (\nabla^2 u_x, \nabla^2 u_y, \nabla^2 u_z),$$

1.3.2 Pressure

In the Navier-Stokes equation, Eq. (1.49), we have a pressure term P. For compressible flow, pressure and velocity can be coupled with the Equation of State. But for incompressible flow, there is no obvious way to couple pressure and velocity. (One can use Poisson's equation as will be seen in subsequent chapters.) So, taking the divergence of the momentum equation and using the continuity equation, we can get a Poisson equation for pressure [16]). The gradient of pressure term also restated as the volumetric stress tensor prevents motion due to normal stresses. We denote the hydrostatic pressure as P_h. To find this, we set $\mathbf{u} = 0$ in the Navier-Stokes equation,

$$\nabla P_h = \mathbf{f} = \rho \mathbf{g}. \tag{1.50}$$

Integrating Eq. (1.50) we get a linear relation,

$$P_h = P\mathbf{g} \cdot \mathbf{x} + P_1,$$

where P_1 is some arbitrary constant. Now the gravity vector is $\mathbf{g} = (0, 0, -g)$. Then

$$P_h = P_1 - g\rho z.$$

Suppose we have a body $\mathcal{D}$ with boundary $\partial\mathcal{D}$ and outward normal $\mathbf{n}$. The following is known as Archimedes principle: The force due to the pressure is where M is the mass of fluid displaced.

Next for moving fluid, in general, the pressure gradient can either be some external pressure gradient driving the motion or a pressure gradient caused by the flow itself. In either case, we can write

$$P = P_h + p',$$

where P_h is the hydrostatic pressure, and p' is what results or causes motion.

Substituting into the Navier-Stokes equations Eq. (1.49) gives

$$\rho\frac{Du}{Dt} = -\nabla p' + \mu\nabla^2\mathbf{u}. \tag{1.51}$$

1.3.3 Reynolds Number

The dimensionless Reynolds number is crucial in foreseeing patterns of fluid flows. The Reynolds number, Re, is used to determine whether the fluid flow is laminar or turbulent. It is one of the main controlling parameters in flows where a numerical model is selected with a set Reynolds number.

The Reynolds number is the ratio of the inertial forces to the viscous forces. This ratio helps to differentiate laminar flows from the turbulent ones.

Inertial forces resist a change in the velocity of an object and are the cause of the fluid movement. These forces are dominant in turbulent flows. If the viscous forces, defined as the resistance to flow, are dominant, the flow is laminar. We suppose the flow has a characteristic speed U and an extrinsic length scale L, externally imposed by geometry. For example, if we look at the flow between two plates, the characteristic speed can be the maximum (or average) speed of the fluid, and a sensible length scale would be the length between the plates.

Let us define the time scale $T = L/U$. Finally, we suppose pressure differences have characteristic magnitude P.

Dividing by ρ in the Navier-Stokes equations we get,

$$\frac{\partial \mathbf{u}}{\partial t} + \mathbf{u} \cdot \nabla \mathbf{u} = -\frac{1}{\rho}\nabla p + \nu \nabla^2 \mathbf{u},$$

where $\nu = \frac{\mu}{\rho}$. Estimating the size of these terms we have

$$\frac{U}{(L/U)}, \quad U \cdot \frac{U}{L}, \quad \frac{1}{\rho}\frac{P}{L}, \quad \nu\frac{U}{L^2}.$$

Dividing by U^2/L, we get

$$1, \quad 1, \quad \frac{P}{\rho U^2}, \quad \frac{\nu}{UL}.$$

Definition 1.18 *Reynolds number: The* Reynolds number *is*

$$Re = \frac{UL}{\nu}, \tag{1.52}$$

Note that the pressure always scales to balance the dominant terms in the equation, and imposes incompressibility, i.e. $\nabla \cdot \mathbf{u} = 0$.

Definition 1.19 *Dynamic similarity: Reynolds' Law of similarity states that when there are two geometrically similar flows, both are essentially equal to each other, as long as they adhere to the same Reynolds number.*

Let us take an example of an actual vehicle and a half scale model with moving air past both vehicles. The Reynolds numbers of both agree when the velocity of the half scale model is doubled. In this state, the proportions of viscous force and inertia force of both cases are equal; hence, the surrounding flows can be defined as similar.

When $Re \ll 1$, the inertia terms are negligible and,

$$P \sim \frac{\rho \nu U}{L} = \frac{\mu U}{L}.$$

So the pressure balances the shear stress. We can approximate the Navier-Stokes equation by dropping the term on the left-hand side and write

$$0 = -\nabla p + \mu \nabla^2 \mathbf{u},$$

where the incompressibility condition is

$$\nabla \cdot \mathbf{u} = 0.$$

These are known as *Stoke's equations* . This is now a linear equation.

When $Re \gg 1$, the viscous terms are negligible. Then, the pressure scales on the momentum flux,

$$P \sim \rho U^2,$$

and on extrinsic scales, we can approximate Navier-Stokes equations by the *Euler equations*

$$\rho \frac{D u}{D t} = -\nabla p \tag{1.53}$$

$$\nabla \cdot \mathbf{u} = 0. \tag{1.54}$$

In this case, the acceleration is proportional to the pressure gradient.

1.4   SCALE INVARIANCE OF NAVIER-STOKES EQUATIONS

The Navier-Stokes equations are invariant under a particular change of time and space scaling. More exactly, assume that, in $\mathbb{R}^3 \times [0, \infty)$, $\mathbf{u}(\mathbf{x}, \mathbf{t})$ and $P(\mathbf{x}, \mathbf{t})$ solve the Navier-Stokes system. Then the same is true for the rescaled functions, that is,

$$\mathbf{u}_\lambda(\mathbf{x}, t) = \lambda \mathbf{u}(\lambda \mathbf{x}, \lambda^2 t) \tag{1.55}$$

$$P_\lambda(\mathbf{x}, t) = \lambda^2 P(\lambda \mathbf{x}, \lambda^2 t) \tag{1.56}$$

The spaces which are invariant under such a scaling are called critical spaces for the Navier-Stokes equations. An example of a critical space for the Navier-Stokes is:

Definition 1.20 *Critical space: A translation invariant Banach space of tempered distributions X is called a critical space for the Navier-Stokes equations if its norm is invariant under the action of the scaling $f(x) \to \lambda f(\lambda x)$ for any $\lambda > 0$.*

The following important theorem is proven in [15].

Theorem 1.1 *Assume an initial condition* $\mathbf{u}\ |_{t=0} = \mathbf{u}_0$ *of the incompressible Navier Stokes equation is scale invariant and locally Hölder continuous in* $\mathbb{R}^3$ *{0} with div* $\mathbf{u} = 0$ *in* $\mathbb{R}^3$*. Then, the Cauchy problem for the Navier-Stokes equations (with continuity equation) satisfying the initial condition, has at least one scale-invariant solution* $\mathbf{u}$ *which is smooth in* $\mathbb{R}^3 \times (0, \infty)$ *and locally Hölder continuous in* $\mathbb{R}^3 \times [0, \infty)$ *{(0, 0)}.*

Here uniqueness may fail for large data. See Chapter 5 for a discussion of smooth functions and Hölder continuous functions.

1.5 COMPLEX POTENTIALS

Every complex analytic function leads to an irrotational, incompressible flow. All such flows relate to an analytic function. The analytic function is the complex potential of the flow.

We call the complex potential function,

$$\Phi = \phi + i\psi. \tag{1.57}$$

Analytic functions give us incompressible, irrotational flows.

Let $\Phi(z)$ be an analytic function on a region Ω. For $z = x + iy$ we write $\Phi(z) = \phi(x, y) + i\psi(x, y)$.

From this we can define a vector field

$$F = \nabla \phi = (\phi_x, \phi_y) = (u, v), \tag{1.58}$$

here u and v are defined by ϕ_x and ϕ_y. From the theory of analytic and harmonic functions, we can make a list of properties of these functions. These are,

1. ϕ and ψ are both harmonic.

2. The level curves of ϕ and ψ are orthogonal.

3. $\Phi' = \phi_x - i\phi_y$.

4. The analytic function Φ gives us an incompressible, irrotational vector field **F**.

We call ϕ a potential function for the vector field **F**. We also call Φ a complex potential function for **F**. The function ψ will be called the stream function of **F** and Φ' is the complex velocity.

Theorem 1.2 *Assuming* $\mathbf{F} = (u, v)$ *is an incompressible, irrotational field on a simply connected region* Ω. *Then, there is an analytic function* Φ *which is a complex potential function for* F, *that is,*

$$\mathbf{F} = \nabla\phi \tag{1.59}$$

Theorem 1.3 *Suppose that* $\Phi = \phi + i\psi$ *is the complex potential for a velocity field* $\mathbf{F}$. *Then, the fluid flows along the level curves of* ψ. *That is,* $\mathbf{F}$ *is everywhere tangent to the level curves of* ψ. *The level curves of* ψ *are called streamlines and* ψ *is called the stream function.*

Example: Stagnation points. Draw the streamlines and identify the stagnation points for the potential $\Phi(z) = z^2$.

Solution: We have $\Phi = (x^2 - y^2) + 2ixy$. So the streamlines are the hyperbolas : $2xy = $ constant. Since $\phi = x^2 - y^2$ increases as $\mid x \mid$ increases and decreases as $\mid y \mid$ increases, the arrows, which point in the direction of increasing ϕ, are as shown in the figure below. The stagnation points are the zeros of $\Phi'(z) = 2z$, i.e. the only stagnation point is at the point $z = 0$. Note. The stagnation points are what we call the critical points of a vector field.

1.5.1 Stagnation Point Flow $u = 0$

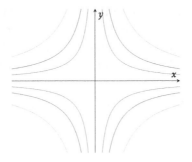

The boundary conditions as $y \to \infty$ give us a picture of what is going on at large y, as shown in the diagram, but near the boundary, the velocity has to begin to vanish.

At the origin, the velocity is zero, and this is a stagnation point. This is also the point of highest pressure. In general, velocity is high at low pressures and low at high pressures. In fluid dynamics, the Euler equations are a set of quasilinear partial differential equations governing adiabatic and inviscid flow. They are named after Leonhard Euler. In particular, they correspond to the Navier–Stokes equations with zero viscosity and zero thermal conductivity. The Euler equations can be applied to incompressible or compressible flow. The incompressible Euler equations consist of Cauchy equations for conservation of mass and balance of momentum, together with the incompressibility condition that the flow velocity is a solenoidal field. The compressible Euler equations consist of equations for conservation of mass, balance of momentum and balance of energy. Here an equation of state to relate pressure to density and temperature is prescribed.

1.6 INVISCID FLOW

1.6.1 Momentum Equation for Inviscid $\nu = 0$ Incompressible Fluid

Consider an arbitrary volume $\mathcal{D}$ with boundary $\partial\mathcal{D}$ and outward pointing normal $\mathbf{n}$.

The total momentum of the fluid in $\mathcal{D}$ can change according to, Momentum flux across the boundary $\partial\mathcal{D}$, surface pressure forces, body forces and viscous surface forces. We will ignore the last one. We can then write the rate of change of the total momentum as

$$\frac{d}{dt}\int_{\mathcal{D}} \rho\mathbf{u}\, dV = -\int_{\partial\mathcal{D}} \rho\mathbf{u}(\mathbf{u}\cdot\mathbf{n})\, dS - \int_{\partial\mathcal{D}} p\mathbf{n}\, \partial S + \int_{\mathcal{D}} \mathbf{f}\, dV. \quad (1.60)$$

The equation can be written as,

$$\frac{d}{dt}\int_{\mathcal{D}} \rho u_i\, dV = -\int_{\partial\mathcal{D}} \rho u_i u_j n_j\, dS - \int_{\partial\mathcal{D}} -p n_i\, dS + \int_{\mathcal{D}} f_i\, dV. \quad (1.61)$$

We can use the divergence theorem to write

$$\int_{\mathcal{D}} \left(\rho\frac{\partial u_i}{\partial t} + \rho\frac{\partial}{\partial x_j}(u_i u_j)\right)\, \partial V = \int_{\mathcal{D}} \left(-\frac{\partial p}{\partial x_i} + f_i\right)\, \partial V. \quad (1.62)$$

Since $\mathcal{D}$ is arbitrary in Eq. (1.62), we must have

$$\rho\frac{\partial u_i}{\partial t} + \rho\frac{\partial u_i}{\partial x_j} + \rho u_i\frac{\partial u_j}{\partial x_j} = -\frac{\partial p}{\partial x_i} + f_i.$$

The last term on the left is the divergence of $\mathbf{u}$, which vanishes by incompressibility, and the remaining terms are just the material derivative of $\mathbf{u}$. So we get

Proposition 1.4 *Euler momentum equation:*

$$\rho\frac{D\mathbf{u}}{Dt} = -\nabla p + \mathbf{f}. \tag{1.63}$$

We can further derive some equations from this.

For conservative forces, we can write $\mathbf{f} = -\nabla\phi$, where ϕ is a. For example, gravity can be written as $\mathbf{f} = \rho\mathbf{g} = \nabla(\rho\mathbf{g}\cdot\mathbf{x})$ for ρ constant. So $\phi = -\rho\mathbf{g}\cdot\mathbf{x} = \rho gz$ if $\mathbf{g} = (0,0,-g)$.

In the case of a steady flow, $\frac{\partial\mathbf{u}}{\partial t}$ vanishes. Then, the momentum equation becomes

$$0 = -\int_{\partial\mathcal{D}} \rho\mathbf{u}(\mathbf{u}\cdot\mathbf{n})\,\partial S - \int_{\partial\mathcal{D}} p\mathbf{n}\,\partial S - \int_{\mathcal{D}} \nabla\phi\,\partial V. \tag{1.64}$$

Thus we have,

Proposition 1.5 *Momentum integral for steady flow:*

$$\int_{\partial\mathcal{D}} (\rho\mathbf{u}(\mathbf{u}\cdot\mathbf{n}) + p\mathbf{n} + \phi\mathbf{n})\,\partial S = 0. \tag{1.65}$$

Recall the vector identity

$$\mathbf{u}\times(\nabla\times\mathbf{u}) = \nabla\left(\frac{1}{2}|\mathbf{u}|^2\right) - \mathbf{u}\cdot\nabla\mathbf{u}.$$

We use this to obtain,

$$\rho\frac{\partial\mathbf{u}}{\partial t} + \rho\nabla\left(\frac{1}{2}|\mathbf{u}|^2\right) - \rho\mathbf{u}\times(\nabla\times\mathbf{u}) = -\nabla p - \nabla\phi. \tag{1.66}$$

and we have,

Proposition 1.6 *Bernoulli's equation:*

$$\frac{1}{2}\rho\frac{\partial|\mathbf{u}|^2}{\partial t} = -\mathbf{u}\cdot\nabla\left(\frac{1}{2}\rho|\mathbf{u}|^2 + p + \phi\right). \qquad (1.67)$$

Note here that high velocity is associated with low pressure and low pressure with high velocity. Bernoulli's principle is a key concept in fluid dynamics that relates pressure, speed and height. Bernoulli's principle states that an increase in the speed of a fluid occurs simultaneously with a decrease in static pressure or with a decrease in the fluid's potential energy. Although Bernoulli proved that the pressure decreases when the flow speed increases, it was Leonhard Euler in 1752 who derived Bernoulli's equation in its usual form. The principle is only applicable for isentropic flows: when the effects of irreversible processes (like turbulence) and non-adiabatic processes (e.g. thermal radiation) are negligible.

In the case where we have a steady flow, we know

$$H = \frac{1}{2}\rho|\mathbf{u}|^2 + p + \phi \qquad (1.68)$$

is constant along streamlines.

If the flow is not steady, we can define the value H and then integrate Bernoulli's equation over a volume $\mathcal{D}$ to obtain

$$\frac{d}{dt}\int_{\mathcal{D}}\frac{1}{2}\rho|\mathbf{u}|^2\,dV = -\int_{\partial\mathcal{D}}H\mathbf{u}\cdot\mathbf{n}\,dS. \qquad (1.69)$$

So H is the transportable energy of the flow.

Definition 1.21 *Consider a pipe associated with two relative heights h_1 and h_2 (Figure 1.3)*

Suppose at the left of the pipe at height h_1, we have a uniform speed v_1, area A_1 and pressure P_1. At height h_2 we have speed v_2, area A_2 and pressure P_2.

By conservation of mass, we have

$$q = v_1 A_1 = v_2 A_2$$

We apply Bernoulli's principle along the central streamline, using the fact that H is constant along streamlines. We have a body force $\rho g z$, different at $z = h_1$ and $z = h_2$. Then, we get

$$\frac{1}{2}\rho v_1^2 + P_1 + \rho g h_1 = \frac{1}{2}\rho v_2^2 + P_2 + \rho g h_2 \qquad (1.70)$$

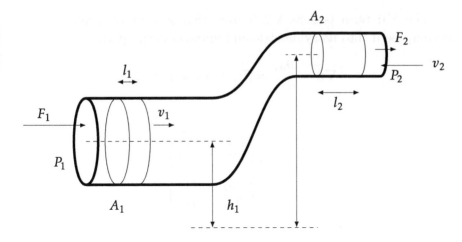

Figure 1.3 Pipe assembly Diagram for Bernoulli Equation

Replacing our v_1 with q/A_1 and v_2 with q/A_2 in Eq. (1.70), we have that

$$\frac{1}{2}\rho\left(\frac{q}{A_1}\right)^2 + P_1 + \rho g h_1 = \frac{1}{2}\rho\left(\frac{q}{A_2}\right)^2 + P_2 + \rho g h_2 \qquad (1.71)$$

which gives upon division by ρ,

$$\left(\frac{1}{A_1^2} - \frac{1}{A_2^2}\right)q^2 = \frac{2(P_2 - P_1)}{\rho} + g(h_2 - h_1)$$

Then we obtain

$$q = \sqrt{\left(\frac{2(P_2 - P_1)}{\rho} + g(h_2 - h_1)\right)\left(\frac{1}{A_1^2} - \frac{1}{A_2^2}\right)^{-1}} \qquad (1.72)$$

1.6.2 The Case of Linear Flows

Considering a point $\mathbf{x}_0 \in \mathbb{R}^3$. Near the point $\mathbf{x}_0$, the flow is made up of three parts: uniform flow, pure strain and pure rotation.

Consider the linear Taylor expansion of $\mathbf{u}$ about $\mathbf{x}_0$,

$$\mathbf{u}(\mathbf{x}) = \mathbf{u}(\mathbf{x}_0) + (\mathbf{x} - \mathbf{x}_0) \cdot \nabla \mathbf{u}(\mathbf{x}_0) + \cdots$$
$$= \mathbf{u}_0 + \mathbf{r} \cdot \nabla \mathbf{u}_0,$$

with $\mathbf{r} = \mathbf{x} - \mathbf{x}_0$ and $\mathbf{u}_0 = \mathbf{u}(\mathbf{x}_0)$.

The $\nabla \mathbf{u}$ term is a rank-2 tensor, that is a matrix, and we can decompose it into its symmetric and antisymmetric parts:

$$\nabla \mathbf{u} = \frac{\partial u_i}{\partial x_j} = E_{ij} + A_{ij} = E + A,$$

where

$$E_{ij} = \frac{1}{2}\left(\frac{\partial u_i}{\partial x_j} + \frac{\partial u_j}{\partial x_i}\right),$$

$$A_{ij} = \frac{1}{2}\left(\frac{\partial u_i}{\partial x_j} - \frac{\partial u_j}{\partial x_i}\right),$$

Since

$$\boldsymbol{\omega} = \nabla \times \mathbf{u}.$$

Then,

$$\boldsymbol{\omega} \times \mathbf{r} = (\nabla \times \mathbf{u}) \times \mathbf{r} = r_j\left(\frac{\partial u_i}{\partial x_j} - \frac{\partial u_j}{\partial x_i}\right) = 2A_{ij}r_j.$$

So we have,

$$\mathbf{u} = \mathbf{u}_0 + E\mathbf{r} + \frac{1}{2}\boldsymbol{\omega} \times \mathbf{r}. \tag{1.73}$$

The first component is uniform flow; the second is the strain field; and the last is the rotation component.

Since we have an incompressible fluid, we have $\nabla \cdot \mathbf{u} = 0$. So E has zero trace,

1.6.3 Vorticity Equation

The vorticity equation of fluid dynamics describes the evolution of the vorticity ω of a particle of a fluid as it moves with its flow; that is, the local rotation of the fluid. The equation is valid in the absence of any concentrated torques and line forces for a compressible, Newtonian fluid. In the case of incompressible flow (i.e., low Mach number) and isotropic fluids, with conservative body forces, the equation becomes the vorticity equation. Diffusion of vorticity is in an analogous form to the heat equation. Vorticity is often generated at rigid boundaries. Circulation, which is a scalar integral quantity, is a macroscopic measure of rotation for a finite area of the fluid, whereas vorticity is a vector field that gives a microscopic measure of the rotation at any point in the fluid.

We consider the Navier-Stokes equation for a viscous fluid,

$$\rho \left(\frac{\partial \mathbf{u}}{\partial t} + \mathbf{u} \cdot \nabla \mathbf{u} \right) = -\nabla p - \nabla \phi + \mu \nabla^2 \mathbf{u}$$

We use the vector identity

$$\mathbf{u} \cdot \nabla \mathbf{u} = \frac{1}{2} \nabla |\mathbf{u}^2| - \mathbf{u} \times \omega,$$

and take the curl to obtain

$$\frac{\partial \omega}{\partial t} - \nabla \times (\mathbf{u} \times \omega) = \nu \nabla^2 \omega,$$

where the curl of the gradient term vanishes. Using the following identity,

$$\nabla \times (\mathbf{u} \times \omega) = (\nabla \cdot \omega)\mathbf{u} + \omega \cdot \nabla \mathbf{u} - (\nabla \cdot \mathbf{u})\omega - \mathbf{u} \cdot \nabla \omega.$$

and since the divergence of the curl vanishes, and $\nabla \cdot \mathbf{u} = 0$ by incompressibility, we get,

$$\frac{\partial \omega}{\partial t} + \mathbf{u} \cdot \nabla \omega - \omega \cdot \nabla \mathbf{u} = \nu \nabla^2 \omega.$$

Now using the definition of the Material derivative, and rearranging terms we obtain,

Proposition 1.7 *Vorticity equation:*

$$\frac{D\omega}{Dt} = \omega \cdot \nabla \mathbf{u} + \nu \nabla^2 \omega. \tag{1.74}$$

Here, the rate of change of vorticity of a fluid particle is caused by $\omega \cdot \nabla \mathbf{u}$, the amplification by stretching or twisting and $\nu \nabla^2 \omega$ (dissipation of vorticity by viscosity). The second term also allows for generation of vorticity at boundaries using the no-slip condition. For an inviscid fluid, we have,

$$\frac{D\omega}{Dt} = \omega \cdot \nabla \mathbf{u}.$$

Taking the dot product with $\boldsymbol{\omega}$, we get

$$\frac{D}{Dt}\left(\frac{1}{2}|\boldsymbol{\omega}|^2\right) = \boldsymbol{\omega}\cdot\nabla u\cdot\boldsymbol{\omega} \tag{1.75}$$

$$= \boldsymbol{\omega}(E+A)\boldsymbol{\omega} \tag{1.76}$$

$$= \omega_i(E_{ij}+A_{ij})\omega_j. \tag{1.77}$$

Since $\omega_i\omega_j$ is symmetric, while A_{ij} is antisymmetric, we are left with,

$$\frac{D}{Dt}\left(\frac{1}{2}|\boldsymbol{\omega}|^2\right) = E_1\omega_1^2 + E_2\omega_2^2 + E_3\omega_3^2. \tag{1.78}$$

If we assume $E_1 > 0$ (since the E_i's sum to 0), let $E_2, E_3 < 0$. So the flow is stretched in the $\mathbf{e}_1$ direction and compressed radially. In the direction of stretching, we have,

$$\frac{D}{Dt}\left(\frac{1}{2}\omega_1^2\right) = E_1\omega_1^2.$$

So the vorticity grows exponentially and this is exactly vorticity amplification by stretching. We have so far assumed the density is constant. If the fluid has a non-uniform density $\rho(\mathbf{x})$, then

$$\frac{D\boldsymbol{\omega}}{Dt} = \boldsymbol{\omega}\cdot\nabla\mathbf{u} + \frac{1}{\rho^2}\nabla\rho\times\nabla p. \tag{1.79}$$

If ρ is constant, then $\frac{D\boldsymbol{\omega}}{Dt} = \boldsymbol{\omega}\cdot\nabla\mathbf{u} = 0$. Thus, the velocity can be described by the gradient of a potential function which satisfies Laplace's equation.

EXERCISES

1.1 Write down the Navier-Stokes equations in cylindrical polar coordinates. Hence, find an exact solution for the shearing azimuthal flow (Taylor-Couette flow) induced between two concentric cylinders of radius r_1 and r_2 when they rotate with angular velocities Ω_1 and Ω_2, respectively. Find an approximation to the Taylor-Couette solution when the gap width between the cylinders is small (in a sense you should define, i.e. small compared to what?) and show that this approximate solution is the same as the exact solution for Couette flow with suitably defined parameters.

Find also an exact solution for the shearing flow (Couette flow) induced by the motion of two parallel planes a distance d apart, say at $z = 0$ and $z = d$, when these move at speeds v_1 and v_2 in the x–direction. (Assume there is no pressure gradient.)

1.2 Starting from the Navier-Stokes equations in the form

$$(\mathbf{u} \cdot \nabla)\mathbf{u} = -\frac{1}{\rho}\nabla p + \nu\nabla^2\mathbf{u}, \quad \nabla.\mathbf{u} = 0,$$

which describe the steady flow of a stream with speed U_∞ past a two-dimensional body of a typical dimension L, derive Prandtl's dimensionless boundary-layer equations

$$u\frac{\partial u}{\partial x} + v\frac{\partial u}{\partial y} = -\frac{dp}{dx} + \frac{\partial^2 u}{\partial y^2}, \quad \frac{\partial u}{\partial x} + \frac{\partial v}{\partial y} = 0$$

in suitable dimensionless boundary layer variables which you should define. Show that $u = v = 0$ for $y = 0$ and $u \to U(x)$ as $y \to \infty$, where $p + \frac{1}{2}U^2 = \text{const}$.

1.3 Show that the scaling $f = \alpha F(\xi)$, $\eta = \xi/\alpha$ leaves the Blasius equation

$$\frac{d^3 f}{d\eta^3} + \frac{1}{2}f\frac{d^2 f}{d\eta^2} = 0$$

unchanged but changes the boundary condition at ∞. Suppose that the solution of the initial-value problem

$$F''' + \frac{1}{2}FF'' = 0$$

with $F(0) = F'(0) = 0$, $F''(0) = 1$ is calculated and it is found that $F'(\infty) = C \ (\simeq (0.332)^{-2/3})$. How can f be found in terms of F and what is the value and physical significance of $f''(0)$?

1.4 Use Karman integral equation to show $\delta = \sqrt{\frac{12\nu x}{U}}$ for the flow profile described as follows:

$$u = \begin{cases} \frac{U}{\delta}y & \text{for } 0 \le y \le \delta \\ U & \text{for } y \ge \delta \end{cases}$$

1.5 Prove that the 3D incompressible Navier-Stokes equations are scale invariant.

1.6 An incompressible fluid of small kinematic viscosity ν flows along a plane wall $y = 0$ with mainstream velocity $U(x)$. Explain how to obtain the two-dimensional Prandtl boundary layer equations

$$u_x \frac{\partial u_x}{\partial x} + u_y \frac{\partial u_x}{\partial y} = \frac{\partial^2 u_x}{\partial y^2} + U\frac{dU}{dx},$$

$$\frac{\partial u_x}{\partial x} + \frac{\partial u_y}{\partial y} = 0.$$

Let ψ denote the stream function. Show that a similar solution of the form

$$\psi = F(x)f(\eta), \text{ with } \eta = \frac{y}{g(x)},$$

is possible if

$$\text{either } U(x) \propto (x - x_0)^m \text{ or } U(x) \propto e^{cx},$$

where x_0, m and c are constants. In the first of these two cases, for what value of m is the boundary layer thickness independent of x?

1.7 Solve the modified diffusion equation in one spatial dimension,

$$\frac{\partial u}{\partial t} = D\frac{\partial^2 u}{\partial x^2} - \alpha u,$$

where x represents the spatial coordinate($-\infty < x < \infty$), and t represents time. At $t = 0$, the initial condition is an impulse at the origin $u(x, t = 0) = \delta(x)$. Use Fourier transforms s to solve the problem.

Hint: If $\tilde{f}(k) = \exp\left(k^2/4a\right)/\sqrt{2a}$, then it's inverse transform is $f(x) = \exp\left(-ax^2\right)$.

1.8 Air at 105 kPa and 37°C flows upward through a 6.0-cm-diameter inclined pipe at a rate of 65 L/s. The pipe diameter is then reduced to 4 cm through a reducer. See Figure 1.4 The pressure change across the reducer is measured by a water density (is 1,000 kg/m³ manometer. The elevation difference between the

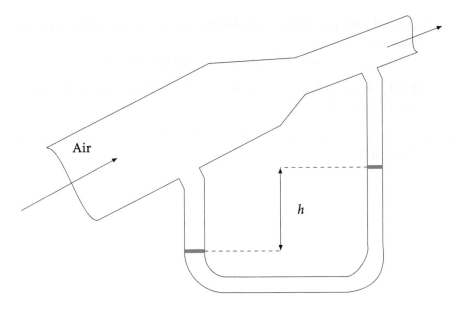

Air

h

Figure 1.4 Application of Bernoulli Principle

two points on the pipe where the two arms of the manometer are attached is 0.20 m. Determine the differential height, h between the fluid levels of the two arms of the manometer. Take that the gas constant of air is $R = 0.287$ kPa· m^3/kg· K.

1.9 An incompressible velocity field is given by

$$u = a(x^2 - y^2)$$

where v is unknown and $w = b$, with $a, b \in \mathbb{R}$. What is the necessary form of the velocity component v?

1.10 Using the velocity field of exercise 9, where we assume $b = 0$ and

$$u = a(x^2 - y^2)$$

$$v = -2axy$$

$$w = 0$$

determine the conditions for a solution of the Navier Stokes momentum equation to exist when the gravity force $f_x = 0$, $f_y = 0$ and $f_z = -g$, where g is the acceleration due to gravity.

1.11 If a velocity potential exists for the velocity field of Exercise 10, that is,

$$u = a(x^2 - y^2); \quad v = -2axy; \quad w = 0.$$

Find this potential and plot its graph. Do stagnation points exist there? Explain.

1.12 Derive the 3D incompressible Navier-Stokes equations in cylindrical coordinates.

## CHAPTER 2

Geometric Algebra

Terry E. Moschandreou and Khoa Nguyen

TVDSB & Western University

2.1 THE GEOMETRIC PRODUCT

The *geometric product* of two vectors $\mathbf{a}$, $\mathbf{b} \in V$ is denoted $\mathbf{ab}$ for some real vector space V.

We can separate the product into two parts, a *symmetric* part and an *anti-symmetric* part.

$$\mathbf{ab} = \frac{1}{2}(\mathbf{ab} + \mathbf{ba}) + \frac{1}{2}(\mathbf{ab} - \mathbf{ba}) \tag{2.1}$$

The **symmetric product** is

$$\frac{1}{2}(\mathbf{ab} + \mathbf{ba}) = \frac{1}{2}(\mathbf{ba} + \mathbf{ab}), \tag{2.2}$$

and the **anti-symmetric product** is

$$\frac{1}{2}(\mathbf{ab} - \mathbf{ba}) = -\frac{1}{2}(\mathbf{ba} - \mathbf{ab}). \tag{2.3}$$

In particular, we have commutativity,

$$\mathbf{ab} = \mathbf{ba} \iff \frac{1}{2}(\mathbf{ab} - \mathbf{ba}) = 0 \tag{2.4}$$

and anti-commutativity,

$$\mathbf{ab} = -\mathbf{ba} \iff \frac{1}{2}(\mathbf{ab} + \mathbf{ba}) = 0. \tag{2.5}$$

In general, the product of $\mathbf{a}$ and $\mathbf{b}$ is non-commutative.

DOI: 10.1201/9781003452256-2

2.1.1 Collinearity and Orthogonality

Note that if **a** and **b** are linearly dependent, then they commute with each other:

$$\mathbf{a} = \alpha\mathbf{b} \implies \mathbf{ab} = \mathbf{ba}, \text{ for } \alpha \in \mathbb{R}.$$

The converse is not always true. For example, consider $\cos(\theta)$ and $\sin(\theta)$. There is a commutative relationship here, $\cos(\theta)\sin(\theta) = \sin(\theta)\cos(\theta))$, but they are not linearly dependent.

Vectors a, b that satisfy

$$\mathbf{a} = \alpha\mathbf{b} \iff \mathbf{ab} = \mathbf{ba}, \text{ for } \alpha \in \mathbb{R},$$

show an equivalence between commutativity and geometric properties.

The following axiom shows an equivalence between commutativity and collinearity, anti-commutativity and orthogonality:

Given vectors **a** and **b**, there exists $\alpha \in \mathbb{R}$ such that

$$\frac{1}{2}(\mathbf{ab} + \mathbf{ba}) = \alpha\mathbf{b}^2.$$

The product absorbs the part of **a** that is a linearly dependent part on **b**.

It follows that

$$\mathbf{a} = \mathbf{a}_{\parallel} + \mathbf{a}_{\perp}$$

for $\mathbf{a}_{\parallel} = \alpha\mathbf{b}$, is a canonical decomposition of **a** with respect to **b** satisfying

$$\frac{1}{2}(\mathbf{ab} + \mathbf{ba}) = \mathbf{a}_{\parallel}\mathbf{b} = \mathbf{ba}_{\parallel}$$

and

$$\frac{1}{2}(\mathbf{ab} - \mathbf{ba}) = \mathbf{a}_{\perp}\mathbf{b} = -\mathbf{ba}_{\perp}.$$

We say that the vectors **a** and **b** are collinear (or parallel) if and only if

$$\mathbf{ab} = \mathbf{ba} \tag{2.6}$$

and **orthogonal** (or perpendicular) if and only if

$$\mathbf{ab} = -\mathbf{ba}. \tag{2.7}$$

Note that, by symmetry,

$$\frac{1}{2}(\mathbf{ab} + \mathbf{ba}) = \alpha\mathbf{b}^2 = \alpha_0\mathbf{a}^2 = \frac{1}{2}(\mathbf{ba} + \mathbf{ab}),$$

for some $\alpha_0 \in \mathbb{R}$.

The geometric algebra G consists of all things that can be generated by products and sums of products of vectors in V.

The product of k orthogonal vectors $A_k = a_1 a_2 \ldots a_k$ is called a **k-vector** (or sometimes a **k-blade**). These determine subspaces of V. The grade of A_k refers to the number of linearly independent vectors required. The grade has the same meaning as dimension, with the additional specificity of 'dimension of a multivector' in the manner defined by geometric algebra. Things in one grade are not in another. Now, $a_\perp b$ is a 2-vector, or a bivector, and it determines a two-dimensional subspace of V, spanned by $a_\perp$ and $\mathbf{b}$. Also, every i-blade is linearly independent of every j-blade, for $i \neq j$.

An arbitrary element $M = \sum_{i=0}^{n} A_i \in G$ is called a **multivector** and consists of a sum of arbitrary k-vectors. For example, the sum of a scalar and a bivector is a multivector. We will look at examples of multivectors that produce rotations, and other transformations, of k-vectors.

Also we can determine the grade of a^2 and b^2. If the geometric product is distributive then we can determine their grade by considering the square of a bivector:

$$(\mathbf{a}_\perp \mathbf{b})^2 = \mathbf{a}_\perp \mathbf{b} \mathbf{a}_\perp \mathbf{b} = -\mathbf{a}_\perp \mathbf{a}_\perp \mathbf{b}^2 = -\mathbf{a}_\perp \mathbf{b}^2 \mathbf{a}_\perp. \tag{2.8}$$

In particular,

$$\mathbf{a}_\perp \mathbf{a}_\perp \mathbf{b}^2 = \mathbf{a}_\perp \mathbf{b}^2 \mathbf{a}_\perp$$

and

$$\mathbf{a}_\| \mathbf{a}_\| \mathbf{b}^2 = \mathbf{a}_\| \mathbf{b}^2 \mathbf{a}_\|$$

show that

$$\mathbf{a} \mathbf{a} \mathbf{b}^2 = \mathbf{a} \mathbf{b}^2 \mathbf{a} \tag{2.9}$$

for arbitrary vectors $\mathbf{a}$ and $\mathbf{b}$. This means that b^2 commutes with all vectors. Elements of the algebra that necessarily commute with all vectors are said to have grade 0 and are called scalars of G, denoted $G_0 \subseteq G$. **Geometric algebra** is precisely the algebra for which $G_0 = \mathbb{R}$.

Since $\mathbf{a}_\| = \alpha \mathbf{b}$ for some scalar α, the symmetric product of any two vectors

$$\frac{1}{2}(\mathbf{ab} + \mathbf{ba}) = \mathbf{a}_\| \mathbf{b} = \alpha \mathbf{b}^2 \tag{2.10}$$

is of grade 0 as well.

2.1.2 The Dot and Wedge Products

For two vectors $\mathbf{a}, \mathbf{b} \in V$, define

$$\mathbf{a} \cdot \mathbf{b} = \frac{1}{2}(\mathbf{ab} + \mathbf{ba}). \tag{2.11}$$

Note that $\mathbf{a} \cdot \mathbf{b} : V \times V \to G_0$ is a map that satisfies the following, for vectors $\mathbf{a}, \mathbf{b}, \mathbf{c} \in V$ and $\alpha \in \mathbb{R}$:

1.
$$\mathbf{a} \cdot \mathbf{b} = \mathbf{b} \cdot \text{asymmetry ()}$$

2.
$$\alpha(\mathbf{a} \cdot \mathbf{b}) = (\alpha\mathbf{a}) \cdot \mathbf{b}\text{linearity ()}$$

3.
$$(\mathbf{a} + \mathbf{b}) \cdot \mathbf{c} = \mathbf{a} \cdot \mathbf{c} + \mathbf{b} \cdot \mathbf{c}\text{distributivity ()}$$

4.
$$\mathbf{a} \cdot \mathbf{b} \in G_0 = \mathbb{R}.$$

Thus, $\mathbf{a} \cdot \mathbf{b}$ is an inner product, called the dot product. This gives us a notion of the length of a vector given as, $\mathbf{a} \cdot \mathbf{a} = \mathbf{a}^2 \in \mathbb{R}$ which is referred to as the square of the magnitude of $\mathbf{a}$.

If $\mathbf{a}^2 \neq 0$, then $\mathbf{a}$ has an inverse

$$\mathbf{a}^{-1} = \mathbf{a}/\mathbf{a}^2 \tag{2.12}$$

satisfying

$$\mathbf{aa}^{-1} = 1.$$

Furthermore, then $\mathbf{a}$ has an associated unit vector

$$\hat{\mathbf{a}} = \mathbf{a}/ \mid \mathbf{a} \mid$$

where $\mid \mathbf{a} \mid = \sqrt{\pm\mathbf{a}^2}$ (allowing $\mathbf{a}^2 < 0$), such that

$$\mathbf{a} = \mid \mathbf{a} \mid \hat{\mathbf{a}}.$$

Next, for two vectors $\mathbf{a}, \mathbf{b} \in V$,

$$\mathbf{a} \wedge \mathbf{b} = \frac{1}{2}(\mathbf{ab} - \mathbf{ba}) \tag{2.13}$$

is an exterior product, which generates higher graded objects, called the wedge product. The exterior product represents the oriented area defined by two vectors **a** and **b** or represents an oriented area in the plane defined by those vectors, also known as a bivector. Now the precise shape does not matter. For example, the bivectors represented by $(1,0) \wedge (0,3) = (3,0) \wedge (0,1)$ are equal because they have the same orientation that is counter-clockwise and the same area (3). Next an important property is that the exterior product is anticommutative.

$$\mathbf{a} \wedge \mathbf{b} = -\mathbf{b} \wedge \mathbf{a}.$$

Using the two previous vectors, note that the order used in the exterior product makes the bivectors either clockwise or counter-clockwise. The exterior product is associative, scalar associative, left distributive, right distributive, anti-symmetric and zero for parallel vectors. ($a \parallel b$ implies $a \wedge b = 0$).

Considering basis vectors, any vector can be written as a linear combination of basis vectors as,

$$\mathbf{a} = a_1\mathbf{e_1} + a_2\mathbf{e_2}$$

$$\mathbf{b} = b_1\mathbf{e_1} + b_2\mathbf{e_2}$$

Taking their exterior product, we get,

$$
\begin{aligned}
\mathbf{a} \wedge \mathbf{b} &= (a_1\mathbf{e_1} + a_2\mathbf{e_2}) \wedge (b_1\mathbf{e_1} + b_2\mathbf{e_2}) \\
&= a_1b_1\mathbf{e_1} \wedge \mathbf{e_1} + a_1b_2\mathbf{e_1} \wedge \mathbf{e_2} + a_2b_1\mathbf{e_2} \wedge \mathbf{e_1} + a_2b_2\mathbf{e_2} \wedge \mathbf{e_2} \\
&= 0 + a_1b_2\mathbf{e_1} \wedge \mathbf{e_2} + a_2b_1\mathbf{e_2} \wedge \mathbf{e_1} + 0 \\
&= a_1b_2\mathbf{e_1} \wedge \mathbf{e_2} - a_2b_1\mathbf{e_1} \wedge \mathbf{e_2} \\
&= (a_1b_2 - a_2b_1)\mathbf{e_1} \wedge \mathbf{e_2}
\end{aligned}
$$

The exterior product of any two vectors is here expressed as just a scalar multiple of $\mathbf{e_1} \wedge \mathbf{e_2}$. Under the geometric product, we get,

$$
\begin{aligned}
\mathbf{ab} &= (a_1\mathbf{e_1} + a_2\mathbf{e_2})(b_1\mathbf{e_1} + b_2\mathbf{e_2}) \\
&= a_1b_1\mathbf{e_1}\mathbf{e_1} + a_1b_2\mathbf{e_1}\mathbf{e_2} + a_2b_1\mathbf{e_2}\mathbf{e_1} + a_2b_2\mathbf{e_2}\mathbf{e_2} \\
&= a_1b_1 + a_1b_2\mathbf{e_1}\mathbf{e_2} + a_2b_1\mathbf{e_2}\mathbf{e_1} + a_2b_2 \\
&= a_1b_1 + a_2b_2 + a_1b_2\mathbf{e_{12}} - a_2b_1\mathbf{e_{12}} \\
&= (a_1b_1 + a_2b_2) + (a_1b_2 - a_2b_1)\mathbf{e_{12}}
\end{aligned}
$$

So here we have a scalar and a bivector.

In summary, the geometric product of two vectors takes the form

$$\mathbf{ab} = \mathbf{a} \cdot \mathbf{b} + \mathbf{a} \wedge \mathbf{b}, \tag{2.14}$$

where $\mathbf{a} \cdot \mathbf{b}$ is a scalar and $\mathbf{a} \wedge \mathbf{b}$ is a bivector.

2.1.3 Geometric Algebra

A geometric algebra G is a vector space equipped with the geometric product, which satisfies the following properties. For $\mathbf{a}, \mathbf{b}, \mathbf{c} \in V$

1.

$$\mathbf{a}(\mathbf{bc}) = (\mathbf{ab})\mathbf{c} = \mathbf{abc} \text{ (associativity)},$$

2.

$$\mathbf{a}(\mathbf{b} + \mathbf{c}) = \mathbf{ab} + \mathbf{ac} \text{ and } (\mathbf{a} + \mathbf{b})\mathbf{c} = \mathbf{ac} + \mathbf{bc} \text{ (distributivity)},$$

3.

there is an $\alpha \in \mathbb{R}$ such that $\dfrac{1}{2}(\mathbf{ab} + \mathbf{ba}) = \alpha \mathbf{b}^2$

4.

$$G_0 = \mathbb{R}$$

Now (3) and (4) are replaced with the following properties. *(contraction property)*

$$\mathbf{a}^2 = g(\mathbf{a}, \mathbf{a}) \in \mathbb{R} \text{ (contraction)},$$

for some inner product g.

2.2 ROTATIONS

We will compare the problem of rotating one vector onto another by use of the matrix product and by use of geometric product.

2.2.1 The Matrix Product

In vector algebra, two-dimensional vectors can be written as,

$$\vec{x} = x_1 \hat{x} + x_2 \hat{y}, \tag{2.15}$$

and are required to be converted to matrix form,

$$\vec{x} = (x_1, x_2),$$

where each column entry denotes a coordinate relative to some implicit coordinate system, so we can use matrix rotation operators.

Given vectors $\vec{a} = (a_1, a_2)$ and $\vec{b} = (b_1, b_2)$, we produce a matrix R that rotates the row vector $\vec{a}$ into $\vec{b}$, i.e.

$$\vec{a}R = \vec{b}. \tag{2.16}$$

Two-dimensional rotations are given by

$$r(\theta) = \begin{bmatrix} \cos(\theta) & \sin(\theta) \\ -\sin(\theta) & \cos(\theta) \end{bmatrix}, \tag{2.17}$$

If $\cos(\theta) = \frac{\vec{a} \cdot \vec{b}}{|\vec{a}||\vec{b}|}$ and $\sin(\theta) = \pm \frac{|\vec{a} \times \vec{b}|}{|\vec{a}||\vec{b}|}$, depending if $\vec{b}$ is θ clockwise or counter-clockwise from $\vec{a}$, then

$$R = \frac{|\vec{b}|}{|\vec{a}|} r$$

is the required form for the rotation. Let $\mathbf{e}$ and $\mathbf{f}$ be an orthonormal basis of a plane. We interpret $\mathbf{i}$ to mean the plane generated by $\mathbf{e}$ and $\mathbf{f}$. With vector algebra we represent planes with normal vectors which only makes sense in 3D. However with GA, we represent a plane by multiplying two orthonormal vectors that generates it and this is true in any dimension. Now the geometric algebra interpretation of $e^{i\theta}$ is a rotation by θ in the plane $\mathbf{i}$, rather than a particular point on a unit circle measured from some zero point.

2.2.2 The Geometric Product

Now we consider the same problem using the geometric product instead of matrix multiplication. We will find a multivector M that rotates $\mathbf{a}$ into $\mathbf{b}$, i.e.

$$\mathbf{a}M = \mathbf{b}. \tag{2.18}$$

If $\mathbf{a}^2 \neq 0$, then $\mathbf{a}$ has an inverse

$$\mathbf{a}^{-1} = \mathbf{a}/\mathbf{a}^2,$$

from which it follows that,

$$M = \mathbf{a}^{-1}\mathbf{b} = \mathbf{a}\mathbf{b}/\mathbf{a}^2.$$

Here this describes vectors related by circular rotations and by hyperbolic rotations.

In a canonical basis, assuming $\mathbf{a}^2 > 0$ and $\mathbf{b}^2 > 0$.

$$\mathbf{a}\mathbf{b} = \mathbf{a} \cdot \mathbf{b} + \mathbf{a} \wedge \mathbf{b} \tag{2.19}$$

$$= \mathbf{a}_{\parallel}\mathbf{b} + \mathbf{a}_{\perp}\mathbf{b} \tag{2.20}$$

$$= (|\,\mathbf{a}\,|\cos(\theta)\hat{\mathbf{b}})\mathbf{b} + (|\,\mathbf{a}\,|\sin(\theta)\hat{\mathbf{a}}_{\perp})\mathbf{b} \tag{2.21}$$

$$= |\,\mathbf{a}\,||\,\mathbf{b}\,|\,(\cos(\theta) + \sin(\theta)\hat{\mathbf{a}}_{\perp}\hat{\mathbf{b}}) \tag{2.22}$$

$$= |\,\mathbf{a}\,||\,\mathbf{b}\,|\,e^{i\theta}, \tag{2.23}$$

where $i = \hat{\mathbf{a}}_{\perp}\hat{\mathbf{b}}$ is a unit bivector and $i^2 = -1$. In addition $\cos(\theta) = \frac{\mathbf{a}\cdot\mathbf{b}}{|\mathbf{a}||\mathbf{b}}$ and $\sin(\theta) = \frac{|\mathbf{a}\wedge\mathbf{b}|}{|\mathbf{a}||\mathbf{b}}$ (which are dependent on $\mathbf{a}^2 > 0$ and $\mathbf{b}^2 > 0$) and determines the projection,

$$\mathbf{a}_{\parallel} = |\,\mathbf{a}\,|\cos(\theta)\hat{\mathbf{b}} \tag{2.24}$$

and rejection,

$$\mathbf{a}_{\perp} = |\,\mathbf{a}\,|\sin(\theta)\hat{\mathbf{a}}_{\perp}. \tag{2.25}$$

M is now,

$$M = \frac{|\,\mathbf{b}\,|}{|\,\mathbf{a}\,|}e^{i\theta},$$

and we have that,

$$\mathbf{a}M = |\,\mathbf{a}\,|\,\hat{\mathbf{a}}M = |\,\mathbf{b}\,|\,\hat{\mathbf{a}}e^{i\theta} = |\,\mathbf{b}\,|\,\hat{\mathbf{b}} = \mathbf{b}.$$

The $e^{i\theta}$ part rotates the unit vector $\hat{\mathbf{a}}$ into $\hat{\mathbf{b}}$.

EXERCISES

2.1 In 2-d two multivectors, A and B are given by,

$$A = a_0 + a_1 e_1 + a_2 e_2 + a_3 e_1 \wedge e_2.$$

$$B = b_0 + b_1 e_1 + b_2 e_2 + b_3 e_1 \wedge e_2$$

where $\{e_1, e_2\}$ are a pair of orthonormal basis vectors. The product of these is,

$$AB = p_0 + p_1 e_1 + p_2 e_2 + p_3 e_1 \wedge e_2.$$

Find formulae for p_0, p_1, p_2 and p_3 and verify that $\langle AB \rangle = \langle BA \rangle$.

2.2 Expand the bivector $a \wedge b$ in terms of geometric products. In particular, prove that it anticommutes with both **a** and **b**, but commutes with all vectors outside the plane.

2.3 Prove that the bivector $a \wedge b$ has magnitude $|a| \, |b| \sin(\theta)$.

2.4 Prove equivalence for the following of the vector cross product in 3-d

$$a \times b = -Ia \wedge b = b \cdot (Ia) = -a \cdot (Ib).$$

Interpret these forms geometrically. Finally prove that,

$$a \times (b \times c) = -a \cdot (b \wedge c) = -(a \cdot bc - a \cdot cb).$$

and the following in terms of the inverse of I,

$$a \cdot (b \times c) = [a, b, c] = a \wedge b \wedge cI^{-1}.$$

2.5 Expand $a \wedge (b \wedge c)$ in terms of geometric products. Determine if the result is antisymmetric on **a** and **b**. Explain the following and give a reason for your answer,

$$a \wedge b \wedge c = \frac{1}{2}(abc - cba).$$

2.6 In 4-D prove that $a \wedge b + c \wedge d$ is a blade iff,

$$a \wedge b \wedge c \wedge d = 0$$

2.7 Using the axioms of geometric algebra prove that the exterior product is associative.

## CHAPTER 3

Compressible Navier–Stokes Equations

Terry E. Moschandreou

TVDSB

3.1 INTRODUCTION

The 3D compressible cylindrical unsteady Navier–Stokes equations are written in an expanded form, for each, u_r, u_θ and u_z:

$$
\frac{\partial u_r}{\partial t} + u_r \frac{\partial u_r}{\partial r} + \frac{u_\theta}{r} \frac{\partial u_r}{\partial \theta} - \frac{u_\theta^2}{r} + u_z \frac{\partial u_r}{\partial z}
$$

$$
- \frac{\mu}{\rho} \left(-\frac{u_r}{r^2} + \frac{\partial^2 u_r}{\partial r^2} + \frac{1}{r} \frac{\partial u_r}{\partial r} + \frac{1}{r^2} \frac{\partial^2 u_r}{\partial \theta^2} - \frac{2}{r^2} \frac{\partial u_\theta}{\partial \theta} + \frac{\partial^2 u_r}{\partial z^2} \right)
$$

$$
- \frac{\mu}{3\rho} \frac{\partial}{\partial r} \left(\frac{\partial u_r}{\partial r} + \frac{1}{r} \frac{\partial u_r}{\partial \theta} + \frac{\partial u_r}{\partial z} \right) + \frac{1}{\rho} \frac{\partial p}{\partial r} - F g_r = 0 \qquad (3.1)
$$

$$
\frac{\partial u_\theta}{\partial t} + u_r \frac{\partial u_\theta}{\partial r} + \frac{u_\theta}{r} \frac{\partial u_\theta}{\partial \theta} + \frac{1}{r} u_r u_\theta + u_z \frac{\partial u_\theta}{\partial z}
$$

$$
- \frac{\mu}{\rho} \left(-\frac{u_\theta}{r^2} + \frac{\partial^2 u_\theta}{\partial r^2} + \frac{1}{r} \frac{\partial u_\theta}{\partial r} + \frac{1}{r^2} \frac{\partial^2 u_\theta}{\partial \theta^2} + \frac{2}{r^2} \frac{\partial u_r}{\partial \theta} + \frac{\partial^2 u_\theta}{\partial z^2} \right)
$$

$$
- \frac{1}{r} \frac{\mu}{3\rho} \frac{\partial}{\partial \theta} \left(\frac{\partial u_\theta}{\partial r} + \frac{1}{r} \frac{\partial u_\theta}{\partial \theta} + \frac{\partial u_\theta}{\partial z} \right) + \frac{1}{\rho r} \frac{\partial p}{\partial \theta} - F g_\theta = 0 \qquad (3.2)
$$

DOI: 10.1201/9781003452256-3

$$\frac{\partial u_z}{\partial t} + u_r \frac{\partial u_z}{\partial r} + \frac{u_\theta}{r} \frac{\partial u_z}{\partial \theta} + u_z \frac{\partial u_z}{\partial z}$$

$$- \frac{\mu}{\rho} \left(\frac{\partial^2 u_z}{\partial r^2} + \frac{1}{r} \frac{\partial u_z}{\partial r} + \frac{1}{r^2} \frac{\partial^2 u_z}{\partial \theta^2} + \frac{\partial^2 u_z}{\partial z^2} \right)$$

$$- \frac{\mu}{3\rho} \frac{\partial}{\partial z} \left(\frac{\partial u_z}{\partial r} + \frac{1}{r} \frac{\partial u_z}{\partial \theta} + \frac{\partial u_z}{\partial z} \right) + \frac{1}{\rho} \frac{\partial p}{\partial z} - Fg_z = 0 \qquad (3.3)$$

where u_r is the radial component of velocity, u_θ is the azimuthal component and u_z is the velocity component in the direction along tube, ρ is density, μ is dynamic viscosity, Fg_r, Fg_θ, Fg_z are body forces on fluid. The total gravity force vector is expressed as $\vec{F_T} = (Fg_r, Fg_\theta, Fg_z)$. The following relationships between starred and non-starred dimensional quantities together with a non-dimensional quantity δ are used:

$$u_r = \frac{1}{\delta} u_r^* \qquad (3.4)$$

$$u_\theta = \frac{1}{\delta} u_\theta^* \qquad (3.5)$$

$$u_z = \frac{1}{\delta} u_z^* \qquad (3.6)$$

$$r = \delta r^* \qquad (3.7)$$

$$\theta = \theta^* \qquad (3.8)$$

$$z = \delta z^*, t = \delta^2 t^* \qquad (3.9)$$

In the above, the scale invariance of the Navier-Stokes equations is evident. A similar approach was adopted in [22] and [19] for the definition of δ, here the density is explicitly defined according to Theorem 3.1.

First we derive Bernoulli's equation for an adiabatic process for a fluid flow. Here we have that,

$$Pv^\gamma = \text{constant}$$

or

$$\frac{P}{\rho^\gamma} = \text{constant} = c_2$$

Hence

$$\int \frac{dP}{\rho} = \int \frac{dp}{(P/c_2)^{1/\gamma}} = (c_2)^{1/\gamma} \int \frac{1}{P^{1/\gamma}} dP = (c_2)^{1/\gamma} \int P^{-1/\gamma} dP$$

$$= (c_2)^{1/\gamma} \left[\frac{P^{-\frac{1}{\gamma}+1}}{\left(-\frac{1}{\gamma}+1\right)} \right] = \frac{(c_2)^{1/\gamma} P^{\left(\frac{\gamma-1}{\gamma}\right)}}{\left(\frac{\gamma-1}{\gamma}\right)} = \frac{\gamma}{\gamma-1} (c_2)^{1/\gamma} (P)^{\left(\frac{\gamma-1}{\gamma}\right)}$$

$$= \left(\frac{\gamma}{\gamma-1}\right) \left(\frac{P}{\rho^\gamma}\right)^{1/\gamma} (P)^{\left(\frac{\gamma-1}{\gamma}\right)}$$

$$= \left(\frac{\gamma}{\gamma-1}\right) \left(\frac{P^{1/\gamma}}{\rho^\gamma \times \frac{1}{\gamma}}\right) (P)^{\left(\frac{\gamma-1}{\gamma}\right)}$$

$$= \left(\frac{\gamma}{\gamma-1}\right) \frac{(P)^{\left(1/\gamma+\frac{\gamma-1}{\gamma}\right)}}{\rho} = \left(\frac{\gamma}{\gamma-1}\right) \frac{P}{\rho}$$

Substituting in Bernoulli's equation for $\int \frac{dP}{\rho}$, we obtain,

$$\left(\frac{\gamma}{\gamma-1}\right) \frac{\rho^\gamma}{\rho} + \frac{V^2}{2} + gz = \text{constant}$$

which is Bernoulli's equation for compressible flow undergoing adiabatic process. Here we have written $P/\rho = c_2\rho^\gamma/\rho$, using adiabatic definition again.

Theorem 3.1 *For compressible Navier–Stokes equations defined on a surface where the radial velocity u_r^* will vanish and axial velocity u_z^* is increasing linearly in the developing flow region in the tube, as $u_z^* = \beta z^*$, $\beta > 0$, the density can be expressed as,*

$$\rho = \rho_0 + \Phi(r, \theta, z, t)^{2/\gamma} F(r, z, t).$$

for arbitrary function F and $\gamma = 1.4$(ratio of specific heats), assuming adiabatic conditon,

$$\rho = (\rho - \rho_0)^\gamma \left(\eta - u_\theta(r, \theta, z, t)^2\right)^{-1}.$$

which is derived from Bernoulli equation above Theorem 3.1 for compressible flow with adiabatic assumption.

Also η is a constant. Moreover, the derivative of u_θ with respect to θ blows up in finite time if and only if ρ is in the form above and Φ and F are both separable, that is,

$$\Phi = F_1(r)F_2(\theta)F_3(z)F_4(t), \quad F = F_a(r)F_b(z)F_c(t),$$

where $F_4(t)$ and $F_c(t)$ are arbitrary functions of the time variable t.

Proof 3.1 *Substitute $\rho = (\rho - \rho_0)^\gamma \left(\eta - u_\theta(r,\theta,z,t)^2\right)^{-1}$ and the blowup form $b_2 = u_\theta = \frac{\phi}{A_1 t + B_1}$ into the left side of Eq. (3.30) using continuity equation, Eq. (3.24). Simplification using Maple software results in an equation where the blowup conditions is set as $A_1 = -B_1/t$ and the resulting pde has two factors of which one has a solution of the form $\rho = \rho_0 + \Phi(r,\theta,z,t)^{2/\gamma}F(r,z,t)$ and subsequently substitution of this form of ρ into the other factor of the PDE, gives a solution which is fully separable for Φ and F. This proves the one direction of the theorem. Furthermore, the function $\Phi = G_2 - \frac{2rB_1\theta}{t}$, for arbitrary function G, using the right-hand side of Eq. (3.30) implies that $G_2 = 0$. This allows us to determine F and its factors. The opposite direction of the theorem is the focus of Chapter 3, that is if the density is expressed as above then there is blowup in the first derivative of u_θ.*

$$\rho = \begin{cases} -i\rho_1 \sqrt{-\frac{\sin(\alpha\theta)}{\alpha\theta}} & \alpha\theta \in (-\pi, 0), \\ \rho_1 \sqrt{\frac{\sin(\alpha\theta)}{\alpha\theta}} & \alpha\theta \in (0, \pi) \end{cases}$$

where $i = \sqrt{-1}$, $\alpha > 0$, ρ_1 is a function derived below, and further below in this chapter a geometric proof is shown that the sign of the radial velocity u_r will change on a surface given by $z^* = \sqrt{(r^{*2} - \theta^{*2})}$. It can also be observed that the density as defined above is real valued on $(-\pi, \pi)$.

The density equation used for compressible flow is,

$$\delta = \frac{\frac{\partial \rho_1(r,\theta,z,t)}{\partial \theta} + z\frac{\partial \rho_1(r,\theta,z,t)}{\partial z} + \rho_{avg}}{r\left(\frac{\partial \rho_1(r,\theta,z,t)}{\partial r} + \frac{\rho_1(r,\theta,z,t)}{r}\right)}$$

with general solution,

$$\rho_1(r,\theta,z,t) = \frac{\rho_{avg}}{\delta} + \frac{F_\rho\left(\frac{\theta\delta + \ln(r)}{\delta}, zr^{-\frac{1}{\delta}}, t\right)}{r}$$

where ρ_{avg} is an averaged density over space and time in the flow.

A specific physical form of solution derived immediately from the general one is given as

$$\rho_1 = \frac{\rho_{avg}}{\delta} + \sum_i \xi_i \exp\left(-\left(\frac{\theta_i \delta + \ln(r)}{\delta}\right)^2\right) r^{\frac{1}{\delta}-1} \frac{1}{z} \exp(-ct)$$

Here at the inlet of the pipe at $z = 0$ we obtain a Dirac Delta type function and the density of vapor or gas is decreasing down the pipe due to frictional losses, hence pressure losses. The choice of weight coefficients ξ_i and variable values θ_i are so that we have high pressure at the center of the tube and low pressure near the wall. The values of $\delta \in (0, \infty)$. Here $\delta \to 0$ and $r^* \to \infty$ such that δr^* is finite and large. At the end of the chapter, for computational simplicity we consider $\theta_i = 0$ and look at the one term which is maximum in the series above.

Replacing Eqs. (2.11) and (2.12) in [22], we use new Eqs. (3.4)–(3.9) above, multiplying scale invariant Eqs. (3.1)–(3.3) by Cartesian unit vectors $\vec{e_{r*}} = (1,0,0)$, $2\vec{e_{\theta*}} = (0,2,0)$ and $\vec{k} = (0,0,1)$ respectively and adding modified equations for Eqs. (3.1)–(3.3) giving the following equations, for the resulting composite vector $\vec{L_1} = \frac{1}{\delta}u_{r*}^*\vec{e_{r*}} + \frac{2}{\delta}u_{\theta*}^*\vec{e_{\theta*}} + \frac{1}{\delta}u_{z*}^*\vec{k}$,

$$\delta^3\left(\frac{\partial \vec{L_1}}{\partial t} + \frac{u_{r*}^*}{\delta^2}\frac{\partial \vec{L_1}}{\partial r^*} + \frac{u_{\theta*}^*}{\delta^2 r^*}\frac{\partial \vec{L_1}}{\partial \theta^*} + \frac{1}{\delta^2}u_{z*}^*\frac{\partial \vec{L_1}}{\partial z^*} - \frac{1}{\delta^3 r^*}u_{\theta*}^{*2}\vec{e_{r*}}\right.$$

$$+ \frac{1}{\delta^3 r^*}u_{r*}^* u_{\theta*}^* \vec{e_{\theta*}} - \frac{\mu}{\rho}\left(-\delta^{-2}\frac{\vec{L_1}}{r^{*2}} + \delta^{-2}\frac{\partial^2 \vec{L_1}}{\partial r^{*2}} + \delta^{-2}\frac{1}{r^*}\frac{\partial \vec{L_1}}{\partial r^*}\right.$$

$$+ \delta^{-2}\frac{1}{r^{*2}}\frac{\partial^2 \vec{L_1}}{\partial \theta^{*2}} + \frac{2}{\delta^2 r^{*2}}\left(\frac{2}{\delta}\frac{\partial u_{r*}^*}{\partial \theta^*}\vec{e_{\theta*}} - \frac{1}{\delta}\frac{\partial u_{\theta*}^*}{\partial \theta^*}\vec{e_{r*}}\right) + \delta^{-2}\frac{\partial^2 \vec{L_1}}{\partial z^{*2}}\right)$$

$$- \frac{1}{\delta^3}\frac{\mu}{3\rho}\frac{\partial}{\partial r^*}\left(\frac{\partial u_{r*}^*}{\partial r^*} + \frac{1}{r^*}\frac{\partial u_{r*}^*}{\partial \theta^*} + \frac{\partial u_{r*}^*}{\partial z^*}\right)\vec{e_{r*}}$$

$$- \frac{1}{\delta^3 r^*}\frac{\mu}{3\rho}\frac{\partial}{\partial \theta^*}\left(\frac{\partial u_{\theta*}^*}{\partial r^*} + \frac{1}{r^*}\frac{\partial u_{\theta*}^*}{\partial \theta^*} + \frac{\partial u_{\theta*}^*}{\partial z^*}\right)\vec{e_{\theta*}}$$

$$- \frac{\mu}{3\delta^3 \rho}\frac{\partial}{\partial z^*}\left(\frac{\partial u_{z*}^*}{\partial r^*} + \frac{1}{r^*}\frac{\partial u_{z*}^*}{\partial \theta^*} + \frac{\partial u_{z*}^*}{\partial z^*}\right)\vec{k}$$

$$\left.+ \frac{1}{\delta\rho}\frac{\partial p}{\partial r^*}\vec{e_{r*}} + \frac{1}{\delta r^*\rho}\frac{\partial p}{\partial \theta^*}\vec{e_{\theta*}} + \frac{1}{\delta\rho}\frac{\partial p}{\partial z^*}\vec{k} - \vec{F_T}\right) = 0 \qquad (3.10)$$

$$\delta^3 \left(\frac{\partial \vec{L}_1}{\partial t} + \frac{\vec{L}_1}{\delta} \frac{\partial \vec{L}_1}{\partial r^*} + \frac{\vec{L}_1}{\delta r^*} \frac{\partial \vec{L}_1}{\partial \theta^*} + \frac{\vec{L}_1}{\delta} \frac{\partial \vec{L}_1}{\partial z^*} - \frac{1}{\delta^3 r^*} u_{\theta^*}^{*2} \vec{e}_{r^*} \right.$$

$$+ \frac{1}{\delta^3 r^*} u_{r^*}^* u_{\theta^*}^* \vec{e}_{\theta^*} - \frac{\mu}{\rho} \left(-\delta^{-2} \frac{\vec{L}_1}{r^{*2}} + \delta^{-2} \frac{\partial^2 \vec{L}_1}{\partial r^{*2}} + \delta^{-2} \frac{1}{r^*} \frac{\partial \vec{L}_1}{\partial r^*} \right.$$

$$+ \delta^{-2} \frac{1}{r^{*2}} \frac{\partial^2 \vec{L}_1}{\partial \theta^{*2}} + \frac{2}{\delta^2 r^{*2}} \left(\frac{2}{\delta} \frac{\partial u_{r^*}^*}{\partial \theta^*} \vec{e}_{\theta^*} - \frac{1}{\delta} \frac{\partial u_{\theta^*}^*}{\partial \theta^*} \vec{e}_{r^*} \right) + \delta^{-2} \frac{\partial^2 \vec{L}_1}{\partial z^{*2}} \right)$$

$$- \frac{\mu}{3\delta^3 \rho} \frac{\partial}{\partial r^*} \left(\frac{\partial u_{r^*}^*}{\partial r^*} + \frac{1}{r^*} \frac{\partial u_{r^*}^*}{\partial \theta^*} + \frac{\partial u_{r^*}^*}{\partial z^*} \right) \vec{e}_{r^*}$$

$$- \frac{1}{\delta^3 r^*} \frac{\mu}{3\rho} \frac{\partial}{\partial \theta^*} \left(\frac{\partial u_{\theta^*}^*}{\partial r^*} + \frac{1}{r^*} \frac{\partial u_{\theta^*}^*}{\partial \theta^*} + \frac{\partial u_{\theta^*}^*}{\partial z^*} \right) \vec{e}_{\theta^*}$$

$$- \frac{\mu}{3\delta^3 \rho} \frac{\partial}{\partial z^*} \left(\frac{\partial u_{z^*}^*}{\partial r^*} + \frac{1}{r^*} \frac{\partial u_{z^*}^*}{\partial \theta^*} + \frac{\partial u_{z^*}^*}{\partial z^*} \right) \vec{k}$$

$$\left. + \frac{1}{\delta \rho} \frac{\partial p}{\partial r^*} \vec{e}_{r^*} + \frac{1}{\delta r^* \rho} \frac{\partial p}{\partial \theta^*} \vec{e}_{\theta^*} + \frac{1}{\delta \rho} \frac{\partial p}{\partial z^*} \vec{k} - \vec{F}_T \right) = 0 \qquad (3.11)$$

expanding previous two equations leads to, due to scale invariance of Navier–Stokes equations, $\vec{L}_2 = u_{r^*}^* \vec{e}_{r^*} + 2u_{\theta^*}^* \vec{e}_{\theta^*} + u_{z^*}^* \vec{k}$. Here the δ drops out of the equation.

3.2 A SOLUTION PROCEDURE FOR δ ARBITRARILY SMALL IN QUANTITY

Multiplication of Eq. (3.11) by ρ and Eq. (3.12) below by $\vec{L}_1$, addition of the resulting equations [19,22], and using the ordinary product rule of differential multivariable calculus a form as in the after Eq. (3.12) is obtained whereby $\vec{a}$ is given by $\vec{a} = \rho \vec{L}_2$.

The is

$$\frac{\partial \rho}{\partial t^*} + u_{r^*}^* \frac{\partial \rho}{\partial r^*} + \frac{u_{\theta^*}^*}{r^*} \frac{\partial \rho}{\partial \theta^*} + u_{z^*}^* \frac{\partial \rho}{\partial z^*}$$

$$= -\rho \left(\frac{u_{r^*}^*}{r^*} + \frac{\partial u_{r^*}^*}{\partial r^*} + \frac{1}{r^*} \frac{\partial u_{\theta^*}^*}{\partial \theta^*} + \frac{\partial u_{z^*}^*}{\partial z^*} \right) = -\rho \overline{L} \qquad (3.12)$$

The Momentum equation is,

$$\rho\frac{\partial \vec{a}}{\partial t} + \vec{a}\cdot\nabla\vec{a} + \rho^2\vec{b}\nabla\cdot\vec{b} - \frac{\rho}{r^*}u_{\theta^*}^{*2}\vec{e_{r^*}} + \frac{\rho}{r^*}u_{r^*}^*u_{\theta^*}^*\vec{e_{\theta^*}} =$$

$$2r^{*-2}\left(2\frac{\partial u_{r^*}^*}{\partial\theta^*}\vec{e_{\theta^*}} - \frac{\partial u_{\theta^*}^*}{\partial\theta^*}\vec{e_{r^*}}\right) + \mu\left(\nabla^2\vec{b} + \frac{1}{3}\nabla(\nabla\cdot\vec{b})\right) + \nabla P + \vec{F_T}$$

Taking the geometric product in the previous equation with the inertial vector term,

$$\vec{f} = \vec{a}\cdot\nabla\vec{a} \tag{3.13}$$

where $\vec{b} = \frac{\vec{a}}{\rho}$ is defined, where in the context of geometric algebra, the following scalar and vector grade equations arise,

$$\vec{f}\cdot\left(\rho^2\frac{\partial\vec{b}}{\partial t} + \rho\vec{b}\frac{\partial\rho}{\partial t}\right) + \|\vec{f}\|^2 + \vec{b}\cdot\vec{f}\rho^2\nabla\cdot\vec{b} - \frac{\rho}{r^*}u_{\theta^*}^{*2}\vec{f}\cdot\vec{e_{r^*}} + \frac{\rho}{r^*}u_{r^*}^*u_{\theta^*}^*\vec{f}\cdot\vec{e_{\theta^*}} =$$

$$2r^{*-2}\left(2\frac{\partial u_{r^*}^*}{\partial\theta^*}\vec{f}\cdot\vec{e_{\theta^*}} - \frac{\partial u_{\theta^*}^*}{\partial\theta^*}\vec{f}\cdot\vec{e_{r^*}}\right) + \mu\vec{f}\cdot\nabla^2\vec{b} + \frac{\mu}{3}\vec{f}\cdot\nabla(\nabla\cdot\vec{b}) + \vec{f}\cdot\nabla P + \vec{f}\cdot\vec{F_T} \tag{3.14}$$

$$\rho^2\frac{\partial\vec{b}}{\partial t} + \rho\vec{b}\frac{\partial\rho}{\partial t} + \vec{b}\rho^2\nabla\cdot\vec{b} - \frac{\rho}{r^*}u_{\theta^*}^{*2}\vec{e_{r^*}} + \frac{\rho}{r^*}u_{r^*}^*u_{\theta^*}^*\vec{e_{\theta^*}}$$

$$= 2r^{*-2}\left(2\frac{\partial u_{r^*}^*}{\partial\theta^*}\vec{e_{\theta^*}} - \frac{\partial u_{\theta^*}^*}{\partial\theta^*}\vec{e_{r^*}}\right) + \mu\left(\nabla^2\vec{b} + \frac{1}{3}\nabla(\nabla\cdot\vec{b})\right) + \nabla P + \vec{F_T} \tag{3.15}$$

The geometric product of two vectors [23] and (current book Chapter 2 on Geometric Algebra), is defined by $\vec{A}\vec{B} = \vec{A}\cdot\vec{B} + \vec{A}\times\vec{B}$. Taking the divergence of Eq. (3.15) results in

$$\left[\rho^2\frac{\partial}{\partial t}\left(\nabla\cdot\vec{b}\right) + \rho\frac{\partial\rho}{\partial t}\nabla\cdot\vec{b} + \vec{b}\cdot\nabla(\rho\frac{\partial\rho}{\partial t})\right] + \nabla\cdot(\vec{b}\rho^2\nabla\cdot\vec{b}) + \text{Div}$$

$$= \nabla\cdot\left(\mu\nabla^2(\vec{b})\right) + \Psi\frac{\mu}{3}\nabla\cdot\left(\nabla(\nabla\cdot\vec{b})\right) + \nabla\cdot(\nabla P) + \nabla\cdot\vec{F_T} \tag{3.16}$$

with the divergence of the following non-linear terms,

$$\text{Div} = \nabla\cdot\left(-\frac{\rho}{r^*}u_{\theta^*}^{*2}\vec{e_{r^*}} + 2\frac{\rho}{r^*}u_{r^*}^*u_{\theta^*}^*\vec{e_{\theta^*}}\right) =$$

$$-\frac{\partial}{\partial r^*}(\frac{\rho}{r^*})u_{\theta^*}^{*2} - 2\frac{\rho}{r^*}u_{\theta^*}^*\frac{\partial u_{\theta^*}^*}{\partial r^*} + 2\frac{\partial}{\partial\theta^*}(\frac{\rho}{r^{*2}})u_{r^*}^*u_{\theta^*}^* + 2\frac{\rho}{r^{*2}}u_{r^*}^*\frac{\partial u_{\theta^*}^*}{\partial\theta^*} + 2\frac{\rho}{r^{*2}}u_{\theta^*}^*\frac{\partial u_{r^*}^*}{\partial\theta^*}$$

$$
\text{Div} = \begin{cases}
-\frac{2\rho}{r^*} u_{\theta*}^* \frac{\partial u_{\theta*}^*}{\partial r^*} + \frac{2\rho}{r^{*2}} u_{r*}^* \frac{\partial u_{\theta*}^*}{\partial \theta^*} - l_1(\theta^*) \frac{2}{r^{*2}} u_{r*}^* u_{\theta*}^* \\
\quad +2\frac{\rho}{r^{*2}} u_{r*}^* \frac{\partial u_{\theta*}^*}{\partial \theta^*} + 2\frac{\rho}{r^{*2}} u_{\theta*}^* \frac{\partial u_{r*}^*}{\partial \theta^*}, \qquad \theta^* \in (-\pi, 0), \\[4pt]
-\frac{2\rho}{r^*} u_{\theta*}^* \frac{\partial u_{\theta*}^*}{\partial r^*} + \frac{2\rho}{r^{*2}} u_{r*}^* \frac{\partial u_{\theta*}^*}{\partial \theta^*} - il_1(\theta^*) \frac{2}{r^{*2}} u_{r*}^* u_{\theta*}^* \\
\quad +2\frac{\rho}{r^{*2}} u_{r*}^* \frac{\partial u_{\theta*}^*}{\partial \theta^*} + 2\frac{\rho}{r^{*2}} u_{\theta*}^* \frac{\partial u_{r*}^*}{\partial \theta^*}, \qquad \theta^* \in (0, \pi)
\end{cases}
$$

where $l_1(\theta^*) = i/2 \, \dfrac{\cos(\theta^* \alpha)\alpha\,\theta^* + \sin(\theta^* \alpha)}{\alpha\,\theta^{*2}} \dfrac{1}{\sqrt{\mp \frac{\sin(\theta^* \alpha)}{\theta^* \alpha}}}$,

and which becomes

$$
\text{Div} = \begin{cases}
-\frac{2\rho}{r^*} u_{\theta*}^* \left[\nabla \times \vec{b} \right]_z - l_1(\theta^*) \frac{2}{r^{*2}} u_{r*}^* u_{\theta*}^* \\
\quad +2\frac{\rho}{r^{*2}} u_{r*}^* \frac{\partial u_{\theta*}^*}{\partial \theta^*} + 2\frac{\rho}{r^{*2}} u_{\theta*}^{*2}, \qquad \theta^* \in (-\pi, 0), \\[4pt]
-\frac{2\rho}{r^*} u_{\theta*}^* \left[\nabla \times \vec{b} \right]_z - il_1(\theta^*) \frac{2}{r^{*2}} u_{r*}^* u_{\theta*}^* \\
\quad +2\frac{\rho}{r^{*2}} u_{r*}^* \frac{\partial u_{\theta*}^*}{\partial \theta^*} + 2\frac{\rho}{r^{*2}} u_{\theta*}^{*2}, \qquad \theta^* \in (0, \pi)
\end{cases}
$$

where vorticity in the z^* direction appears. It will be assumed for now that the vorticity is linear in r^* and separable in r^* from the remaining independent variables. It can be generalized to a form $r^* F(r^*, \theta^*, z^*, t^*)$ with the restriction that F is increasing in r^* toward the wall of the tube. For the moment, Div $= 0$ will be set without justification and $\delta \approx 0$(but not zero) which imply,

$$
u_{r*}^* = \begin{cases}
\dfrac{2u_{\theta*}^* \sin(\alpha\theta^*)\theta^* \left(\omega_3 r^* - u_{\theta*}^*\right)}{2\sin(\alpha\theta^*)\left(\frac{\partial}{\partial\theta^*} u_{\theta*}^*\right)\theta^* - u_{\theta*}^* \left(\cos(\alpha\theta^*)\alpha\theta^* + \sin(\alpha\theta^*)\right)} & \theta^* \in (-\pi, 0), \\[16pt]
\dfrac{2u_{\theta*}^* \sin(\alpha\theta^*)\theta^* \left(\omega_3 r^* - u_{\theta*}^*\right)}{2\sin(\alpha\theta^*)\left(\frac{\partial}{\partial\theta^*} u_{\theta*}^*\right)\theta^* - u_{\theta*}^* \left(\cos(\alpha\theta^*)\alpha\theta^* + \sin(\alpha\theta^*)\right)} & \theta^* \in (0, \pi),
\end{cases}
$$

The fact that the vorticity is twice the angular velocity ($\vec{\omega} = 2\vec{\omega_A}$) is used; it can be seen that the angular velocity will be high near the wall due to viscous friction and negligible near $r^* = 0$. The angular velocity of a fluid particle in 3D is,

$$
\vec{\omega_A} = \frac{\vec{r} \times \vec{u}}{|\vec{r}|^2}
$$

and as a result, the vorticity ω_3 is calculated as,

$$\omega_3 = \frac{2}{|\vec{r}|^2}\left(r^* u^*_{\theta*} - \theta^* u^*_{r*}\right),$$

Substitution into the formula for u^*_{r*} above and solving for u^*_{r*} results in,

$$u^*_{r*} = \left(2u^{*2}_{\theta*}\,\theta^* \sin(\alpha\theta^*)\left(r^{*2} - \theta^{*2} - z^{*2}\right)\right) \times$$

$$\left(2\theta^* \sin(\alpha\theta^*)(r^{*2} + \theta^{*2} + z^{*2})\left(\frac{\partial u^*_{\theta*}}{\partial\theta^*}\right) + \left(((4r^* + 1)\theta^{*2} + r^{*2} + z^{*2})\sin(\alpha\theta^*)\right.\right.$$

$$\left.\left. + \alpha\theta^* \cos(\alpha\theta^*)(r^{*2} + \theta^{*2} + z^{*2}))u^*_{\theta*}\right)^{-1}$$

The radial velocity, u^*_{r*}, depends on the terms $u^*_{\theta*}$ and hyperbolic part, $\left(r^{*2} - \theta^{*2} - z^{*2}\right)$. The sign of the radial velocity will change on the surface given by $z^* = \sqrt{\left(r^{*2} - \theta^{*2}\right)}$

3.3 CHARACTERIZATION OF THE SIGN OF THE VORTICITY

The cylindrical function $(z^*)^2 = (r^*)^2 - (\theta^*)^2$ defines interior and exterior regions where the radial velocity will be positive and conversely negative; the radial velocity will vanish on regions which belong to the cylindrical surface outlined by the cylindrical function. Graphing modalities encounter difficulties due to plotting software protocols for handling square root functions. This cylindrical function has a natural representation, which can be found by warping the cylindrical coordinate system from $(r*, \theta*)$ to (ϕ, ξ) which is outlined by the substitution $\{r* = \phi\cosh\xi, \theta* = \phi\}$. This transformation has the surface Jacobian outlined by:

$$J^{\alpha}_{\alpha'} = \begin{bmatrix} \cosh\xi & \phi\sinh\xi \\ 1 & 0 \end{bmatrix} \tag{3.17}$$

In this natural representation, the radial velocity null-surface is outlined by:

$$\mathbf{x} = \begin{bmatrix} \phi\cosh\xi\cos\phi \\ \phi\cosh\xi\sin\phi \\ \phi\sinh\xi \end{bmatrix}, \xi \in [0, \xi_{\max}]\,,\; \phi \in [0, \phi_{\max}] \tag{3.18}$$

where ξ_{max} is specified to be coincident with a cylinder of radius $r^* = r_{max}$ oriented along the z-axis at the value $z^* = z_{max}$. Since it is known that the conditions specify the equalities $z_{max} = \phi_{max} \sinh \xi_{max}$ and $r_{max} = \phi_{max} \cosh \xi_{max}$, the value of ξ_{max} can be given by

$$\tanh \xi_{max} = \frac{z_{max}}{r_{max}} \;,\; \phi^2_{max} = r^2_{max} - z^2_{max} \qquad (3.19)$$

This image can be plotted continuously (See Figure 3.1a–d). An important property of this surface is that taking the cylindrical

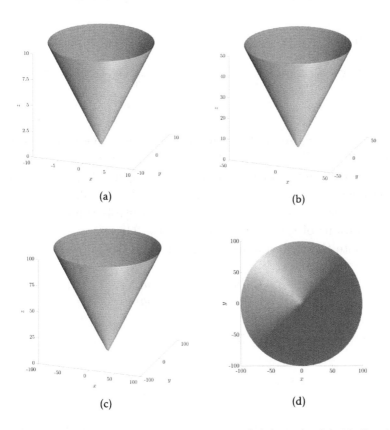

(a)

(b)

(c)

(d)

Figure 3.1 Surface where u_r vanishes, $z = |r|$ (cone). (a) Null-velocity surface ($r = 0..10$). (b) Null-velocity surface ($r = 0..50$), (c) Null-velocity surface ($r = 0..100$) Front view. (d) Null-velocity surface ($r = 0..100$) Top view.

representation of the function, in the large regimes of r, the function reduces to a cone. At infinity for z^*, we approach a perfect circle as shown in Figure 3.1.

3.4   NON-LINEAR FURTHER REDUCTION

The other nonlinear terms in Eq. (3.15), ie, $2r^{*-2}\left(2\frac{\partial u_{r^*}^*}{\partial\theta^*}\vec{e_{\theta^*}} - \frac{\partial u_{\theta^*}^*}{\partial\theta^*}\vec{e_{r^*}}\right)$ have a divergence equal to,

$$-\left(4\frac{\frac{\partial^2}{\partial\theta^{*2}}u_{r^*}^*\left(r^*,\theta^*,z^*,t^*\right)}{r^{*3}} + 4\frac{\frac{\partial}{\partial\theta^*}u_{\theta^*}^*\left(r^*,\theta^*,z^*,t^*\right)}{r^{*3}} - 2\frac{\frac{\partial^2}{\partial\theta^*\partial r^*}u_{\theta^*}^*\left(r^*,\theta^*,z^*,t^*\right)}{r^{*2}}\right)$$

which is equal to

$$\frac{\partial}{\partial\theta^*}\left(4r^{*-2}[\nabla\times\vec{b}]_z^* - 2r^{*-2}\frac{\partial u_{\theta^*}^*}{\partial r^*} - 8r^{*-2}\frac{u_{\theta^*}^*}{r^*}\right) = \frac{\partial}{\partial\theta^*}\left(4r^{*-2}\omega_3\left(r^*,\theta^*,z^*,t^*\right)\right.$$
$$\left. -\frac{r^{*-2}}{2}\omega_3\left(r^*,\theta^*,z^*,t^*\right) - 4r^{*-2}\omega_3\left(r^*,\theta^*,z^*,t^*\right)\right)$$

The value of r^* is chosen to be large as mentioned previously for δ small in value. This implies that the aforementioned expression is negligible and can be omitted from the subsequent analysis. This is true since ω_3 is assumed to be linear in r^* as will be further clarified below. It is known that the three-dimensional vorticity controls the breakdown of smooth solutions of the 3D Euler equations. [24]

We denote the following in Eq. (2.14)

$$\Psi = 2r^{*-2}\left(2\frac{\partial u_{r^*}^*}{\partial\theta^*}\vec{f}\cdot\vec{e_{\theta^*}} - \frac{\partial u_{\theta^*}^*}{\partial\theta^*}\vec{f}\cdot\vec{e_{r^*}}\right)$$

and in Eq. (2.15) we denote by ψ

$$\psi = \nabla\cdot2r^{*-2}\left(2\frac{\partial u_{r^*}^*}{\partial\theta^*}\vec{e_{\theta^*}} - \frac{\partial u_{\theta^*}^*}{\partial\theta^*}\vec{e_{r^*}}\right)$$

As it will be shown in this work, for r^* not very large so that the flow is confined to a central core in the tube, the term ψ is significant but not unbounded on $t^* \in (0, \infty)$. It is only when $r^* \to \infty$ that solutions breakdown outside the central core in the vicinity of the wall of the tube. So ψ dependent on ω_3 does control the breakdown of smooth solutions of the 3D compressible equations as will be seen in this chapter.

The expression involving Ψ in Eq. (3.14) will vanish further below when we take a dot product in the z^* direction of flow, or $\vec{k}$ direction downstream in tube. We therefore do not include it after Eq. (3.22) below. Upon multiplication of Eq. (3.16) by,

$$H = \frac{\rho \vec{b} \cdot \vec{f}}{\frac{\partial \rho}{\partial t}} \tag{3.20}$$

the resulting equation is

$$\left[\rho^2 H \frac{\partial}{\partial t}(\nabla \cdot \vec{b}) + \rho^2 \vec{b} \cdot \vec{f} \nabla \cdot \vec{b} + H \vec{b} \cdot \nabla(\rho \frac{\partial \rho}{\partial t})\right] + H \nabla \cdot (\vec{b} \rho^2 \nabla \cdot \vec{b}) =$$
$$\mu H \nabla \cdot \nabla^2 \vec{b} + \frac{\mu}{3} H \nabla \cdot (\nabla(\nabla \cdot \vec{b})) + H \nabla \cdot (\nabla P) + H \nabla \cdot \rho \vec{F_T} \tag{3.21}$$

which results upon using Eq. (3.14) in,

$$\rho^2 H \frac{\partial}{\partial t}(\nabla \cdot \vec{b}) - \vec{f} \cdot \left(\rho^2 \frac{\partial \vec{b}}{\partial t} + \rho \vec{b} \frac{\partial \rho}{\partial t}\right) - \|\vec{f}\|^2 - \vec{f} \cdot \nabla^2 \vec{b} - \frac{\mu}{3} \vec{f} \cdot \nabla(\nabla \cdot \vec{b})$$
$$- \vec{f} \cdot \nabla P + H \vec{b} \cdot \nabla(\rho \frac{\partial \rho}{\partial t}) + H \nabla \cdot (\vec{b} \rho^2 \nabla \cdot \vec{b}) = -\mu H (\nabla \cdot \nabla^2 \vec{b})$$
$$- \frac{\mu}{3} H \nabla \cdot (\nabla(\nabla \cdot \vec{b})) + H \nabla \cdot (\nabla P) + H \nabla \cdot \rho \vec{F_T} + \Psi \tag{3.22}$$

The continuity equation is written in terms of $\vec{b}$ as,

$$\frac{\partial \rho}{\partial t} + \rho \nabla \cdot \vec{b} + \vec{b} \cdot \nabla \rho = 0 \tag{3.23}$$

and

$$\nabla \cdot \vec{b} = -\frac{1}{\rho}\frac{\partial \rho}{\partial t} - \frac{1}{\rho}\nabla \rho \cdot \vec{b} \tag{3.24}$$

where the following compact expression is given,

$$Y^* = \nabla \cdot \vec{b} \tag{3.25}$$

Multiplying by $(\vec{f} \cdot \vec{f})^{-1}$ in Eq. (3.22) and, using properties of third derivatives involving the gradient and in particular the fact that the

Laplacian of the divergence of a vector field is equivalent to the divergence of the Laplacian of a vector field, leads to the following form,

$$W^* \frac{\partial Y^*}{\partial t} - G(\rho, \frac{\partial \rho}{\partial t})W^* - F(\rho, \frac{\partial \rho}{\partial t})\vec{b}\vec{f}(1 + \vec{f} \cdot \nabla P)$$

$$- \rho^{-2}V(\mu)W^*\nabla^2(Y^*) - \vec{b}\vec{f}\frac{\partial \rho}{\partial t}\frac{\rho^{-1}}{\|\vec{f}\|^2}\vec{f} \cdot \left(\frac{\partial \vec{b}}{\partial t}\right)$$

$$+ \Omega - \vec{b}\vec{f}\frac{\partial \rho}{\partial t}\frac{\rho^{-3}}{\|\vec{f}\|^2}\mu\vec{f} \cdot \nabla^2\vec{b} - \vec{b}\vec{f}\frac{\partial \rho}{\partial t}\frac{\rho^{-3}}{\|\vec{f}\|^2}\frac{\mu}{3}\vec{f} \cdot \nabla(\nabla \cdot \vec{b})$$

$$- \rho^{-2}\vec{b}\vec{f}\frac{1}{\|\vec{f}\|^2}\vec{b} \cdot \vec{f} \, \nabla \cdot \left(\vec{b}\rho^2\nabla \cdot \vec{b} + \nabla P + \rho\vec{F_T}\right) = 0 \qquad (3.26)$$

where $\Omega = \rho^{-2}H\vec{b} \cdot \nabla(\rho\frac{\partial \rho}{\partial t})$ in Eq. (3.26) and,

$$W^* = \left(\frac{\vec{f} \cdot \vec{b}}{\|\vec{f}\|^2}\vec{f}\right)\vec{b} = \left(\frac{\vec{f} \cdot \vec{b}}{\|\vec{f}\|^2}\vec{f}\right) \cdot \vec{b} + \left(\frac{\vec{f} \cdot \vec{b}}{\|\vec{f}\|^2}\vec{f}\right) \times \vec{b} = \xi + \vec{f} \times \left(\frac{\vec{f} \cdot \vec{b}}{\|\vec{f}\|^2}\vec{b}\right)$$

$$(3.27)$$

This involves the vector projection of $\vec{b}$ onto $\vec{f}$ which is written in the conventional form,

$$proj_{\mathbf{f}}\mathbf{b} = \frac{\mathbf{f} \cdot \mathbf{b}}{\|\mathbf{f}\|^2}\mathbf{f} \qquad (3.28)$$

Eq. (3.26) can be written compactly as

$$\frac{\partial Y^*}{\partial t} - G(\rho, \frac{\partial \rho}{\partial t}) - \rho^{-2}V(\mu)\nabla^2 Y^* - \rho^{-2}\nabla^2 P$$
$$= \frac{U_{\vec{f}}\left[\mathbf{Q}(\rho, \frac{\partial \rho}{\partial t}, \vec{b}, \frac{\partial \vec{b}}{\partial t}, \nabla P, \vec{F_T}) + \vec{f}\right]}{U_{\vec{f}}\vec{b}} \qquad (3.29)$$

where $U_{\vec{f}}\vec{\xi}$ is the scalar projection for $\vec{b}$, $G = \frac{1}{\rho^2}\left(\frac{\partial \rho}{\partial t}\right)^2$, $\mathbf{Q}$ (a differential operator defined by Eqs. (3.15) and (3.26) and hence for a constant positive function α,

$$\frac{\partial Y^*}{\partial t} - G(\rho, \frac{\partial \rho}{\partial t}) - \rho^{-2}V(\mu)\nabla^2 Y^* - \rho^{-2}\nabla^2 P$$
$$= \frac{\|\mathbf{Q}(\rho, \frac{\partial \rho}{\partial t}, \vec{b}, \frac{\partial \vec{b}}{\partial t}, \nabla P, \vec{F_T}) + \vec{f}\|}{\|\vec{b}\|} = \alpha \geq 0 \qquad (3.30)$$

with solution in terms of a function $\mathcal{B}$,

$$Y^* = \nabla \cdot \vec{b} = \mathcal{B}(\alpha, r, \theta, z, t) \tag{3.31}$$

At this stage of the analysis, we introduce the vorticity equation for compressible flow,

$$\frac{\partial \vec{\omega}}{\partial t} + (\vec{a} \cdot \nabla)\vec{\omega} = (\vec{\omega} \cdot \nabla)\vec{a} - \vec{\omega}(\nabla \cdot \vec{a}) + \frac{\nabla \rho}{\rho^2} \times \nabla P + \nabla \times (\frac{\nabla \cdot \tau}{\rho}) + \nabla \times (\frac{F}{\rho})$$

We consider the third component of the vorticity equation in z^*. It is assumed that the vorticity is an exponential function of z^* and t^*, ie $\omega = r^* G(\theta^*) e^{-\tanh(\alpha z^*)} e^{-\tanh(\alpha t^*)}$, for some general function of θ^*. Recall that $a_1 = \rho b_1 = 0$ on surface $r^{*2} - \theta^{*2} - z^{*2} = 0$, $a_2 = \rho b_2 = \frac{1}{2r^*}\omega(r^*, \theta^*, z^*, t^*)(r^{*2} + \theta^{*2} + z^{*2})$. This can be non-zero except at the center of the tube where there the flow is irrotational, also $G(\theta^*) \neq 0$ further away from the center of tube. Substitution of this form of $\vec{\omega}$ into the vorticity equation and letting a_3^* be an increasing linear function in the form βz^*, results in $\nabla \cdot \vec{a}$ being solved for and expanded in a series for small α and large β, resulting in,

$$\nabla \cdot \vec{b} = O(1/r^*) \tag{3.32}$$

It is worthy to note that ρ is decreasing down the tube and a_3^* is increasing, and this is physically true for gas pipe flows.

Here we note that due to the appearance of the z^*-direction of vorticity in the cylindrical Navier–Stokes equations that this expression's form can control the breakdown of smooth solutions for 3-D cylindrical compressible equations as we will see below in this chapter. The compressible Navier-Stokes equations have regular solutions that blow up in finite time. The remaining part of this chapter is to prove this using integral calculus methods. It will also be seen that if ρ is independent of t^* then there are classical global solutions in t^*.

Recall from the very onset definition of ρ that it is changing exponentially in time. It may be proven that $\mathcal{B}$ is written in terms of a non-homogeneous Green's function in spatial and time variables. Next we have from Eq. (3.30),

$$\left[\frac{\partial Y^*}{\partial t} - G(\rho, \frac{\partial \rho}{\partial t}) - \rho^{-2} V(\mu) \nabla^2 Y^* - \rho^{-2} \nabla^2 P \right] \vec{f} \cdot \vec{b} =$$
$$\vec{f} \cdot \left[\mathbf{Q}(\rho, \frac{\partial \rho}{\partial t}, \vec{b}, \frac{\partial \vec{b}}{\partial t}, \nabla P, \vec{F_T}) + \vec{f} \right] \tag{3.33}$$

where $\vec{f}$ drops out on both sides to obtain

$$\left[\vec{b}\frac{\partial Y^*}{\partial t} - \vec{b}G(\rho, \frac{\partial\rho}{\partial t^*}) - \vec{b}\rho^{-2}V(\mu)\nabla^2 Y^* - \vec{b}\rho^{-2}\nabla^2 P\right] =$$
$$\left[Q(\rho, \frac{\partial\rho}{\partial t}, \vec{b}, \frac{\partial\vec{b}}{\partial t}, \nabla P, \vec{F_T}) + \vec{f}\right]$$

Using Eq. (3.32) ie. $\nabla \cdot \vec{b} = O(1/r^*)$ and substituting in the left side of Eq. (3.34) and taking the limit as $r^* \to \infty$ for ρ as defined in this work, the left side of (3.34) vanishes with the exception of term, $-\vec{b}G(\rho, \frac{\partial\rho}{\partial t^*})$ (cancels with exact term that is part of Q), and we obtain the following for Q,

$$\left[-2\frac{\partial\vec{b}}{\partial t} - \frac{1}{\rho}\vec{b}\frac{\partial\rho}{\partial t} - F(\rho, \frac{\partial\rho}{\partial t})(\vec{f_b} + \vec{F_T})\right] - \rho^{-2}\vec{b}\nabla \cdot \vec{F_T} = 0 \quad (3.34)$$

where $F(\rho, \frac{\partial\rho}{\partial t}) = \rho^{-3}\frac{\partial\rho}{\partial t}$. Ω in Eq. (3.26) vanishes due to assumption on rate of change of density with respect to t and $c = 1/\delta$. Also, $F(\rho, \frac{\partial\rho}{\partial t})\vec{f} = \rho^{-3}\frac{\partial\rho}{\partial t}\rho^2\vec{b} \cdot \nabla\vec{b} = \rho^{-1}\frac{\partial\rho}{\partial t}\vec{b} \cdot \nabla\vec{b} = c\vec{f_b}$.

We obtain the following upon taking the curl of Eq. (3.34),

$$-2\frac{\partial}{\partial t^*}\nabla \times \vec{b} - c\nabla \times \vec{f_b} - \nabla \times \left(\rho^{-2}\vec{b}\nabla \cdot \vec{F_T}\right) = 0. \quad (3.35)$$

Multiply Eq. (3.34) by the normal vector $\cos(\theta)\vec{a}$ which is the normal component of $\vec{a}$ at wall of moving control volume (CV) in the z^* direction,(direction of flow downstream in tube)

$$\cos(\theta)\vec{a} \cdot \left[-2\frac{\partial}{\partial t^*}\nabla \times \vec{b} - c\nabla \times \vec{f_b} - \nabla \times \left(\rho^{-2}\vec{b}\nabla \cdot \vec{F_T}\right)\right] = 0 \quad (3.36)$$

3.5 STOKES THEOREM APPLIED TO DYNAMIC SURFACES

Recalling Divergence theorem and Stoke's theorem, for general $\mathbf{F}$,

$$\iiint_V (\nabla \cdot \mathbf{F})dV = \oiint_{S(V)} \mathbf{F} \cdot \hat{n}dS$$

$$\iint\limits_{S} (\nabla \times \mathbf{F}) \cdot d\mathbf{S} = \oint_{C} \mathbf{F}(\mathbf{r}) \cdot d\mathbf{r} \tag{3.37}$$

where C is the contour of a circle in control volume of tube and S consists of all surfaces of control volume.

Defining the following vector field,

$$\vec{W} = -\frac{\partial}{\partial t} \nabla \times \vec{b} - \frac{1}{2} \nabla \times \left(\rho^{-2} \vec{b} \nabla \cdot \vec{F_T} \right), \tag{3.38}$$

$$\oiint\limits_{S(V)} \mathbf{W} \cdot \hat{n} dS = \iiint\limits_{V} (\nabla \cdot \mathbf{W}) dV$$

$$= \frac{c}{2} \oint_{C} \mathbf{f_b}(\mathbf{r}) \cdot d\mathbf{r}$$

where $\hat{n} = \cos(\theta)\vec{a}$, and Stoke's theorem has been used. Applying Stoke's theorem to $\vec{W}$, hence $\vec{b}$. It is evident that the derivative of the

$$\frac{d}{dt^*} \oint_{C} \vec{b} \cdot d\mathbf{r} = \oint_{C} \frac{\partial \vec{b}}{\partial t^*} \cdot d\mathbf{r} + \oint_{C} \vec{b} \cdot d\vec{b} = \oint_{C} \frac{\partial \vec{b}}{\partial t^*} \cdot d\mathbf{r} + \oint_{C} d \mid \vec{b} \mid^2 / 2.$$

The last integral in the previous series of equalities is an integral of a perfect differential around a closed path and is therefore equal to zero. To use Stokes theorem for dynamic surfaces that change with time, it can be seen that using the Calculus of Moving Surfaces (CMS) [25,53], the only requirement is that the paraboloid surface considered here has the same boundary as a disk which is coincident to the boundary of the paraboloid. Under this observation, the boundary is stationary, and so the surface integral over a dynamic surface can be reduced down to the line integral around the stationary path, and is thus equivalent to the path integral around a closed disk coincident with the mouth of the paraboloid. Therefore, the dynamic nature of the paraboloids studied in this chapter is not a point of concern when applying Stokes Equation to obtain a line integral. Proceeding we obtain,

$$\frac{c}{2} \oint_{C} \vec{f_b}(\mathbf{r}) \cdot \vec{T} ds = - \oint_{C} \left(\frac{\partial \vec{b}}{\partial t^*} + \frac{1}{2} \rho^{-2} \vec{b} \nabla \cdot \vec{F_T} \right) \cdot d\mathbf{r} \tag{3.39}$$

where $\vec{T}$ is unit tangent vector to closed curve C and ds is arc length,

$$\oint_C (\frac{c}{2}\vec{f}_b(\mathbf{r}) + \frac{\partial \vec{b}}{\partial t^*} + \frac{1}{2}\rho^{-2}\vec{b}\nabla \cdot \vec{F}_T) \cdot d\mathbf{r} = 0 \qquad (3.40)$$

The third term in the parenthesis in Eq. (3.40) is integrated by parts for line integral and we obtain the following,

$$\oint_C (\frac{c}{2}\vec{f}_b(\mathbf{r}) + \frac{\partial \vec{b}}{\partial t^*} - \frac{1}{2}\vec{F}_T\nabla \cdot (\rho^{-2}\vec{b})) \cdot d\mathbf{r} = 0 \qquad (3.41)$$

Parametrizing the circle as $r = g(\theta)$ in polar coordinates it can be proven that the line integral in Eq. (3.41) is,

$$\oint_C (c\frac{f_{b_1}}{2} + \frac{\partial b_1}{\partial t^*} - \frac{1}{2}F_{T_1}\nabla\cdot(\rho^{-2}\vec{b}))\cdot d\theta - \oint_C (c\frac{f_{b_2}}{2} + \frac{\partial b_2}{\partial t^*} - \frac{1}{2}F_{T_2}\nabla\cdot(\rho^{-2}\vec{b}))\cdot dr = 0 \qquad (3.42)$$

The normal form of Green's theorem can be used for the line integral in Eq. (3.42) setting first,

$$M = \frac{\partial b_1}{\partial t^*} + cb_1\frac{\partial b_1}{\partial r^*} + c\frac{b_2\frac{\partial b_1}{\partial \theta}}{r^*} - \frac{1}{2}F_{T_1}\nabla \cdot (\rho^{-2}\vec{b}) \qquad (3.43)$$

$$N = \frac{\partial b_2}{\partial t^*} + cb_1\frac{\partial b_2}{\partial r^*} + c\frac{b_2\frac{\partial b_2}{\partial \theta}}{r^*} - \frac{1}{2}F_{T_2}\nabla \cdot (\rho^{-2}\vec{b}) \qquad (3.44)$$

The line integral in Eq. (3.42) is equal to the following:

$$\iint_R \left(\frac{\partial M}{\partial r} + \frac{\partial N}{\partial \theta}\right) dr d\theta \qquad (3.45)$$

where M and N are given by Eqs. (3.43) and (3.44) respectively and R is the open disk with boundary C.

A fluid particle undergoing circular motion along boundary C is changing with respect to θ and t. It follows that,

$$\frac{\partial u_\theta}{\partial \theta} = \frac{\partial u_\theta}{\partial t}\frac{dt}{d\theta} \qquad (3.46)$$

So then since $\frac{d\theta}{dt}$ is the instantaneous angular speed (with respect to time) we write this as $|\mathbf{u}| = b_2$

Hence the force $F_{T_2} = m\frac{\partial u_\theta}{\partial \theta}u_\theta$ where m is the mass of the fluid particle.

3.6 ANALYSIS FOR HUNTER SAXTON EQUATION

The derivative of Eq. (3.44) with respect to θ is the Hunter Saxton equation if and only if the derivative of $\frac{1}{2}F_{T_2}\nabla \cdot (\rho^{-2}\vec{b})$ is equal to $\frac{1}{2}\frac{\partial b_2}{\partial \theta}^2$. This is the auxiliary PDE necessary to obtain the Hunter Saxton equation. The very last term on the right-hand side of Eq. (3.44) is $-\frac{1}{2}F_{T_2}\nabla \cdot (\rho^{-2}\vec{b})$. F_{T_2} and is moved to the right-hand side of the equation set to zero. The left side is differentiated with respect to θ and set to $\frac{1}{2}\frac{\partial b_2}{\partial \theta}^2$ and the resulting equation is the Hunter Saxton PDE.

So there is this auxiliary PDE to solve for ρ in terms of b_2 which is possible explicitly with the use of Maple. See Appendix A.1 for the Maple program.

So then we have the following theorem:

Theorem 3.2 *Hunter Saxton For the Compressible 3D Navier–Stokes equations, if the density $\rho(r,\theta,z,t)$ does not blow up in finite time at say $t = T_f < \infty$ then there will be finite time blowup for $b_2(r,\theta,z,t)$ at $t = T_f < \infty$. This follows due to the reciprocal adiabatic condition for gas flow between pressure and velocity and hence between density and velocity. It can be verified in the program in Appendix A.1 that ρb_2 is constant with respect to time.*

Thus,

$$\frac{\partial}{\partial \theta}\left(\frac{\partial b_2}{\partial t^*} + b_2\frac{\partial b_2}{\partial \theta}\right) - \frac{1}{2}\left(\frac{\partial b_2}{\partial \theta}\right)^2 = 0 \tag{3.47}$$

Equation (3.47) is known to have finite time singularities. See reference [26] for the finite time blowup of the Hunter-Saxton equation on the line.

EXERCISES

3.1 Compressed air ($\gamma = 1.4$) is supplied from a reservoir to a pipe, 1 cm in diameter and 5 m long. It is estimated that the average friction factor, f, of the flow in the pipe is 0.02. At the end of this long pipe is a short nozzle whose opening to the atmosphere has one-half of the cross-sectional area of the pipe. Assuming that frictional effects in the nozzle can be neglected find the following related to conditions when the flow through the pipe/nozzle combination is choked: (1) The Mach number of the flow entering the pipe. (2) The ratio of the pressure in the reservoir to the pressure in the exit (throat) from the nozzle.

3.2 Burger's equation in non-conservative form is,

$$\frac{\partial u}{\partial t} + u\frac{\partial u}{\partial x} = \nu\frac{\partial^2 u}{\partial x^2}$$

Put this equation in conservative form by considering the special case of a steady state $\frac{\partial}{\partial t} \equiv 0$. Show that the conservative form of the equation can be integrated and conclude that it has the same behavior in the far field as the kinematic wave equation. Discuss the type of shock wave you obtain what happens as $\nu \to 0$.

3.3 Consider frictionless, steady flow of a compressible fluid in an infinitesimal stream tube (Figure 3.2). a. Show by the continuity and momentum theorems and equations that,

$$\frac{d\rho}{\rho} + \frac{dA}{A} + \frac{dV}{V} = 0.$$

$$dp + \rho V\,dV + \rho g\,dz = 0.$$

b. Determine the integrated forms of these equations for an incompressible fluid.

c. Derive the appropriate equations for unsteady frictionless, compressible flow, in a stream tube of cross-sectional area which depends on both space and time.

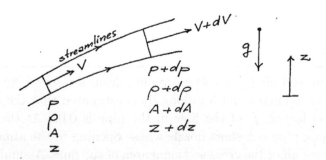

Figure 3.2 Stream tube diagram.

## CHAPTER 4

Hydrodynamic Stability and Maple

Terry E. Moschandreou

TVDSB

4.1 INTRODUCTION

We examine velocity, shear and inertia flows. Such flows are important in the study of fluid dynamics. The transition problem from laminar to turbulence is crucial in fluid dynamics, and hydrodynamic stability is a means to analyze such flows.

4.2 RAYLEIGH–TAYLOR INSTABILITY

The Rayleigh–Taylor instability, or RT instability, is an instability of an interface between two fluids of different densities which occurs when the lighter fluid is pushing the heavier fluid. Examples include the behavior of water suspended above oil in the gravity of Earth, mushroom clouds like those from volcanic eruptions and atmospheric nuclear explosions, supernova explosions where expanding core gas is accelerated into denser shell gas, instabilities in plasma fusion reactors and fusion with inertial confinement.

Water suspended above a surface separating oil is an everyday example of Rayleigh–Taylor instability and can be modeled by two completely different layers of immiscible fluid, the denser fluid on top of the less dense one and both subject to Earth's gravity. The equilibrium here is unstable to any perturbation or disturbance of the interface: if

DOI: 10.1201/9781003452256-4

a parcel of heavier fluid is displaced downward with an equal volume of lighter fluid displaced upward, the potential energy of the configuration is lower than the initial state. The disturbance will thus grow and lead to a further release of potential energy, as the denser material moves down under the gravitational field, and the less dense material is displaced upwards. This was the setup as studied by Lord Rayleigh. The important insight by G. I. Taylor was his realization that this situation is equivalent to the situation when the fluids are accelerated, with the less dense fluid accelerating into the denser fluid.

As the RT instability develops, the initial perturbations progress from a linear growth phase into a non-linear growth phase, eventually developing plumes flowing upwards and spikes falling downwards. In the linear phase, the fluid movement can be closely approximated by linear equations, and the amplitude of perturbations can grow exponentially with time. In the non-linear phase, perturbation amplitudes are too large for a linear approximation, and non-linear equations are required to describe fluid motions. The difference in the fluid densities divided by their sum is defined as the Atwood number, A. For A close to 0, RT instability flows take the form of symmetric "fingers" of fluid; for A close to 1, the much lighter fluid "below" the heavier fluid takes the form of larger bubble -like forms.

In this section, we investigate the stability of a surface between two fluids:

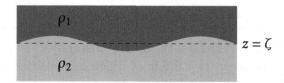

We assume the two fluids have densities ρ_1 and ρ_2 and are separated by a smooth interface, and we perform a normal mode analysis for the stability of fluid given a disturbance.

4.2.1 Normal Mode Analysis

For our linear stability analysis, we begin with an initially stationary flow and assume there are small perturbations to the velocity and the physical variables. The density and pressure are only functions of z and we consider a small disturbance in terms of a periodic wave along a plane at $z = \zeta$. For a single mode disturbance (in say $x - z$ plane) the amplitude A of a disturbance is described by:

$$A(x, z, t) = A_k(z, t)\exp(ikx).,$$

where k is the wavenumber. Seeking solutions in time for $A_k(z, t)$ also in an exponential form:

$$A_k(z, t) = A_k(z)\exp(nt),$$

we have,

$$A(x, z, t) = A_k(z)\exp(ikx + nt),$$

We have an eigenvalue problem here where specific boundary conditions are employed. Here n is the eigenvalue corresponding to the wavenumber k. Superimposed modes can be obtained from single modes which would then have independent solutions with a corresponding eigenvalue for each wavenumber. The case of a positive eigenvalue results in a positive exponential and an unstable mode and a negative eigenvalue results in a stable mode. For our velocities, density and pressure, we have,

$$u = u(x, z, t) = u_k(z)\exp(ikx + nt),$$

$$w = w(x, z, t) = w_k(z)\exp(ikx + nt),$$

$$\delta\rho(x, z, t) = \delta\rho_k(z)\exp(ikx + nt),$$

$$\delta P = \delta P(x, z, t) = \delta P_k(z)\exp(ikx + nt)$$

where δ is used to represent a small perturbation in the flow quantities. Now using assumptions for two-dimensional flow with incompressibility, velocities are expressed as an initial constant added to a small perturbation. With inviscid flow, initial velocities are zero, the momentum equation is linearized and we assume an isothermal type condition where we neglect the Energy equation. This results in a reduced form of the Navier–Stokes equations,

$$\frac{\partial u}{\partial x} + \frac{\partial w}{\partial z} = 0,$$

$$\rho\frac{\partial u}{\partial t} = -\frac{\partial\delta P}{\partial x},$$

$$\rho \frac{\partial w}{\partial t} = -\delta \rho g - \frac{\partial \delta P}{\partial z},$$

$$\frac{\partial \delta \rho}{\partial t} = -w \frac{\partial \rho}{\partial z},$$

Now substituting into the four amplitude relations gives,

$$iku + \frac{\partial w}{\partial z} = 0,$$

$$n\rho u = -ik\delta P,$$

$$n\rho w = -\delta \rho g - \frac{\partial \delta P}{\partial z},$$

$$n\delta \rho = -w \frac{\partial \rho}{\partial z}.$$

Multiplying the second equation by ik and substituting $-\frac{dw}{dz}$ for iku from the first equation immediately above, we observe that,

$$-n\rho \frac{dw}{dz} = k^2 \delta P,$$

and rearranging we obtain,

$$\delta P = -\frac{n}{k^2} \rho \frac{dw}{dz}.,$$

and substituting into the third equation,

$$n\rho w = -\delta \rho g + \frac{n}{k^2} \frac{d}{dz} \left(\rho \frac{dw}{dz} \right).$$

Next solving the last equation for $\delta \rho$ we get,

$$\delta \rho = -w \frac{1}{n} \frac{d\rho}{dz},$$

and substituting into $n\rho w = -\delta \rho g + \frac{n}{k^2} \frac{d}{dz} \left(\rho \frac{dw}{dz} \right)$, and multiplying through by k^2/n gives,

$$\frac{d}{dz} \left(\rho \frac{dw}{dz} \right) - \rho k^2 w = -wg \frac{k^2}{n^2} \frac{d\rho}{dz},$$

The previous equation is the governing ODE to be next used for the instability between the two fluids.

4.3 THE RAYLEIGH-TAYLOR INSTABILITY FOR TWO INCOMPRESSIBLE FLUIDS

Recalling the diagram of two incompressible fluids with differing densities, separated by an interface with a perturbation imposed, for a layer of fluid above ρ_1 or below ρ_2 the interface ζ, the density is constant and the equation $\frac{d}{dz}\left(\rho\frac{dw}{dz}\right) - \rho k^2 w = -wg\frac{k^2}{n^2}\frac{d\rho}{dz}$, reduces to,

$$\frac{d^2w}{dz^2} - k^2 w = 0, \tag{4.1}$$

We use boundary conditions that the velocity is matched at the interface for the two solutions and the velocity is zero at large distances above and below the interface surface, so that we have,

$$w_1 = w_0 e^{kz}, \quad (z < \zeta),$$

$$w_2 = w_0 e^{-kz}, \quad (z > \zeta).$$

Recall,

$$\frac{d}{dz}\left(\rho\frac{dw}{dz}\right) - \rho k^2 w = -wg\frac{k^2}{n^2}\frac{d\rho}{dz}.$$

and applying this equation at the interface $z = \zeta$, multiplying by dz and integrating, results in,

$$\int d\left(\rho\frac{dw}{dz}\right) - \int \rho k^2 w dz = -\int wg\frac{k^2}{n^2}d\rho.$$

Integrating about an infinitesimal distance $dz \approx 0$, about the interface $z = \zeta$, the second term vanishes, and the other two terms can be equated as,

$$\Delta\left(\rho\frac{dw}{dz}\right) = -wg\frac{k^2}{n^2}\,\Delta\rho,$$

$$-\rho_2 kw - \rho_1 kw = -wg\frac{k^2}{n^2}a\left(\rho_2 - \rho_1\right),$$

$$-\left(\rho_2 + \rho_1\right) = -g\frac{k}{n^2}\left(\rho_2 - \rho_1\right),$$

Finally solving for the eigenvalue n we obtain,

$$n = \sqrt{gk\frac{\rho_2 - \rho_1}{\rho_2 + \rho_1}} \tag{4.2}$$

Thus, the system is unstable if the heavy fluid is above the lighter fluid ($\rho_2 > \rho_1$), because the eigenvalue is real, and stable if the light fluid is above the heavy fluid, ($\rho_2 < \rho_1$) because the eigenvalue is imaginary. The quantity $(\rho_2 - \rho_1)(\rho_2 + \rho_1)^{-1}$ is called the Atwood number (A), and a more compact form for the eigenvalue is then $n = gkA$. For a positive Atwood number, the interface is unstable, and for a negative value, it is stable.

4.4 RAYLEIGH-BENARD CONVECTION

RBC is a buoyancy -driven flow in a container with a temperature gradient. As the fluid at the bottom heats up, its density decreases, so buoyant forces push the less-dense fluid up toward the cooler end of the container. Meanwhile, the cooler fluid at the top is denser, so it sinks and displaces the warmer fluid. There is a continuous density gradient, and this cyclical motion creates Benard convection cells. Raleigh-Benard convection is one type of natural convection, since it is not propelled by an outside force. Researchers use the Raleigh number to measure the "amount of convection" in a flow. This number helps researchers compare results between experiments conducted with differently sized containers and different fluids. Fluids with a Raleigh number below 1708 are not convective; heat moves through the fluid only by conduction among the molecules. The molecules are not moving yet. At Ra = 1708, laminar convection begins. Laminar means that the layers of fluid in the convection cells are not interfering with each other; the flow is smooth. At high Rayleigh numbers, Rayleigh-Benard convection is a turbulent flow , which means that the fluid layers are interfering with each other. The next system we want to study is:

The experimental setup uses a layer of liquid, e.g. water, between two parallel planes. The height of the layer is small compared to the horizontal dimension. At first, the temperature of the bottom plane is the same as the top plane. The liquid will then tend toward an equilibrium, where its temperature is the same as its surroundings. (Once there, the liquid is perfectly uniform: to an observer it would appear the same from any position. This equilibrium is also asymptotically stable: after a local, temporary perturbation of the outside temperature, it will go back to its uniform state, in line with the second law of thermodynamics).

Then, the temperature of the bottom plane is increased slightly yielding a flow of thermal energy conducted through the liquid. The system will begin to have a structure of thermal conductivity: the temperature, and the density and pressure with it, will vary linearly between the bottom and top plane. A uniform linear gradient of temperature will be established. Conduction is when heat diffuses from the bottom to the top, with no fluid motion and convection occurs when the fluid at the bottom heats up, expands, becomes lighter, and proceeds to the top.

To understand the system mathematically, we must deal with the case where we have density variations throughout the fluid. Again, recall the Navier-Stokes equations

$$\rho\left(\frac{\partial \mathbf{u}}{\partial t} + \mathbf{u} \cdot \nabla \mathbf{u}\right) = -\frac{1}{\rho}\nabla P - g(1 - \alpha(T - T_0))\hat{z} + \nu\nabla^2\mathbf{u}, \quad \nabla \cdot \mathbf{u} = 0.$$

where the kinematic viscosity is $\nu = \mu/\rho$. We are interested in the heat flow between the two plates. The lower plate is located at $z = 0$ and has temperature T_0. The upper plate sits at $z = d$ and has the temperature $T_0 - \Delta T$. There is a time-independent solution to the governing equations when $\mathbf{u} = 0$. Here there is conduction when there is no fluid movement. The situation is not stable when perturbations occur. The solution would become unstable due to the fluid moving. We would obtain an instability that looks like convection rolls or Benard cells. Let's look at the vorticity. Taking the curl of the Navier–Stokes equations gives,

$$\frac{\partial \omega}{\partial t} + \mathbf{u} \cdot \nabla \omega = \frac{1}{\rho^2}\nabla\rho \times \nabla P + \omega \cdot \nabla \mathbf{u} + \nu\nabla^2\mathbf{u} \qquad (4.3)$$

If the direction of the gradient of the pressure is in a direction that is different from the gradient of density then vorticity will be created. We assumed that the heavier fluid sits on top of the lighter fluid. Therefore, the potential energy gained by the fluid is,

$$P.E. = gd^4 \rho_0 \alpha \, \Delta T.$$

For the fluid to turn over viscous forces must be overcome, if it turns over at a speed of U, then the viscous force per volume is $\mu \nabla^2 \mathbf{u} \approx \mu \frac{U}{d^2}$. The work to turn the fluid over is,

$$W.D. \approx \mu U d^2.$$

Here the gain in potential energy is sufficient to overcome the work done by viscous forces when,

$$\frac{gd^2 \alpha \, \Delta T}{\nu U}. \gg 1$$

The heat equation is,

$$\frac{\partial T}{\partial t} = \frac{\kappa}{c_P} \nabla^2 T \tag{4.4}$$

Using the heat equation, the time scale where heat diffuses is $\tau_d f \approx \frac{d^2 c_P}{\kappa}$. If the fluid turns over, it takes a time $t \approx d/U$. If we equate these times scales we obtain $U \approx \kappa / c_P d$. Here we introduce the Rayleigh number,

$$R_a = \frac{g \alpha c_P d^3 \, \Delta T}{\nu \kappa} \gg 1$$

There will exist an instability if and only if $R_a \gg 1$.

The time-independent solution to the Navier–Stokes equations when the velocity is zero is,

$$P = P_0 - \rho_0 g \left(z - \frac{\alpha \, \Delta T}{2d} z^2 \right)$$

$$T = T_0 = \frac{\Delta T z}{d}$$

As mentioned previously there is no convection here.

Let us perturb these solutions. The perturbations will be the usual $\delta T(\mathbf{x}, t)$ and $\delta P(\mathbf{x}, t)$ respectively.

We take these as small quantities so as to linearize the equations of motion.

The Navier–Stokes equation becomes,

$$\frac{\partial \mathbf{u}}{\partial t} = -\frac{1}{\rho_0} \delta \nabla P + \alpha g \delta T \hat{\mathbf{z}} + \nu \nabla^2 \mathbf{u} \tag{4.5}$$

The temperature equation becomes,

$$\frac{\partial \delta T}{\partial t} - \frac{\Delta T}{d} u_z = \frac{\kappa}{c_P} \nabla^2 \delta T \tag{4.6}$$

The velocity perturbations satisfy the continuity equation.

So the perturbations with this feature will introduce convection rolls.

In general, equations are hard to solve, and we want to make approximations. A common approximation is the Boussinesq approximation which we have addressed.

Thus, what we do is that we assume that the density is constant except in the buoyancy force. Here we take the limit $g \to \infty$ but $\rho' \to 0$ with $g\rho'$ remaining finite.

Recall that our density is to be given by a function of temperature T. We will use a leading order approximation

$$\rho(\mathbf{x}, t) = \rho_0(1 - \alpha(T(\mathbf{x}, t) - T_0)).$$

We can now return to our original problem. We have a fluid of viscosity ν and thermal diffusivity κ. There are two planes a distance d apart, with the top held at temperature $T_0 - \Delta T$ and bottom at T_0. We make the Boussinesq approximation.

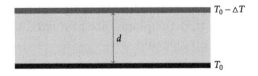

We can see the physics in these equations if we write them in terms of dimensionless variables. We introduce new variables where we rescale both space and time as

$$\tilde{t} = \frac{\kappa t}{c_p d^2} \qquad \tilde{\mathbf{x}} = \frac{\mathbf{x}}{d} \qquad \tilde{\mathbf{u}} = \frac{c_p d}{\kappa}\mathbf{u}$$

$$\tilde{T} = \frac{\delta T}{\Delta T} \qquad \tilde{P} = \frac{c_p^2 d^2 \delta P}{\kappa^2 \rho_0}$$

In terms of these new variables, our equations of temperature and motion become:

$$\frac{\partial \tilde{T}}{\partial \tilde{t}} - \tilde{u}_z = \tilde{\nabla}^2 \tilde{T} \tag{4.7}$$

$$\frac{\partial \tilde{\mathbf{u}}}{\partial \tilde{t}} = -\tilde{\nabla}\tilde{P} + R_a P_r \tilde{T}\hat{\mathbf{z}} + P_r \tilde{\nabla}^2 \tilde{\mathbf{u}} \tag{4.8}$$

Here the equations of motion depend on two dimensionless constants: the Rayleigh number and Prandtl number

$$Ra = \frac{g\alpha c_P \Delta T d^3}{\nu \kappa}$$

$$Pr = \frac{c_P \nu}{\kappa},$$

These are the two parameters that control the behavior of the system. In particular, the Prandtl number measures the competition between viscous and diffusion forces. Different fluids have different Prandtl numbers. In a gas, $\frac{\nu}{\kappa} \sim 1$, since both are affected by the mean free path of a particle. It has the value $P_r = 0.7$ for air and $P_r \approx 7.5$ for water at room temperature.

We now solve Eq. (4.8) subject to special boundary conditions. Here we set $\tilde{u}_z = 0, \frac{\partial \tilde{u}_z}{\partial \tilde{z}} = 0 \; \frac{\partial^2 \tilde{u}_z}{\partial \tilde{z}^2}$ at $\tilde{z} = 0$ and $\tilde{z} = 1$ and this means there is no flow through the planes, top and bottom. In addition to this Dirichlet conditions are prescribed for the temperature, $\tilde{T} = 0$ at $\tilde{z} = 0$ and $\tilde{z} = 1$ which means that the planes are at a fixed temperature, the lower one hot and the top one cold.

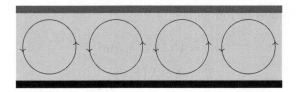

It is convenient to assume the top and bottom surfaces are stress-free: $\frac{\partial \tilde{u}_x}{\partial \tilde{z}} = \frac{\partial \tilde{u}_y}{\partial \tilde{z}} = 0$, which implies

$$\frac{\partial}{\partial \tilde{z}} \left(\frac{\partial \tilde{u}_x}{\partial \tilde{x}} + \frac{\partial \tilde{u}_y}{\partial \tilde{y}} \right) = 0.$$

This then gives us $\frac{\partial^2 \tilde{u}_z}{\partial \tilde{z}^2} = 0$. It is true that these conditions are correct for a free surface, but they are mathematically convenient, and we will impose them. Also $\tilde{\nabla} \cdot \tilde{\mathbf{u}} = 0$.

We now take the curl of the second equation, Eq. (4.8) which gives,

$$\frac{\partial \tilde{\omega}}{\partial \tilde{t}} = R_a P_r \tilde{\nabla} \tilde{T} \times \hat{\mathbf{z}} + P_r \tilde{\nabla} \tilde{\omega}^2 \tag{4.9}$$

where $\tilde{\omega} = \tilde{\nabla} \times \tilde{\mathbf{u}}$.

Now taking another curl, that is of Eq. (4.9) we obtain,

$$\frac{\partial \tilde{\nabla}^2 \tilde{\mathbf{u}}}{\partial \tilde{t}} = R_a P_r \left(\hat{\mathbf{z}} \tilde{\nabla}^2 \tilde{T} - \tilde{\nabla} \left(\frac{\partial \tilde{T}}{\partial \tilde{z}} \right) \right) + P_r \tilde{\nabla}^4 \tilde{\mathbf{u}} \tag{4.10}$$

where the identity with incompressible flow has been used: $\tilde{\nabla} \times \left(\tilde{\nabla} \times \tilde{\mathbf{u}} \right) = -\tilde{\nabla}^2 \tilde{\mathbf{u}}$.

This is fourth order in spatial derivatives. We are interested in the z component which is,

$$\frac{\partial \tilde{\nabla}^2 \tilde{u}_z}{\partial \tilde{t}} = R_a P_r \left(\frac{\partial^2 \tilde{T}}{\partial \tilde{x}^2} + \frac{\partial^2 \tilde{T}}{\partial \tilde{y}^2} \right) + P_r \tilde{\nabla}^4 \tilde{u}_z \tag{4.11}$$

Together with the first equation of Eq. (4.7) we attempt to solve this PDE using separation of variables.

We write our solution as,

$$\tilde{u}_z = V(\tilde{z})X(\tilde{x}, \tilde{y})e^{\gamma \tilde{t}}$$
$$\tilde{T} = \theta(\tilde{z})X(\tilde{x}, \tilde{y})e^{\gamma \tilde{t}}.$$

Solutions with $\lambda > 0$, will be unstable. With these ansatz, the temperature equation Eq. (4.7) becomes,

$$\frac{1}{\theta}\left[\frac{d^2\theta}{d\tilde{z}^2} - \lambda\theta + V\right] = -\frac{1}{X}\left[\frac{\partial^2 X}{\partial \tilde{x}^2} + \frac{\partial^2 X}{\partial \tilde{y}^2}\right] \qquad (4.12)$$

The left side depends on $\tilde{z}$ and the right-hand side depends only on $\tilde{x}$ and $\tilde{y}$, so that both sides are a constant, and this must be positive for the solution to be bounded.

We name it Λ^2.

The technique of separation of variables yields two solutions; the first is,

$$X(\tilde{x}, \tilde{y}) = e^{i(\Lambda_x \tilde{x} + \Lambda_y \tilde{y})} \qquad (4.13)$$

where $\Lambda^2 = \Lambda_x^2 + \Lambda_y^2$. Oscillations have wavelength $\gamma = 2\pi d/\Lambda$.

We also have that,

4.5   SOLUTION OF RAYLEIGH BENARD CONVECTION WITH MAPLE

$$L1 = \frac{\partial^3 u_z}{\partial x^2 \partial t} + \frac{\partial^3 u_z}{\partial y^2 \partial t} + \frac{\partial^3 u_z}{\partial z^2 \partial t} - Ra\,Pr\left(\frac{\partial^2 T}{\partial x^2} + \frac{\partial^2 T}{\partial y^2} + \frac{\partial^2 T}{\partial z^2} - \right.$$
$$\left. Pr\left(\frac{\partial^4 u_z}{\partial x^4} + 2\frac{\partial^4 u_z}{\partial y^2 \partial x^2} + 2\frac{\partial^4 u_z}{\partial z^2 \partial x^2} + \frac{\partial^4 u_z}{\partial y^4} + 2\frac{\partial^4 u_z}{\partial z^2 \partial y^2} + \frac{\partial^4 u_z}{\partial z^4}\right)\right) = 0$$

$$(4.14)$$

$$P = \frac{\partial^2 u_z}{\partial x^2} + \frac{\partial^2 u_z}{\partial y^2} + \frac{\partial^2 u_z}{\partial z^2} \qquad (4.15)$$

$$L2 = \frac{\partial T}{\partial t} - u_z - \frac{\partial^2 T}{\partial x^2} - \frac{\partial^2 T}{\partial y^2} - \frac{\partial^2 T}{\partial z^2} = 0 \qquad (4.16)$$

with maple code

$$A1 := subs(uz(x, y, z, t) = V(z)X(x, y)\exp(\gamma t), T(x, y, z, t)$$
$$= \theta(z)X(x, y)\exp(\gamma t), L1) \qquad (4.17)$$

$$- \left(-V \left(\frac{\partial^4 X}{\partial x^4} \right) P_r{}^2 Ra - 2V \left(\frac{\partial^4 X}{\partial y^2 \partial x^2} \right) P_r{}^2 Ra - V \left(\frac{\partial^4 X}{\partial y^4} \right) P_r{}^2 Ra \right.$$

$$- X \left(\frac{d^4}{dz^4} V \right) P_r{}^2 Ra + \left(-2 \left(\frac{d^2 V}{dz^2} \right) P_r{}^2 Ra + \theta P_r Ra - \gamma V \right) \frac{\partial^2 X}{\partial x^2}$$

$$+ \left(-2 \left(\frac{d^2 V}{dz^2} \right) P_r{}^2 Ra + \theta P_r Ra - \gamma V \right) \frac{\partial^2 X}{\partial y^2}$$

$$\left. - X \left(- \left(\frac{d^2 \theta}{dz^2} \right) P_r Ra + \left(\frac{d^2 V}{dz^2} \right) \gamma \right) \right) e^{\gamma t} = 0 \tag{4.18}$$

and again using Maple,

$$A2 := subs(uz(x,y,z,t) = V(z)X(x,y)\exp(\gamma t),$$
$$T(x,y,z,t) = \theta(z)X(x,y)\exp(\gamma t), L2) \tag{4.19}$$

$$\theta X \gamma - V X - \theta \left(\frac{\partial^2 X}{\partial x^2} \right) - \theta \left(\frac{\partial^2 X}{\partial y^2} \right) - \left(\frac{d^2 \theta}{dz^2} \right) X = 0 \tag{4.20}$$

Dividing by θX the previous differential equation is separable,

$$- \frac{\frac{\partial^2}{\partial x^2} X(x,y)}{X(x,y)} - \frac{\frac{\partial^2}{\partial y^2} X(x,y)}{X(x,y)} = -\Lambda^2 = 0 \tag{4.21}$$

$$\frac{d^2}{dx^2} F_1(x) = -c_1 F_1(x) \tag{4.22}$$

$$\frac{d^2}{dy^2} F_2(y) = F_2(y) c_1 + \Lambda^2 F_2(y) \tag{4.23}$$

$$A_2 = \gamma - \frac{V(z)}{\theta(z)} - \frac{\frac{d^2}{dz^2} \theta(z)}{\theta(z)} = -\Lambda^2 \tag{4.24}$$

with Maple code,

$$A_1 = subs(X(x,y) = \left(C_1 \sin\left(\sqrt{c_1} x\right) + C_2 \cos\left(\sqrt{c_1} x\right) \right)$$
$$\left(C_1 \sin\left(\sqrt{-\Lambda^2 - c_1} y\right) + C_2 \cos\left(\sqrt{-\Lambda^2 - c_1} y\right) \right), A_1) \tag{4.25}$$

gives us the following equation,

$$\mathcal{A}_1 = \left(\left(2\Lambda^2 P_r^2 Ra + \gamma\right)\frac{\mathrm{d}^2 V}{\mathrm{d}z^2} - \Lambda^2\theta P_r Ra + \Lambda^2\left(\Lambda^2 P_r^2 Ra + \gamma\right)V + Ra\, P_r\left(\left(\frac{\mathrm{d}^4 V}{\mathrm{d}z^4}\right)P_r - \frac{\mathrm{d}^2\theta}{\mathrm{d}z^2}\right)\right)e^{\gamma t}$$
$$\left(C_1\sin\left(\sqrt{-\Lambda^2 - c_1 y}\right) + C_2\cos\left(\sqrt{-\Lambda^2 - c_1 y}\right)\right)\left(C_1\sin\left(\sqrt{c_1}x\right) + C_2\cos\left(\sqrt{c_1}x\right)\right) = 0 \tag{4.26}$$

Next using the dsolve command in Maple, we have,

$$Q = \mathrm{dsolve}(\mathcal{A}_1, \mathcal{A}_2); \tag{4.27}$$

With solutions for V and θ as functions of z given as,

$$V(z) = C_6\, e^{-1/2\, z\sqrt{\dfrac{2\gamma P_r^2 Ra^2 - 2\gamma Ra - 2\sqrt{4\Lambda^4 P_r^4 Ra^2 + 4\Lambda^2 P_r^4 Ra^2\gamma + P_r^4 Ra^2\gamma^2 + 4\Lambda^2 P_r^4 Ra^2\gamma + P_r^4 Ra^2\gamma^2 + 4\Lambda^2 P_r^2 Ra\gamma - 4 P_r^3 Ra^2\gamma + 2 P_r^2 Ra\gamma^2 + \gamma^2 Ra}}{P_r Ra}}} +$$

$$C_5\, e^{1/2\, z\sqrt{\dfrac{2\gamma P_r^2 Ra^2 - 2\gamma Ra - 2\sqrt{4\Lambda^4 P_r^4 Ra^2 + 4\Lambda^2 P_r^4 Ra^2\gamma + P_r^4 Ra^2\gamma^2 + 4\Lambda^2 P_r^4 Ra^2\gamma + P_r^4 Ra^2\gamma^2 + 4\Lambda^2 P_r^2 Ra\gamma - 4 P_r^3 Ra^2\gamma + 2 P_r^2 Ra\gamma^2 + \gamma^2 Ra}}{P_r Ra}}} +$$

$$C_4\, e^{-1/2\, z\sqrt{\dfrac{2\gamma P_r^2 Ra^2 - 2\gamma Ra + 2\sqrt{4\Lambda^4 P_r^4 Ra^2 + 4\Lambda^2 P_r^4 Ra^2\gamma + P_r^4 Ra^2\gamma^2 + 4\Lambda^2 P_r^4 Ra^2\gamma + P_r^4 Ra^2\gamma^2 + 4\Lambda^2 P_r^2 Ra\gamma - 4 P_r^3 Ra^2\gamma + 2 P_r^2 Ra\gamma^2 + \gamma^2 Ra}}{P_r Ra}}} +$$

$$C_3\, e^{1/2\, z\sqrt{\dfrac{2\gamma P_r^2 Ra^2 - 2\gamma Ra + 2\sqrt{4\Lambda^4 P_r^4 Ra^2 + 4\Lambda^2 P_r^4 Ra^2\gamma + P_r^4 Ra^2\gamma^2 + 4\Lambda^2 P_r^4 Ra^2\gamma + P_r^4 Ra^2\gamma^2 + 4\Lambda^2 P_r^2 Ra\gamma - 4 P_r^3 Ra^2\gamma + 2 P_r^2 Ra\gamma^2 + \gamma^2 Ra}}{P_r Ra}}} +$$

$$C_1\sin\left(\Lambda z\right) + C_2\cos\left(\Lambda z\right) \tag{4.28}$$

θ is,

$$Z_1 = \frac{1}{2}\frac{1}{P_r R_a(2\Lambda^2+\gamma)}\Bigg((2\Lambda^2+\gamma)\left(-\sqrt{R_a^2(2\Lambda^2+\gamma)^2P_r^4-4P_r^3R_a^2+2R_a\gamma(2\Lambda^2+\gamma)P_r^2+\gamma^2}+\right.$$

$$\left.(R_aP_r^2+1)\gamma+2\Lambda^2P_r^2R_a\right)C_3\,e^{-1/2\frac{\sqrt{2}\sqrt{R_a\left(R_a\gamma P_r^2-\gamma-\sqrt{R_a^2(2\Lambda^2+\gamma)^2P_r^4-4P_r^3R_a^2+2R_a\gamma(2\Lambda^2+\gamma)P_r^2+\gamma^2}\right)}}{P_r R_a}z}+$$

$$C_4(2\Lambda^2+\gamma)\left(-\sqrt{R_a^2(2\Lambda^2+\gamma)^2P_r^4-4P_r^3R_a^2+2R_a\gamma(2\Lambda^2+\gamma)P_r^2+\gamma^2}+\right.$$

$$\left.(R_aP_r^2+1)\gamma+2\Lambda^2P_r^2R_a\right)^{1/2}e^{-1/2\frac{\sqrt{2}\sqrt{R_a\left(R_a\gamma P_r^2-\gamma-\sqrt{R_a^2(2\Lambda^2+\gamma)^2P_r^4-4P_r^3R_a^2+2R_a\gamma(2\Lambda^2+\gamma)P_r^2+\gamma^2}\right)}}{P_r R_a}z}+ \tag{4.29}$$

$$Z_2 = C_5(2\Lambda^2+\gamma)\left(\sqrt{R_a^2(2\Lambda^2+\gamma)^2P_r^4-4P_r^3R_a^2+2R_a\gamma(2\Lambda^2+\gamma)P_r^2+\gamma^2}+\right.$$

$$\left.(R_aP_r^2+1)\gamma+2\Lambda^2P_r^2R_a\right)e^{-1/2\frac{\sqrt{2}\sqrt{R_a\left(R_a\gamma P_r^2-\gamma+\sqrt{R_a^2(2\Lambda^2+\gamma)^2P_r^4-4P_r^3R_a^2+2R_a\gamma(2\Lambda^2+\gamma)P_r^2+\gamma^2}\right)}}{P_r R_a}z}+$$

$$C_6(2\Lambda^2+\gamma)\left(\sqrt{R_a^2(2\Lambda^2+\gamma)^2P_r^4-4P_r^3R_a^2+2R_a\gamma(2\Lambda^2+\gamma)P_r^2+\gamma^2}+\right.$$

$$\left.(R_aP_r^2+1)\gamma+2\Lambda^2P_r^2R_a\right)^{1/2}e^{1/2\frac{\sqrt{2}\sqrt{R_a\left(R_a\gamma P_r^2-\gamma+\sqrt{R_a^2(2\Lambda^2+\gamma)^2P_r^4-4P_r^3R_a^2+2R_a\gamma(2\Lambda^2+\gamma)P_r^2+\gamma^2}\right)}}{P_r R_a}z}+$$

$$2P_rR_a\left(C_2\cos(\Lambda z)+C_1\sin(\Lambda z)\right) \tag{4.30}$$

Finally we have $\theta(z) = Z_1 + Z_2$ as the temperature profile solution.

4.6 CLASSICAL KELVIN-HELMHOLTZ INSTABILITY

Recalling the case of Rayleigh–Taylor instability, but this time for Classical Kelvin-Helmholtz instability we have a non-static situation where there is some horizontal flow in both layers with densities ρ_1 and ρ_2, and they may be of different velocities.

We use velocity potentials

$$\text{upper half above interface } : \mathbf{v_2} = \nabla\Phi_2 \quad \text{for} z > \zeta$$

$$\text{lower half below interface} : \mathbf{v_1} = \nabla\Phi_1 \quad \text{for} z < \zeta$$

Using the Navier–Stokes equation and using the identity $(\mathbf{v} \cdot \nabla) = \nabla(\mathbf{v}^2/2) - \nabla \times \mathbf{v}$ together with potential flow assumption, and thus $\nabla \times \mathbf{v} = 0$, we have the following equation,

$$\nabla\left(\frac{\partial\Phi}{\partial t}\right) + \nabla\left(\frac{1}{2}\mathbf{v}^2\right) + \nabla\left(\frac{P}{\rho}\right) = \mathbf{g}.$$

Since the gravitational acceleration is vertical in the negative $\mathbf{k}$ direction, we have Bernoulli's theorem.

We write the velocity potentials as,

$$\Phi_1 = U_1 x + \phi_1$$

$$\Phi_2 = U_2 x + \phi_2$$

where ϕ_1, and ϕ_2 are infinitesimal perturbations respectively in the upper and lower half with respect to the interface.

Describing the interface we let $z = \zeta(x,t)$ which describes the z-location of the interface.

The velocity of the interface in the z-direction must match the fluid velocities in the z-direction of the two phases.

For the upper part, we have that,

$$\frac{\partial\zeta}{\partial x}\frac{\partial\Phi_1}{\partial x} + \frac{\partial\zeta}{\partial t} = \frac{\partial\Phi_1}{\partial z}$$

which leads to,

$$\frac{\partial \zeta}{\partial t} + \frac{\partial \zeta}{\partial x}\left(U_1 + \frac{\partial \phi}{\partial x}\right) = \frac{\partial \phi_1}{\partial z}$$

To leading order the second term of the partial derivative of ϕ_1 wrt to x can be omitted.

For the lower part, we obtain a similar result, and the two equations are,

$$\frac{\partial \zeta}{\partial t} + \frac{\partial \zeta}{\partial x}U_1 = \frac{\partial \phi_1}{\partial z}$$

$$\frac{\partial \zeta}{\partial t} + \frac{\partial \zeta}{\partial x}U_2 = \frac{\partial \phi_2}{\partial z}$$

Using the Bernoulli Equation first for the upper part, we can match the perturbed interface case to the unperturbed case.

We have that,

$$\frac{\partial \phi_1}{\partial t} + \frac{1}{2}\left(U_1 + \frac{\partial \phi_1}{\partial x}\right)^2 + g\zeta + \frac{P_1}{\rho_1} = \frac{1}{2}U_1^2 + \frac{P_u}{\rho_1}$$

For the lower half, a similar equation is obtained with the same initial pressure and P_1 and P_2 will equal each other and P_u is the pressure of the unperturbed interface.

Equating the two pressures P_1 and P_2 gives,

$$\rho_1\left(\frac{\partial \phi_1}{\partial t} + U_1\frac{\partial \phi_1}{\partial x} + g\zeta\right) = \rho_2\left(\frac{\partial \phi_2}{\partial t} + U_2\frac{\partial \phi_2}{\partial x} + g\zeta\right).$$

A similar eigenmode analysis is done as in the Rayleigh-Taylor instability problem done previously,

Consider the symmetrical ansatz,

$$\phi_1 = \phi_1(z)\exp\left[i\left(kx - nt\right)\right].$$

The Laplace equation is satisfied by both ϕ_1 and ϕ_2 and using the far field conditions for the unperturbed state, that is $z \to \infty$ implies $\phi_1 \to 0$ and $z \to -\infty$ implies $\phi_2 \to 0$, Thus we have that the solutions for ϕ_1, ϕ_2 and ζ are given by the following for a single Fourier mode,

$$\phi_1 = \phi_a\exp(kz)\exp\left[i(kx - nt)\right].$$

$$\phi_2 = \phi_b\exp(-kz)\exp\left[i(kx - nt)\right].$$

$$\zeta = \zeta_{ab}\exp\left[i(kx - nt)\right]$$

Substituting the above expressions with corresponding mode amplitudes gives the following three algebraic equations:

$$-in\zeta_{ab} + iU_1k\zeta_{ab} = k\phi_a.$$

$$-in\zeta_{ab} + iU_2k\zeta_{ab} = -k\phi_b.$$

$$\rho_1\left(-in\phi_a + iU_1k\phi_a + g\zeta_{ab}\right) = \rho_2\left(-in\phi_b + iU_2k\phi_b + g\zeta_{ab}\right)$$

The far field now has velocity with $\phi_1 \to 0$ as $z \to \infty$ and $\phi_2 \to 0$ as $z \to -\infty$.

Non-trivial solutions are obtained with $\zeta_{ab} \neq 0$ when the following is true,

$$n^2(\rho_1 + \rho_2) - 2nk\left(\rho_1 U_1 + \rho_2 U_2\right) + k^2\left(\rho_1 U_1^2 + \rho_2 U_2^2\right) + kg\left(\rho_2 - \rho_1\right) = 0 \tag{4.31}$$

The dispersion relation is as found before when the fluid is stationary and $U_1 = U_2 = 0$ For Kelvin-Helmholtz instability for zero gravitational field that is pure shear, it follows that,

$$n_{1/2} = \frac{k\left(\rho_1 U_1 + \rho_2 U_2\right)}{\rho_1 + \rho_2} \pm \frac{\sqrt{\rho_1\rho_2}}{\rho_1 + \rho_2}\left| U_1 - U_2\right|.$$

EXERCISES

4.1 Explain what the difference is between *laminar* and *turbulent* flow. Hydrodynamic stability theory aims to explain transition to turbulence by linearising the Navier-Stokes equations. Explain how this is done, and derive the Orr-Sommerfeld equation for two-dimensional stream function disturbances to a plane parallel flow $U(y)$ of the form $\psi = \phi(y)\exp[i\alpha(x - ct)]$, and give suitable boundary conditions on ϕ for plane Poiseuille flow. Hint: The linearized equation for a two-dimensional disturbance can be written in the form,

$$\frac{\partial}{\partial t}\triangle\psi + U\frac{\partial}{\partial x}\triangle\psi - U''\frac{\partial\psi}{\partial x} = \frac{1}{R_e}\triangle^2\psi.$$

where $\triangle\psi$ is the $y-$ component of the vorticity and R_e is the Reynolds number.

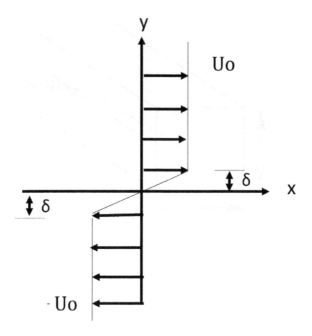

Figure 4.1 Shear layer.

4.2 Consider the piecewise-linear-profile model for a shear layer shown in Figure 4.1. Examine the stability of this profile to infinitesimal perturbations on the basis of linear inviscid stability theory. Determine the wavenumber of the perturbation corresponding to the maximum growth rate.

4.3 Consider the pressure-driven flow of a Newtonian fluid in a long channel with square cross- section (with side = a; see Figure 4.2). The length of the channel is large compared to its side a.

  a. Write down the governing equations and boundary conditions for steady, fully developed , uni-directional flow in this channel.

  b. Next derive the linearized governing equations for the stability of the shown uni-directional base flow.

  c. Write down the normal mode form for perturbations of all components of velocities and pressure, and derive the linearized equations for the normal modes.

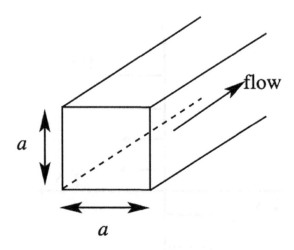

Figure 4.2 Flow in a square channel.

4.4 Consider the set of coupled ODEs,

$$\frac{dx}{dt} = x\left(\alpha_1 - \beta_1 x - y\right)$$

$$\frac{dy}{dt} = y\left(\alpha_2 - \beta_2 y - x\right)$$

where α_i and β_i, ($i=1,2$) are constant parameters.

a. Determine the four possible steady states for this system of ODE.

b. By using a linear stability analysis about the four steady states, determine the nature of the four steady states (that is if there is a stable node, unstable node, unstable saddle point, etc.), and explain what the criteria are in terms of the parameters α_i and $\beta_i(i=1,2)$.

4.5 Consider an unperturbed liquid cylinder that is infinitely long and surrounded by air at atmospheric pressure P_{atm} of radius a. Next axisymmetric perturbations are imposed on the interface in such a way that the perturbed interface position $r = a + \zeta(z)$ is independent of the θ coordinate in the cylindrical coordinate system.

a. Derive an expression for the unit normal $\mathbf{n}$ pointing into the air for the perturbed interface.

b. Derive an expression for the curvature $\nabla \cdot \mathbf{n}$ for the perturbed interface.

Hint: $\nabla = \mathbf{e_r}\frac{\partial}{\partial r} + \mathbf{e}_\theta \frac{1}{r}\frac{\partial}{\partial \theta} + \mathbf{e_z}\frac{\partial}{\partial z}$

$\nabla \cdot \mathbf{n} = \frac{1}{r}\frac{\partial(rn_r)}{\partial r} + \frac{1}{r}\frac{\partial n_\theta}{\partial \theta} + \frac{\partial n_z}{\partial z}$

## CHAPTER 5

Mathematics Preliminaries

Terry E. Moschandreou
TVDSB

5.1 INTRODUCTION

A major goal of Real Analysis is to teach one to understand mathematical proofs as well as to be able to formulate them and write them. Two processes are here involved. One is to discover the ideas required or necessary for the proof and secondly to express these ideas in mathematical language. Important concepts in real analysis are introduced in this chapter that are particularly important in fluid mechanics.

5.2 MATHEMATICS PRELIMINARIES

We introduce the important idea of a norm in a vector space.

Definition 5.1 *A norm on a vector space V is a function $||\cdot|| : V \to [0, \infty)$ satisfying the following three properties:*

1. *(Definiteness) $||v|| = 0$ if and only if $v = 0$,*

2. *(Homogeneity) $||\lambda v|| = |\lambda|||v||$ for all $v \in V$ and $\lambda \in \mathbb{Z}$,*

3. *(Triangle inequality) $||v_1 + v_2|| \leq ||v_1|| + ||v_2||$ for all $v_1, v_2 \in V$.*

A seminorm is a function $||\cdot|| : V \to [0, \infty)$ which satisfies (2) and (3) but not always (1), and a vector space with a norm is called a normed space.

                            DOI: 10.1201/9781003452256-5

This is consistent with the definition of a metric $d : X \times X \to [0, \infty)$, which satisfies the following:

1. $d(x, y) = 0$ if and only if $x = y$,

2. $d(x, y) = d(y, x)$ for all $x, y \in X$,

3. $d(x, y) + d(y, z) \geq d(x, z)$ for all $x, y, z \in X$.

Proposition 5.1 *Let $|| \cdot ||$ be a norm on a vector space V. Then*

$$d(v, w) = ||v - w||$$

defines a metric on V, which is the metric induced by the norm.

Proof 5.1 *Property (1) of the norm implies property (1) of metrics, since*

$$d(v, w) = ||v - w|| = 0 \iff v - w = 0 \iff v = w.$$

For property (2) of the metric,

$$||v - w|| = ||(-1)(w - v)|| = |-1| \cdot ||w - v|| = ||w - v||,$$

by using property (2) of the norm. And finally, (3) of the metric is implied by property (3) of the norm due to $(x - y) + (y - z) = (x - z)$.

Example 5.1 *The Euclidean norm on $\mathbb{R}^n$ or $\mathbb{C}^n$, given by*

$$||x||_2 = \left(\sum_{i=1}^{n} |x_i|^2 \right)^{1/2},$$

is indeed a norm. We also define

$$||x||_\infty = \max_{1 \leq i \leq n} |x_i|$$

and more generally (for $1 \leq p < \infty$)

$$||x||_p = \left(\sum_{i=1}^{n} |x_i|^p \right)^{1/p}.$$

Let us consider infinite dimensional vector spaces. We have the following definition,

Definition 5.2 *Let X be a metric space. The vector space $C_\infty(X)$ is defined as*

$$C_\infty(X) = \{f : X \to \mathbb{C} : f \text{ continuous and bounded}\}.$$

Now $C_\infty([0,1])$ is $C([0,1])$, since all continuous functions on $[0,1]$ are bounded.

Proposition 5.2 *For a metric space X, we define a norm on the vector space $C_\infty(X)$ as*

$$||u||_\infty = \sup_{x \in X} |u(x)|.$$

Proof 5.2 *Properties (1) and (2) of a norm follow from the definitions, next property (3) follows, if $u, v \in C_\infty(X)$, then for any $x \in X$,*

$$|u(x) + v(x)| \leq |u(x)| + |v(x)| \leq max(|u(x)| + |v(x)|) \leq max|u(x)| + max|v(x)|$$

by the triangle inequality for $\mathbb{C}$, and thus a maximum of $||u|| + ||v||$. Thus, we have

$$|u(x) + v(x)| \leq ||u||_\infty + ||v||_\infty \quad \forall x \in X \text{ which implies} ||u + v||_\infty$$
$$= \sup_x |u(x) + v(x)| \leq ||u||_\infty + ||v||_\infty.$$

Let us define convergence in the norm, that is, $u_n \to u$ in $C_\infty(X)$ if

$$\lim_{n \to \infty} ||u_n - u||_\infty = 0,$$

and using $\epsilon - N$ definition we have,

$$\forall \epsilon > 0, \; \exists N \in \mathbb{N} : \forall n \geq N, \forall x \in X, |u_n(x) - u(x)| < \epsilon.,$$

This is the definition of **uniform convergence** on X. Some examples of normed vector spaces are,

Example 5.2 *The ℓ^p space is the space of (infinite) sequences*

$$\ell^p = \{\{a_j\}_{j=1}^\infty : ||a||_p < \infty\},$$

where we define the ℓ^p norm

$$\|a\|_p = \begin{cases} \left(\sum_{j=1}^{\infty} |a_j|^p\right)^{1/p} & 1 \le p < \infty \\ \sup_{1 \le j \le \infty} |a_j| & p = \infty. \end{cases}$$

Example 5.3 *The sequence $\left\{\frac{1}{j^p}\right\}_{j=1}^{\infty}$ is in ℓ^p for all $p > 1$ but not in ℓ^1 where the p-series test has been used.*

The sequence $\{a_j\}_{j=1}^{\infty}$ is a Cauchy sequence if for all $\epsilon > 0$ there exists N s.t.$\|a_m - a_n\| \le \epsilon$ for $m, n \ge N$.

Cauchy sequences always converge. Completeness means that if every Cauchy sequence that converges to a limit and the limit is an element of the metric space then the metric space is called complete.

Definition 5.3 *A normed space is a Banach space if it is complete with respect to the metric induced by the norm.*

Example 5.4 *For any $n \in \mathbb{Z}_{\ge 0}$, $\mathbb{R}^n$ and $\mathbb{C}^n$ are complete with respect to any of the $\| \cdot \|_p$ norms.*

Theorem 5.1 *For any metric space X, the space of bounded, continuous functions on X is complete, and $C_{\infty}(X)$ is a Banach space.*

Proof 5.3 *Take a Cauchy sequence, and introduce a candidate for the limit, and show that (1) this is in the space and (2) convergence occurs.*

For the Cauchy sequence $\{u_n\}$, we show that it is bounded under the norm $C_{\infty}(X)$. Note that there exists some positive integer N_0 such that for all $n, m \ge N_0$,

$$\|u_n - u_m\|_{\infty} < 1.$$

for example.
So for all $n \ge N_0$,

$$\|u_n\|_{\infty} \le \|u_n - u_{N_0}\|_{\infty} + \|u_{N_0}\|_{\infty} < 1 + \|u_{N_0}\|_{\infty}$$

by the triangle inequality, and thus for all $n \in \mathbb{N}$,

$$\|u_n\|_{\infty} \le \|u_1\|_{\infty} + \cdots + \|u_{N_0}\|_{\infty} + 1$$

Now for all $n \geq N_0$, $\|u_n\|_\infty$ is bounded as well as for $n < N_0$.
   For $x \in X$,

$$|u_n(x) - u_m(x)| \leq \sup_x |u_n(x) - u_m(x)| = \|u_n - u_m\|_\infty,$$

and since $\{u_n\}$ is a Cauchy sequence, for any $x \in X$, the sequence of complex numbers $\{u_n(x)\}$ evaluated at 'x' is Cauchy. Now the space of complex numbers is a complete metric space, so for all $x \in X$, $u_n(x)$ converges to some limit. Defining the candidate function as,

$$u(x) = \lim_{n \to \infty} u_n(x).$$

Show that this is in $C_\infty(X)$ and that we have convergence under the **uniform convergence** *norm. We have that,*

$$|u(x)| = \lim_{n \to \infty} |u_n(x)|$$

The right-hand side is bounded by the $\|\cdot\|_\infty$ norm, and thus a constant C. Thus,

$$\sup_{x \in X} |u(x)| \leq C,$$

and u is a bounded function. Finally, we show continuity and convergence. Fix $\epsilon > 0$; since $\{u_n\}$ is Cauchy, there exists some N such that for all $n, m \geq N$, we have $\|u_n - u_m\|_\infty < \frac{\epsilon}{2}$. Now for any $x \in X$,

$$|u_n(x) - u_m(x)| \leq \|u_n - u_m\|_\infty < \frac{\epsilon}{2},$$

so taking the limit as $m \to \infty$, we have that for all $n \geq N$,

$$|u_n(x) - u(x)| \leq \frac{\epsilon}{2}$$

Therefore, $\sup_x |u_n(x) - u(x)| \leq \frac{\epsilon}{2} < \epsilon$, and hence $\|u_n - u\|_\infty \to 0$. Next $\|u_n - u\|_\infty \to 0$, implies $u_n \to u$ uniformly on X, and the uniform limit of a sequence of continuous functions is continuous. Therefore, the candidate u is in $C_\infty(X)$ and is the limit of the sequence denoted by u_n, hence $C_\infty(X)$ is complete and is a Banach space.

Exercise: Prove l^p is a Banach space.

$$c_0 = \{a \in \ell^\infty : \lim_{j \to \infty} a_j = 0\}$$

is a Banach space.

5.3   THE LEBESGUE INTEGRAL

The Lebesgue integral for simple functions is, for any simple function ϕ written in the canonical form $\phi = \sum_{j=1}^{n} a_j \chi_{A_j}$ for disjoint sets A_j, we have $\int_E \phi = \sum_{j=1}^{n} a_j m(A_j)$. For a nonnegative measurable function f, there exist a sequence of simple functions that increase pointwise to f. Thus, we can try to define the Lebesgue integral as the limit of the integrals of simple functions.

Definition 5.4 *For a space E that is measurable, the set of all functions $f : E \to \overline{\mathbb{R}}$ such that f is measurable is denoted by $L_{mclass}(E)$.*

Definition 5.5 *Let $f \in L_{mclass}(E)$. Then the Lebesgue integral of f is*

$$\int_E f = \sup \left\{ \int_E \phi : \phi \in L_{mclass}(E) \ simple, \phi \leq f \right\}.$$

Proposition 5.3 *Let $E \subset \mathbb{R}$ be a set with $m(E) = 0$. Then for all $f \in L_{mclass}(E)$, we have $\int_E f = 0$.*

Proof 5.4 *Let $f \in L_{mclass}(E)$. If ϕ is a simple function in the form $\sum_{j=1}^{n} a_j \chi(A_j)$ with $\phi \leq f$, then $m(A_j) \leq m(A) = 0$, so in the sum $\sum_{j=1}^{n} a_j m(A_j)$. So, $\int_E \phi = 0$, and the supremum over all simple functions ϕ is also zero.*

Proposition 5.4 *If $\phi \in L_{mclass}(E)$ is a simple function, then the two definitions of $\int_E f$, for simple functions and general nonnegative measurable functions, agree. If $f, g \in L_{mclass}(E)$, $c \in [0, \infty)$ is a nonnegative real number, and $f \leq g$, then we have $\int_E cf = c \int_E f$ and $\int_E f \leq \int_E g$. Finally, if $f \in L_{mclass}(E)$ and $F \subset E$, then $\int_F f \leq \int_E f$.*

Proposition 5.5 *If $f, g \in L_{mclass}(E)$, and $f \leq g$ almost everywhere on E, then $\int_E f \leq \int_E g$.*

Proof 5.5 *Since $f_1 \leq f_2 \leq \cdots$, we know that $\int_E f_1 \leq \int_E f_2 \leq \cdots$. Thus, $\{\int_E f_n\}$ is a nonnegative increasing sequence of nonnegative numbers, which means that the limit $\lim_{n \to \infty} \int_E f_n$ exists in $[0, \infty]$. Also, because $\lim_{n \to \infty} f_n(x) = f(x)$ for all x, we know that $f_n \leq f$ for* *all n, which means that $\int_E f$ (which is also some number in $[0, \infty]$) must satisfy*

$$\int_E f_n \leq \int_E f \implies \lim_{n\to\infty} \int_E f_n \leq \int_E f.$$

It is sufficient to prove the reverse inequality (ie. $\int_E f \leq \int_E \lim_{n\to\infty} \int_E f_n$), and this can be done by showing that $\int_E \phi \leq \int_E \lim_{n\to\infty} \int_E f_n$ for every simple function $\phi \leq f$ (Note: eventually f_n will be larger than ϕ).

Let $\epsilon \in (0, 1)$. If $\phi = \sum_{j=1}^m a_j \chi_{A_j}$ is an arbitrary simple function with $\phi \leq f$, then we can define the set

$$E_n = \{x \in E : f_n(x) \geq (1 - \epsilon)\phi(x)\}.$$

Since $(1 - \epsilon)\phi(x) < f(x)$ for all x and there is strict equality now that ϵ is positive), and $\lim_{n\to\infty} f_n(x) = f(x)$, every x must be in some E_n. Therefore, we have

$$\bigcup_{n=1}^{\infty} E_n = E.$$

Because $f_1 \leq f_2 \leq \cdots$, we know that $E_1 \subset E_2 \subset \cdots$. So now we see that,

$$\int_E f_n \geq \int_{E_n} f_n \geq \int_{E_n} (1-\epsilon)\phi = (1-\epsilon) \int_{E_n} \phi = (1-\epsilon) \sum_{j=1}^m a_j m(A_j \cap E_n)$$

since inequality is valid on E_n, and the $A_j \cap E_n$. are measurable and disjoint. And now, since E_n increases to E, and thus $E_1 \cap A_j \subset E_2 \cap A_j \subset \cdots$ increases to A_j, then as $n \to \infty$, $m(A_j \cap E_n) \to m(A_j)$. Therefore, taking limits on both sides we find that for all $\epsilon \in (0, 1)$,

$$\lim_{n\to\infty} \int_E f_n \geq \lim_{n\to\infty} (1 - \epsilon) \sum_{j=1}^m a_j m(A_j \cap E_n) = (1 - \epsilon) \sum_{j=1}^m a_j m(A_j) = (1 - \epsilon) \int_E \phi.$$

Taking $\epsilon \to 0$ gives us $\int_E \phi \leq \lim_{n\to\infty} \int_E f_n$,

Theorem 5.2 *Monotone Convergence Theorem: If $\{f_n\}$ is a sequence of nonnegative measurable functions (in $L_{mclass}(E)$) such that $f_1 \leq f_2 \leq \cdots$ pointwise on E, and $f_n \to f$ pointwise on E for some f (in $L_{mclass}(E)$) since the pointwise limit of measurable functions is measurable, then*

$$\lim_{n\to\infty} \int_E f_n = \int_E f.$$

Let us introduce the following corollaries,

Corollary 5.1 *Let $f \in L_{mclass}(E)$, and let $\{\phi_n\}_n$ be a sequence of simple functions such that $0 \le \phi_1 \le \phi_2 \le \cdots \le f$, with $\phi_n \to f$ pointwise. Then $\int_E f = \lim_{n \to \infty} \int_E \phi_n$.*

Here we can take any pointwise increasing sequence of simple functions and determine the limit, instead of computing the supremum explicitly using MCT.

Corollary 5.2 *If $f, g \in L_{mclass}(E)$, then $\int_E (f + g) = \int_E f + \int_E g$.*

Theorem 5.3 *Let $\{f_n\}_n$ be a sequence in $L_{mclass}(E)$. Then*

$$\int_E \sum_n f_n = \sum_n \int_E f_n.$$

Proof 5.6 *By, Corollary 5.2 shows that for each N, we have*

$$\int_E \sum_{n=1}^{N} f_n = \sum_{n=1}^{N} \int_E f_n.$$

Now because

$$\sum_{n=1}^{1} f_n \le \sum_{n=1}^{2} f_n \le \cdots,$$

and by definition of the infinite sum, we have pointwise convergence $\sum_{n=1}^{N} f_n \to \sum_{n=1}^{\infty} f_n$ as $N \to \infty$, the Monotone Convergence Theorem tells us that

$$\int_E \sum_{n=1}^{\infty} f_n = \lim_{N \to \infty} \int_E \sum_{n=1}^{N} f_n = \lim_{N \to \infty} \sum_{n=1}^{N} \int_E f_n = \sum_{n=1}^{\infty} \int_E f_n.$$

as desired.

The following theorem is used in several places in the text in other chapters.

Theorem 5.4 *Let $f \in L_{mclass}(E)$. Then $\int_E f = 0$ if and only if $f = 0$ almost everywhere on E.*

Proof 5.7 *If $f = 0$ almost everywhere, then $f \leq 0$ almost everywhere, implying that $\int_E f \leq \int_E 0 = 0$, so the integral is less than or equal to zero. For the other direction, we define*

$$F_n = \left\{ x \in E : f(x) > \frac{1}{n} \right\}, \quad F = \{ x \in E : f(x) > 0 \}.$$

We know that $F = \bigcup_{n=1}^{\infty} F_n$ since whenever $f(x) > 0$, we have $f(x) > \frac{1}{n}$ for some n large enough, and also $F_1 \subset F_2 \subset \cdots,$. Now we compute the following

$$0 \leq \frac{1}{n} m(F_n) = \int_{F_n} \frac{1}{n} \leq \int_{F_n} f,$$

so that $\frac{1}{n} m(F_n) = 0$ implies $m(F_n)$ for all n, and thus by continuity of measure

$$m(F) = m \left(\bigcup_{n=1}^{\infty} F_n \right) = \lim_{n \to \infty} \sum_{n=0}^{\infty} m(F_n) = 0.$$

as desired.

With fewer assumptions of MCT we have,

Theorem 5.5 *If $\{f_n\}_n$ is a sequence in $L_{mclass}(E)$ such that $f_1(x) \leq f_2(x) \leq \cdots$ for almost all $x \in E$ and $\lim_{n \to \infty} f_n(x) = f(x)$, then $\int_E f = \lim_{n \to \infty} \int_E f_n$.*

Proof 5.8 *Let F be the set of $x \in E$ where we have two assumptions above. So, $m(E \setminus F) = 0$, so $f - \chi_F f = 0$ and $f_n - \chi_F f_n = 0$ almost everywhere for all n. MCT implies,*

$$\int_E f = \int_E f \chi_F = \int_F f = \lim_{n \to \infty} \int_F f_n,$$

where the first equality holds due to the functions $f, f\chi_F$ being equal almost everywhere, and the third equality holds due to $\{f_n\}$ satisfying the assumptions of MCT on F. We then simplify this to

$$= \lim_{n \to \infty} \int_F f_n = \lim_{n \to \infty} \int_E f_n,$$

because $E \setminus F$ has measure zero and thus any integral over the region has measure zero.

Therefore sets of measure zero do not contribute to the Lebesgue integral. An equivalent to the MCT is Fatou's lemma,

Theorem 5.6 *Fatou's lemma: Let $\{f_n\}_n$ be a sequence in $L_{mclass}(E)$. Then*

$$\int_E \liminf_{n\to\infty} f_n(x) \le \liminf_{n\to\infty} \int_E f_n(x).$$

The liminf of a sequence is defined pointwise,

$$\liminf_{n\to\infty} a_n = \sup_{n\ge 1}\left[\inf_{k\ge n} a_k\right],$$

Proof 5.9 *It is true that, since we are taking the infimum over a smaller set,*

$$\liminf_{n\to\infty} f_n(x) = \sup_{n\ge 1}\left[\inf_{k\ge n} f_k(x)\right],$$

and inside the brackets the expression is increasing in n. So the supremum is actually a limit of a pointwise increasing sequence of functions:

$$= \lim_{n\to\infty}\left[\inf_{k\ge n} f_k(x)\right].$$

Now by MCT we have,

$$\int_E \liminf f_n = \lim_{n\to\infty}\int_E \left(\inf_{k\ge n} f_k\right),$$

and for all $j \ge n$ and for all $x \in E$, we know that $\inf_{k\ge n} f_n(x) \le f_j(x)$, so for all $j \ge n$, we have a fixed bound

$$\int_E \inf_{k\ge n} f_k \le \int_E f_j,$$

and we can take the infimum over all j on the right-hand side and have a true inequality,

$$\int_E \inf_{k\ge n} f_k \le \inf_{j\ge n}\int_E f_j.$$

Substituting into the Monotone Convergence Theorem equality gives,

$$\int_E \liminf f_n = \lim_{n\to\infty}\int_E \left(\inf_{k\ge n} f_k\right) \le \lim_{n\to\infty}\left[\inf_{j\ge n}\int_E f_j\right] = \liminf \int_E f_n.$$

As for functions that can be infinite, we have the following important result,

Theorem 5.7 *Let $f \in L_{mclass}(E)$, and let that $\int_E f < \infty$. Then the set $\{x \in E : f(x) = \infty\}$ is a set of measure zero.*

Proof 5.10 *Define the set $F = \{x \in E : f(x) = \infty\}$. We know that for all n, we have $n\chi_F \leq f\chi_F$, so integrating both sides yields*

$$nm(F) \leq \int_E f\chi_F \leq \int_E f < \infty.$$

Therefore, for all n, $m(F) \leq \frac{1}{n}\int_E f$, which goes to 0 as $n \to \infty$, so we must have $m(F) = 0$.

It is true that **every** Riemann integrable function on a closed and bounded interval is Lebesgue integrable and that these two integrals will agree, and this way we completely characterize the functions which are Riemann integrable, they must be continuous almost everywhere.

Definition 5.6 *Let $f : E \to \mathbb{C}$ be a measurable function. For any $1 \leq p < \infty$, we define the L^p norm*

$$\|f\|_{L^p(E)} = \left(\int_E |f|^p\right)^{1/p}.$$

Furthermore, we define the L^∞ norm or essential supremum of f as

$$\|f\|_{L^\infty(E)} = \inf\{M > 0 : m(\{x \in E : |f(x)| > M\}) = 0\}.$$

This is used in Chapter 8 on the subject of the periodic Navier Stokes equations.

Proposition 5.6 *If $f : E \to \mathbb{C}$ is measurable, then $|f(x)| \leq \|f\|_{L^\infty(E)}$ almost everywhere on E. Also, if $E = [a, b]$ is a closed interval and $f \in C([a, b])$, then $\|f\|_{L^\infty([a,b])} = \|f\|_\infty$ is the usual sup norm on bounded continuous functions.*

Theorem 5.8 *Holder's inequality for L^p spaces: If $1 \leq p \leq \infty$ and $\frac{1}{p} + \frac{1}{q} = 1$, and $f, g : E \to \mathbb{C}$ are measurable functions, then*

$$\int_E |fg| \leq \|f\|_{L^p(E)}\|g\|_{L^q(E)}.$$

Theorem 5.9 *Minkowski's inequality for L^p spaces: If $1 \le p \le \infty$ and $f, g : E \to \mathbb{C}$ are two measurable functions, then $\|f + g\|_{L^p(E)} \le \|f\|_{L^p(E)} + \|g\|_{L^p(E)}$.*

A similar result also holds for $L^\infty(E)$.

Definition 5.7 *For any $1 \le p \le \infty$, define the L^p space*

$$L^p(E) = \{f : E \to \mathbb{C} : f \text{ measurable and } \|f\|_p < \infty\},$$

where we consider elements f, g of $L^p(E)$ to be equivalent if $f = g$ almost everywhere.

This last condition to make the L^p norms, and our space is a space of equivalence classes instead of functions:

$$[f] = \{g : E \to \mathbb{C} : \|g\|_p < \infty \text{ and } g = f \text{ a.e.}\}.$$

Theorem 5.10 *The space $L^p(E)$ with pointwise addition and natural scalar multiplication operations is a vector space, and it is a normed vector space under $\| \cdot \|_p$.*

Proposition 5.7 *Let $E \subset \mathbb{R}$ be measurable. Then $f \in L^p(E)$ if and only if (n runs through positive integers)*

$$\lim_{n \to \infty} \int_{[-n,n] \cap E} |f|^p < \infty.$$

Proof 5.11 *We can rewrite our sequence as*

$$\left\{ \int_{[-n,n] \cap E} |f|^p \right\}_n = \int_E \chi_{[-n,n]} |f|^p.$$

Since we know that $\{\chi_{[-n,n]} |f|^p\}$ is a pointwise increasing sequence of measurable functions, and for all $x \in E$ we have

$$\lim_{n \to \infty} \chi_{[-n,n]}(x) |f(x)|^p = |f(x)|^p.$$

Thus, by the Monotone Convergence Theorem,

$$\int_E |f|^p = \lim_{n \to \infty} \int_E \chi_{[-n,n]} |f|^p = \lim_{n \to \infty} \int_{[-n,n] \cap E} |f|^p,$$

and thus, the two quantities are finite for exactly the same set of fs.

Corollary 5.3 *If $f : \mathbb{R} \to \mathbb{C}$ is a measurable function, and there exists some $C \geq 0$ and $q > 1$ so that for almost every $x \in \mathbb{R}$, we have*

$$|f(x)| \geq C(1 + |x|)^{-q},$$

then $f \in L^p(\mathbb{R})$ for all $p \geq 1$.

Exercise.

Proposition 5.8 *Let $a < b$ and $1 \leq p < \infty$ so that $f \in L^p([a,b])$, and take some $\epsilon > 0$. Then there exists some $g \in C([a,b])$ such that $g(a) = g(b) = 0$, so that $\|f - g\|_p < \epsilon$.*

In other words, the space of continuous functions $C([a,b])$ is dense in $L^p([a,b])$, and it's a proper subset because we can find elements in L^p that are not continuous.

Let $s \in \mathbb{R}$, the homogeneous Sobolev space is,

$$\dot{H}^s\left(\mathbb{T}^3\right) := \{f = \sum_{\mathbf{k} \in \mathbb{Z}^3} a_k e^{i\mathbf{k}\cdot\mathbf{x}}; a_0 = 0 \text{ and } \sum_{k \neq 0} |k|^{2s}|a_k|^2 < \infty\} \quad (5.1)$$

with associated norm,

$$\|f\|_{\dot{H}^s} := \left(\sum_{k \neq 0} |k|^{2s}|a_k|^2\right)^{1/2}$$

The inhomogeneous Sobolev Space is,

$$H^s\left(\mathbb{T}^3\right) := \{f = \sum_{\mathbf{k} \in \mathbb{Z}^3} a_k e^{i\mathbf{k}\cdot\mathbf{x}}; a_0 \neq 0 \text{ and } \sum_{k} |k|^{2s}|a_k|^2 < \infty\} \quad (5.2)$$

with associated norm,

$$\|f\|_{H^s} := \left(\sum_{k \in \mathbb{Z}} |k|^{2s}|a_k|^2\right)^{1/2}$$

The particular inhomogeneous Sobolev space $H^{\frac{1}{2}}\left(\mathbb{T}^3\right)$ is a scale invariant space for (PNS). See Chapter 8 on PNS.

The following two theorems are used in Chapter 8 on the subject of the periodic Navier Stokes equations.

Theorem 5.11 *Prekopa-Leindler: Let $0 < \lambda < 1$ and let f, g, and h be nonnegative integrable functions on $\mathbb{R}^n$ satisfying,*

$$h((1-\lambda)x + \lambda y) \geq f(x)^{1-\lambda} g(y)^\lambda,$$

for all $x, y \in \mathbb{R}^n$. Then

$$\int_{\mathbb{R}^n} h(x) dx \geq \left(\int_{\mathbb{R}^n} f(x) dx \right)^{1-\lambda} \left(\int_{\mathbb{R}^n} g(x) dx \right)^\lambda$$

Theorem 5.12 *Gagliardo-Nirenberg: Let $1 \leq q \leq \infty$ and $j, k \in \mathbb{N}$, $j < k$, and either,*

$$\begin{cases} r = 1 \\ \frac{j}{k} \leq \theta \leq 1 \end{cases} \qquad \begin{cases} 1 < r < \infty \\ k - j - \frac{n}{r} = 0, 1, 2, \ldots \\ \frac{j}{k} \leq \theta < 1 \end{cases}$$

If we set $\frac{1}{p} = \frac{j}{n} + \theta \left(\frac{1}{r} - \frac{k}{n} \right) + \frac{1-\theta}{q}$, then there exists constant C independent of u such that

$$||\nabla^j u||_p \leq C ||\nabla^k u||_r^\theta ||u||_q^{1-\theta}, \quad for \quad all \quad u \in L^q(\mathbb{R}^n) \cap W^{k,r}(\mathbb{R}^n)$$

EXERCISES

5.1 (Define l^p space). Prove l^p is a Banach space.

5.2 Prove that
$$c_0 = \{a \in \ell^\infty : \lim_{j \to \infty} a_j = 0\}$$
is a Banach space.

5.3 Prove the following proposition, If $\phi \in L_{mclass}(E)$ is a simple function, then the two definitions of $\int_E f$, for simple functions and general nonnegative measurable functions, agree. If $f, g \in L_{mclass}(E)$, $c \in [0, \infty)$ is a nonnegative real number, and $f \leq g$, then we have $\int_E cf = c \int_E f$ and $\int_E f \leq \int_E g$. Finally, if $f \in L_{mclass}(E)$ and $F \subset E$, then $\int_F f \leq \int_E f$.

5.4 Prove the following proposition,

If $f, g \in L_{mclass}(E)$, and $f \leq g$ almost everywhere on E, then $\int_E f \leq \int_E g$.

5.5 Prove the following theorem, The space $L^p(E)$ with pointwise addition and natural scalar multiplication operations is a vector space, and it is a normed vector space under $||\cdot||_p$.

5.6 Prove the following proposition,

Let $E \subset \mathbb{R}$ be measurable. Then $f \in L^p(E)$ if and only if (n runs through positive integers)

$$\lim_{n \to \infty} \int_{[-n,n] \cap E} |f|^p < \infty.$$

5.7 Prove that,

Let $a < b$ and $1 \le p < \infty$ so that $f \in L^p([a,b])$, and take some $\epsilon > 0$. Then there exists some $g \in C([a,b])$ such that $g(a) = g(b) = 0$, so that $||f - g||_p < \epsilon$.

5.8 Prove the theorem, Let $0 < \lambda < 1$ and let f, g, and h be nonnegative integrable functions on $\mathbb{R}^n$ satisfying,

$$h((1 - \lambda)x + \lambda y) \ge f(x)^{1-\lambda} g(y)^\lambda,$$

for all $x, y \in \mathbb{R}^n$. Then

$$\int_{\mathbb{R}^n} h(x)dx \ge \left(\int_{\mathbb{R}^n} f(x)dx \right)^{1-\lambda} \left(\int_{\mathbb{R}^n} g(x)dx \right)^\lambda.$$

5.9 Show that, If $f : \mathbb{R} \to \mathbb{C}$ is a measurable function, and there exists some $C \ge 0$ and $q > 1$ so that for almost every $x \in \mathbb{R}$, we have

$$|f(x)| \ge C(1 + |x|)^{-q},$$

then $f \in L^p(\mathbb{R})$ for all $p \ge 1$.

5.10 Let $C^1([a,b])$ be the space of continuous functions on $[a,b]$ whose derivative (one-sided derivative at the endpoints) exists and is continuous on $[a,b]$.

Suppose $u \in C([a,b]) \cap C^1((a,b))$ and u' can be extended continuously to $[a,b]$. Show that $u \in C^1([a,b])$.

Show that $||u|| = \sup_{x \in [a,b]} |u(x)| + \sup_{x \in [a,b]} |u'(x)|$ is a norm on $C^1([a,b])$ which makes $C^1([a,b])$ into a Banach space. [Hint for completeness: if $\{u_n\} \subset C^1([a, b])$ satisfies $u_n \to u$ uniformly and $u'_n \to v$ uniformly, take limits in the equation $u_n(x) - u_n(a) = \int_a^x u'_n(s)\, ds$ to show that $u \in C^1([a, b])$ and $u' = v$.]

5.11 If $0 < A \le 1$, a function $u \in C([a, b])$ is said to satisfy a Holder condition of order A (or to be Holder continuous of order A) if

$$\sup_{x \ne y} \frac{|u(x) - u(y)|}{|x - y|^A} < \infty.$$

Denote by $\wedge^A([a, b])$ this set of functions.

Show that

$$\|u\|_A = \sup_{x \in [a,b]} |u(x)| + \sup_{x \ne y} \frac{|u(x) - u(y)|}{|x - y|^A}$$

is a norm which makes $\wedge^A$ into a Banach space. Show that $C^1([a, b]) \subset \wedge^A([a, b])$ and that the inclusion map is continuous with respect to the norms defined above (i.e., the norm on $C^1([a, b])$ defined in exercise 11 and that on $\wedge^A([a, b])$ defined.).

5.12 Prove that the dual norm to the ℓ^2 norm on $\mathbf{F}^n$ is again the ℓ^2 norm, and show that the ℓ^2 norm is the only norm on $\mathbf{F}^n$ with this property.

## CHAPTER 6

Simplified Periodic Navier–Stokes (PNS) and Rayleigh-Plesset (RP) Equations

Keith Afas and Terry E. Moschandreou
Western University & TVDSB

6.1 INTRODUCTION

Fluid mechanics is a field which possesses diversity in approaches, solutions, and versatility. The crown jewels of the field of fluid mechanics - the Navier–Stokes equations (NSEs) - are among some of the most impactful mathematical equations. One phenomenon that arises from the NSEs, vortex or pattern formation, is caused by non-linear interactions in complex fluid systems. It has been seen that by prescribing initial conditions in terms of a Fourier series, it is possible to trigger instabilities from a primary static pattern determined by particular initial conditions. This is amenable to treatment on a 3-D Torus or Lattice of periodic cells [19]; these instabilities can be analyzed as strong indicators to a possible transition to disorder [37].

In general, instabilities can be represented by saddle points of the associated system, and the same may be extended to the Navier–Stokes system. For example, networks of webs of saddle points have

 DOI: 10.1201/9781003452256-6

been shown to exist for the Kuramoto-Sivashinsky equation [38,39] with similar possibilities for the Navier–Stokes equations; Saddle points in general are associated with unstable solutions of PDE systems, the prototypical example of local instability analysis given in earlier computational studies [40]. Various instabilities may occur that eventually develop into turbulence with vortices and eddies, and in some cases lead to Spatio-Temporal Chaos (STC) [41]. Moving saddle points play an important role in turbulent dynamics and strong turbulent motion is connected and related to saddle point analysis. Recent work on Incompressible flows, velocity-pressure correlation, and global regularity can be seen in earlier studies [42], as well as time-dependent flows of the two-dimensional NSEs on a Torus [43]. In addition, while the principles forming the NSEs rely on intuitive conservation principles from Continuum Mechanics, such as the conservation of mass and momentum, the resultant NSEs describing the general motion of fluids are immensely complex. NSEs often do not admit analytic solutions without perturbation analysis or asymptotic approximations. As such, their existence and uniqueness remain elusive as the motivation of the famed Millenium Problem, proposed by Fefferman in 2008 [44]. While the NSEs are generally accepted as the prime governing equations in fluid analysis, there has been a push throughout centuries to propose alternative models that address specific fluid processes in a manner that they consistently admit analytical results. These alternative models may be shown to either draw from the NSEs, or be phenomenologically derived; in the latter case, phenomenological models may be validated to mimic numerical predictions from the NSEs with small deviations, though not being obtained from the NSEs. Consequentially, it is difficult to determine if phenomenological models encompass conservation principles implicitly.

One field of research that enjoys such alternative models is the field governing matter transitions, such as those that study dynamics of fluid inclusions, informally referred to as "bubbles" [45]. Bubbles have a rich history in fluid mechanics and soft condensed matter physics, as well as in fields of mathematics such as differential geometry. Their attention from several fields of research originates due to the versatility of their geometries, and the simplicity of their physics principles. One of the earliest studies on bubbles was performed in the early 19th century, and culminated in the Young–Laplace Equation [46]:

$$R = \frac{2\gamma}{\Delta P} \tag{6.1}$$

The equation, relating the steady-state radius R of a homogeneous spherical bubble to its surface tension γ and pressure differential with its exterior ΔP, was well formulated by Thomas Young and Pierre Simon de Laplace; In 1830, the connection to differential geometry was made by Carl Friedrich Gauss realizing the spherical radius as the membrane's curvature, establishing the equation in its celebrated form today. The Young–Laplace-Gauss equation culminated in the realization that bubbles were affected by surface tension and pressure which locally affected its curvature. This connection to differential geometry emphasized that bubbles are in themselves, not only phenomenon relating to the NSEs, but rather distinct geometrical entities simultaneously. The realization of the relation between the Young-Laplace-Gauss equation to the NSEs occurred through the work of Lord Rayleigh and Milton Plesset in the 20th century culminating in the Rayleigh-Plesset equation. This equation extends the Young-Laplace-Gauss equation from being able to predict steady-state radius, to being able to predict a time-varying bubble radius; therefore, the non-linear Rayleigh-Plesset equation is most often used for cavitation dynamics. The equation was obtained in the 20th century through a thorough application of the NSEs, assuming a cavity with a pressure difference, an external liquid with a pre-specified density and viscosity, and incorporating boundary conditions [47]. Using these observations, an approximation to the NSEs resulted in the following non-linear ODE for the spherical radius $R(t)$ of a cavitation bubble:

$$\rho_L \overbrace{\left(R\partial_t^2 R + \frac{3}{2}\left(\partial_t R\right)^2 \right)}^{\text{Momentum Term}} + \overbrace{\frac{4\nu_L \rho_L}{R}\partial_t R}^{\text{Viscosity Term}} + \overbrace{\frac{2\gamma}{R} + \Delta P}^{\text{Young-Laplace}} = 0 \tag{6.2}$$

In the steady-state case where $\partial_t = 0$, the Young-Laplace Equation emerges. While the Rayleigh-Plesset equation accomplishes much in understanding cavitation dynamics, one of the critical consequences is the relation to differential geometry is lost since the equation is derived from the NSEs. What is lost is the application to non-spherical surfaces, or even non-closed geometries; this is since the dynamics of such surfaces are matters pertaining to differential geometry, and not exclusively to the NSEs. In addition, it is difficult to see how the equation might be generalized further to more complex fluid mechanics matters; an example is the spontaneous emergence of eddies and vortices present in 2D incompressible fluids. This has been commented in earlier studies [48], however the additional presence of cavitation dynamics is difficult to relate to the field of vortex dynamics which itself is a phenomenological sub-branch of fluid mechanics.

One possible solution is to provide an alternative derivation of the Rayleigh-Plesset equation from a starting point other than the NSEs. Such a derivation would simply require a framework that is geometrical in nature, accommodates for kinetic processes, and has physical conservation principles embedded in its framework, or at the very least, implementable as constraints. One such framework, developed in 2019, has the potential to predict the evolution of surfaces, subject to an energy functional, and also permits the implementation of constraints requiring a conservation of surface area and/or volume [49]. This framework was formulated under the assumption of invariance under the Calculus of Moving Surfaces (CMS), and as such is especially well suited for considering the time-evolution of irregular geometries which may be intractable by modern physics. Not only would a derivation of the Rayleigh-Plesset Equation using this CMS variational framework open up the possibility of a way to treat fluids outside of the NSEs, but would also allow for the analysis of cavitation and other processes within incompressible eddies and vortices.

6.2 ONSET OF TURBULENCE: EDDIES AND VORTICIES IN INCOMPRESSIBLE FLUIDS

In this section, a general presentation of a solution of the incompressible NSEs will be given, and the existence of instabilities leading to the formation of eddies and vortices in incompressible fluids will be demonstrated by showing there exist saddle points, and possibly limit cycles (attractors that are closed orbits). The analytical solution of the NSEs presented in this chapter is expressed as an asymptotic series in terms of its asymptotic variable which can be integrated a second time, term by term with respect to time to obtain time-varying spatial variables in the phase portraits examined. Considering a cubical lattice, the presence of singularities will also be addressed using the concept of rotund surfaces.

6.2.1   3D Incompressible Navier–Stokes Equations

The 3D incompressible unsteady Navier–Stokes Equations NSEs in Cartesian coordinates may be written in the form for the velocity field $\mathbf{u}^* = u^{*i}\vec{\mathbf{e}}_i$, $u^{*i} = \{u_x^*, u_y^*, u_z^*\}$ are:

$$\rho\left(\frac{\partial}{\partial t^*} + u^{*j}\nabla_{*j}\right)u_i^* - \mu\nabla_*^2 u_i^* + \nabla_{*i}P^* = \rho F_i^* \tag{6.3}$$

where ρ is constant density, μ is dynamic viscosity, and $\mathbf{F}^* = F^{*i}\vec{\mathbf{e}}_i$ are body forces on the fluid. The components of the velocity vector, and pressure in $\mathbf{u} = (u)^i\vec{\mathbf{e}}_i$, $\mathbf{P} = (P)^i\vec{\mathbf{e}}_i$, coordinates $\mathbf{x}_i$ and time t, are reparametrized according to the following form utilizing the non-dimensional quantity δ(assumed negative and such that $|\delta|$ is arbitrarily small):

$$u_i^* = \frac{1}{\delta}u_i \ , \ P_i^* = \frac{1}{\delta^2}P_i \ , \ x_i^* = \delta x_i \ , \ t^* = \delta^2 t \tag{6.4}$$

Along with Eq. (6.3), the continuity equation in Cartesian co-ordinates, is given in tensor index notation by:

$$\nabla^i u_i = 0 \tag{6.5}$$

6.2.2   Decomposition of NSE's, Limit Cycles, and Vorticies

Equation (6.3), together with Eq. (6.5) and using the initial condition of $\vec{u}^*(\vec{x}^*, 0) = \vec{\xi}(\vec{x}^*)$ such that $\nabla \cdot \vec{\xi} = 0$ encompass the NSEs along with an incompressible initial condition. Ensuring periodic boundary conditions specified in [44] defined on a cube domain Ω with associated Lattice in $\mathbb{R}^3$ is referred to as the periodic BVP for the NSEs in $\mathbb{R}^3$. It has been shown in earlier NSE work [19] that a solution for Eq. (6.3) with transformation given by Eq. (6.4) exists in the form,

$$\mathbf{u} = (u_x, u_y, u_z) : R^3 \to R^3 \tag{6.6}$$

where u_z in Eq. (6.4) satisfies the following integral equation,

$$\frac{\partial}{\partial t}\left(\frac{\partial u_z}{\partial z}\right) + \frac{1}{u_z^2}\nabla u_z^2 \cdot \frac{\partial \vec{b}}{\partial t} + \left(\frac{1}{\delta} - 1\right)\frac{1}{u_z^2}\left(\frac{\partial u_z}{\partial t}\right)^2 + \frac{1}{u_z^2}\left(u_z\frac{\partial u_z}{\partial z}\frac{\partial u_z}{\partial t}\right)$$

$$+ \frac{\mu}{\rho} \frac{1}{u_z^2} \left(1 - \frac{1}{\delta}\right) \frac{\partial u_z}{\partial t} \nabla^2 u_z + \left(\frac{1}{\delta} - 1\right) \frac{1}{\rho} \frac{1}{u_z^2} \frac{\partial u_z}{\partial t} \frac{\partial P}{\partial z}$$

$$+ \frac{1}{u_z^2} \iint_S \left(\frac{1}{\delta \rho} u_z^2 \nabla_{xy} P + \vec{b} \frac{1}{\rho} u_z \frac{\partial P}{\partial z}\right) \cdot \vec{n} \, dS + \delta^2 \frac{1}{u_z^2} \vec{F_T} \cdot \nabla u_z^2$$

$$- \frac{1}{u_z^2} \delta^3 u_z \frac{\partial u_z}{\partial t} \frac{\partial u_z}{\partial z} F_z + \delta^3 \vec{b} \cdot \nabla \left(u_z F_z\right)$$

$$= \frac{1}{\|\vec{b}\| u_z^2} \int_U \|\frac{\partial u_z}{\partial t} \vec{b} \cdot \left(\vec{b} \otimes \nabla u_z\right)\| dV \tag{6.7}$$

Here, it is assumed that $\vec{b}$ from Eq. (6.7) is given by:

$$\vec{b} = \tfrac{1}{\delta} u_x \vec{i} + \tfrac{1}{\delta} u_y \vec{j} \tag{6.8}$$

and δ is arbitrarily small. The x, y force is $\vec{F_T} = (F_{T_1}, F_{T_2})$ and the z force is F_z. It is worth noting that Poisson's Equation can be written as in [19],

$$\tfrac{1}{\rho} P_{zz} = -\triangle(\tfrac{1}{2} u_z^2) + \left(\tfrac{\partial u_z}{\partial x}\right)^2 + \left(\tfrac{\partial u_z}{\partial y}\right)^2 - u_z \nabla_{xy}^2 u_z - \left(\tfrac{\partial^2}{\partial z \partial t} uz + \tfrac{\partial^3}{\partial z^3} uz\right) \tag{6.9}$$

It was shown in [19] that since,

$$P_{xx} + P_{yy} = \tfrac{\partial^2}{\partial z \partial t} uz \, (x, y, z, t) + \tfrac{\partial^3}{\partial z^3} uz \, (x, y, z, t) \,, \tag{6.10}$$

then,

$$\tfrac{1}{\rho} \triangle P = -\triangle(\tfrac{1}{2} u_z^2) + \left(\tfrac{\partial u_z}{\partial x}\right)^2 + \left(\tfrac{\partial u_z}{\partial y}\right)^2 - u_z \nabla_{xy}^2 u_z \tag{6.11}$$

So solving for P_z in Eq. (6.7), and differentiating with respect to z thus obtaining P_{zz} gives a PDE in terms of u_z and $\vec{b}$ alone. It can be determined that

$$\begin{cases} u_x = C_* \left(\sin\left(4 \, t^*\right) x \delta^2 + \left(\cos\left(4 \, t^*\right) + 2\right) y \delta^2\right) \\ u_y = C_* \left(\left(\cos\left(4 \, t^*\right) - 2\right) x \delta^2 - \sin\left(4 \, t^*\right) y \delta^2\right) \end{cases} \tag{6.12}$$

where C_* is a real valued scaling constant. Here, the velocity $\vec{b}$ is a kinematic benchmark example for testing vortex criteria [48]. Also $\frac{\partial u_z}{\partial z} = -\nabla \cdot \vec{b} \neq 0$. It is an indeterminate form $0/0$ at $\delta = 0$. Solving for u_z in Eq. (6.7) gives a class of solutions G_i. Here the defining equations are in terms of the vorticity components in 3d. Since u_x and u_y are z-independent, the vorticity component equations for a constant vorticity are, $\frac{\partial u_z}{\partial x} = -C$, $\frac{\partial u_z}{\partial y} = C$ and $\frac{\partial u_y}{\partial x} - \frac{\partial u_x}{\partial y} = C$. It has been shown [19] and detailed in Appendix that u_z may be given by,

$$u_z = F_0 + F(x,y)H(z) \times$$
$$G\left(y, 8 \int c_3 y \left(\sin\left(4\,t\right) - \cos\left(4\,t\right)\right) e^{-2\,c_3(\sin(4\,t)-\cos(4\,t))}\, dt + xe^{2\,c_3(\cos(4\,t)-\sin(4\,t))}\right)$$

$$(6.13)$$

where $F_0 \in \mathbb{R}^+$ is a lifting constant and,

$$H = C_1 + C_2\, e^{\frac{z}{\sqrt{c_3}}} + C_3\, e^{-\frac{z}{\sqrt{c_3}}} \tag{6.14}$$

In addition it should be noted that $F(x,y)$ is a general function to be determined, C_1, C_2, C_3 and c_3 are arbitrary constants, and G belongs to a general family of functions. Elements of the family of solutions G have second partial derivative with respect to t equal to zero since by a dimensional argument the jerk vector is zero for arbitrarily small parameter. The full expression for G is shown here to be exactly in the form,

$$G = \int \exp\left(\frac{\left(-\frac{1}{8} - \frac{i}{8}\right) e^{4\,it - 2\,c_3\,\cos(4\,t) + 2\,c_3\,\sin(4\,t)}}{c_3\left(iye^{8\,it} + ix + e^{8\,it}x + y\right)}\right) dt \tag{6.15}$$

where $i = \sqrt{-1}$. In the first part of this paper, the long-run behavior of the trajectories of the associated dynamical system of the equivalent form for the 3d Navier–Stokes equations Eqs. (6.3) and (6.4) - that is Eq. (6.7) - is proven to exist, and it is shown that the trajectories for $n = 3$ converge in time to a closed orbit of period T in each cell of the lattice. The system is also shown to have saddle points and orbits whose interiors are saddle surfaces. Before the proof of the main result, some preliminary definitions and notations are introduced:

Solutions are obtained in nonstarred variables first and then results are shown in Figure 6.1a and b in starred variables. An upper bound for the integral can be expressed as follows. The arithmetic mean-geometric mean Inequality for complex numbers is used here [50]. Let $\phi \in (0, \pi/2)$ and $W_\phi = \{z \in C :| \arg z \,|\leq \phi\}$. Then for all $z_1, \ldots, z_n \in W_\phi$,

$$\prod_{k=1}^{n} |\, z_k\,|^{\frac{1}{n}} \leq \sigma\left(\frac{1}{n}\left[\frac{n}{2}\right], \cos(2\phi)\right) |\,\frac{1}{n}\sum_{k=1}^{n} z_k\,| \tag{6.16}$$

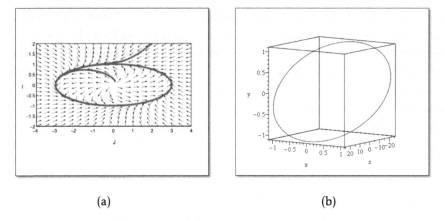

(a)   (b)

Figure 6.1 Example of a periodic orbit in a cube. (a) Elliptical limit set in oblique coordinates. (b) Elliptical limit set in lattice cubical cell.

where

$$\sigma(\mu,a) = \frac{\sqrt{2}(2\mu)^{-\mu}\left(a(2\mu-1)+\sqrt{4\mu(1-\mu)+a^2(1-2\mu)^2}\right)^{\mu}}{\left(2(1-\mu)+a^2(2\mu-1)+a\sqrt{4\mu(1-\mu)+a^2(1-2\mu)^2}\right)^{1/2}}$$

(6.17)

For $n = 2$, there is a summative bound on the product of the two complex functions z_1, z_2 in Eq. (6.16),

$$\begin{cases} z_1 = \left(-\frac{1}{8}-\frac{i}{8}\right)e^{4it-2\,c_3\cos(4t)+2\,c_3\sin(4t)} \\ z_2 = \left[c_3\left(iye^{8it}+ix+e^{8it}x+y\right)\right]^{-1} \end{cases}$$

(6.18)

Here, the terms in Eq. (6.18) appear in Eq. (6.15). The following chain of inequalities results for the modulus of the complex function $\int_a^t \exp w(s)ds$, where $w = z_1 z_2$:

$$\left|\int_a^t e^{w(s)}ds\right| \leq \int_a^t \left|e^{w(s)}\right|ds \leq \int_a^t e^{|w(s)|}ds$$

(6.19)

$$\int_a^t e^{|w(s)|}ds = \int_a^t e^{|z_1 z_2|}ds \leq \int_a^t e^{\zeta|z_2|}ds$$

(6.20)

where $\zeta = \sup_{\mathbb{K}}|z_1|$ where $\mathbb{K} = \{s \in \mathbb{R}|a \leq s \leq t]\}$. The upper bound at the end of the chain above has a closed-form antiderivative. It is shown in this chapter that there potentially exists a limit cycle which converges to an orbit associated with a saddle surface. It has been verified that $\int_a^t e^{\zeta|z_2|}ds$ has existing saddle points which are secondary instabilities emerging from a stationary periodic structure. Integral evaluation is immediate in terms of exponential integral functions, and furthermore this solution can be integrated to obtain $\int\int e^{\zeta|z_2|}dtdt$. This may be obtained by using an asymptotic series expansion for large parameter c_3 found in most symbolic and numerical libraries (ie. Maple 2021's *asympt* command).

In the result of Figure 6.2c and d, $\sqrt{c_3}\delta \approx 1$ due to symmetry of $H(z^*)$ in Eq. (6.14). Here in the limit as $\delta \to 0$, $\sqrt{c_3} \to \infty$. Finally, since an upper bound is established which are the closed orbits it can only be concluded that the field lines attract to the closed orbit from within it starting at the zero vector in $\mathbb{R}^3$ toward a maximum or on the other side towards a minimum. An example of a limit cycle is shown in Figure 6.1.

In Figure 6.1a, a slanted elliptical limit cycle can be seen plotted along coordinates I and J which are oblique coordinates. The embedding of this elliptical limit set in $\mathbb{R}^3$ can be seen in Figure 6.1b.

6.2.3 Convergence & Singularities of Incompressible Eddies and Vortices along Edge of Cube Lattice

An important criterion for proving that a sequence of real numbers converges without knowing its limit is named the Cauchy criterion:

Definition. Let $\{Sn\}_{n=1}^{\infty}$ be a sequence of real numbers. Then $\{Sn\}_{n=1}^{\infty}$ is called a Cauchy sequence if for any $\epsilon > 0$ there exists an $N \in I$ such that

$$| S_m - S_n | < \epsilon \quad (m, n \geq N). \tag{6.21}$$

The Cauchy criterion is used to show that spacecurves $(x^*(t), y^*(t), z^*(t))$ parametrized in t(time) converge to the orbit in Figure 6.2a–d that is a closed periodic orbit. In Maple 2021 this was shown by keeping a length of interval fixed at say 2 units in t and shifting the interval towards very large values. The measure of the interval stays the same but the orbit generated converges (to Figure 6.2a and c). In Figure 6.2b and d, the limit of the integral of solution as c_3 approaches infinity was

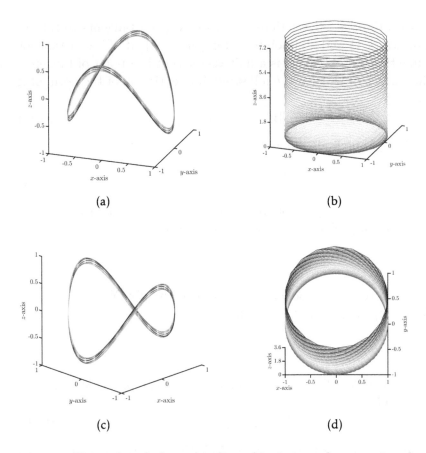

(a)

(b)

(c)

(d)

Figure 6.2 Example of the periodic orbit in a cube associated with Eq. (6.14). (a) Saddle-type orbit (c_3=1280). (b) 2D set-limit cycle for $c_3 \to \infty$. (c) Figure 6.2a rotated. (d) Figure 6.2b rotated.

obtained explicitly to obtain the portrait to show that the attracting set is a two-dimensional set and finally for c_3 arbitrarily small that is an amplitude that becomes arbitrarily large gave way to a swirling orbit towards minus infinity for z^* (Figure not shown). There is the question of singularities occurring along the edges and at corners of a cube lattice. A cube in the lattice can be transformed into a slight rotund ball by slightly rounding its vertices and edges and puffing out its facets. Then the supporting lines and planes of Ω can be regarded as tangents, each one touching the surface of Ω at one and only one point

[51]. A continuous 1-1 relationship exists between tangents to rotund boundaries of Ω and the points of tangency [51]. If $\mathbf{Y}^T$ is a tangent unit functional on Ω and $\mathbf{x}$ is a unit vector that is a point of tangency on the surface of Ω, then the following differentiation of the norm of $\mathbf{x}$ holds,

$$d\|\mathbf{x}\| = \mathbf{Y}^T d\mathbf{x}/\|\mathbf{x}\| \qquad (6.22)$$

Equivalently, one finds differentiating with respect to t,

$$\frac{d\|\mathbf{x}\|}{dt} = \mathbf{Y}^T \frac{d\mathbf{x}}{dt}/\|\mathbf{x}\| \qquad (6.23)$$

where T is the transpose of the vector $\mathbf{Y}$. Recalling G in Eq. (6.15), and the ensuing velocity function u_z, it is clear that,

$$\mathbf{x_3} = \int u_z dt \qquad (6.24)$$

where $\mathbf{x_3}$ is the z- component of $\mathbf{x}$. Substituting for $\mathbf{Y}_3 = u_z$ and $\mathbf{x_3}$ in Eq. (6.24) into Eq. (6.23), it follows that,

$$u_z^2 = \frac{1}{2}\frac{d\|\mathbf{x}\|^2}{dt} \qquad (6.25)$$

Let $\|N\| = \|\mathbf{x}\|^2$. The Energy equality for Incompressible NSE's as used in [19,52] is:

$$\frac{d\|\mathbf{u}\|^2}{dt} = -2\nu\|\nabla\mathbf{u}\|^2 \qquad (6.26)$$

where ν is the viscosity of the fluid. Substituting u_x, u_y, and u_z, into Eq. (6.26) and retaining only the integrand,

$$2\,u_x\,(x,y,t)\,\tfrac{\partial}{\partial t}u_x\,(x,y,t) + 2\,u_y\,(x,y,t)\,\tfrac{\partial}{\partial t}u_y\,(x,y,t) + F\,(x,y)\,\tfrac{\partial^2}{\partial t^2}N\,(x,y,t)$$
$$= -4\nu\left(4\,(\sin(4\,t))^2 + \tfrac{F(x,y)C_2^2 \frac{\partial}{\partial t}N(x,y,t)}{c_3}\left(e^{\frac{z}{\sqrt{c_3}}}\right)^2\right)$$

$$(6.27)$$

This can be simplified to the following form:

$$(2\sin(4\,t)\,x + 2\,(\cos(4\,t)+2)\,y)\,(4\cos(4\,t)\,x - 4\sin(4\,t)\,y)$$
$$+ (2\,(\cos(4\,t)-2)\,x - 2\sin(4\,t)\,y)\,(-4\sin(4\,t)\,x - 4\cos(4\,t)\,y)$$
$$+ F\,(x,y)\,\tfrac{\partial^2}{\partial t^2}N\,(x,y,t) = -4\nu\left(4\,(\sin(4\,t))^2 + \tfrac{F(x,y)C_2^2 \frac{\partial}{\partial t}N(x,y,t)}{c_3}\left(e^{\frac{z}{\sqrt{c_3}}}\right)^2\right)$$

$$(6.28)$$

Solving for $N(\mathbf{x}, t)$, differentiating it with respect to t and then taking the square root gives u_z. Using this function together with Eqs. (6.13 and 6.15), and upper bound $\int_a^t e^{\zeta|z_2|}$ the solution for u_z obtained, gives two forms of expression for u_z for the rotund surface Ω. For the first form given by Eq. (6.15) and it's upper bound using the initial condition by setting $t = 0$ and equating u_z to $A \cos(m(x - 8\pi)) \cos(n(y - 8\pi)) \sin(p(z - 8\pi))$, where A is the amplitude of the sinusoid and a representative shift for values of u_z in the cube with center $(8\pi, 8\pi, 8\pi)$ or $\mathbf{0}$. The dimensions of each cube of the lattice are $(2\pi)^3$. The general form of solution in Eq. (6.13) upon using the initial condition leads to obtaining $F(x, y)$. Recall F_0 is any positive constant. Substituting this $F(x, y)$ into the solution of Eqs. (6.26) and (6.28) gives a PDE with an analytical solution. Again by use of initial condition the integration unknown function $F_1(x, y)$ obtained by integrating the PDE in Eq. (6.28) can be determined readily. A final expression was thus obtained for u_z and ultimately $\int u_z dt$ with orbits shown in Figure 6.2a–d. Due to the periodic Lattice similar results can be obtained for any cube of $(2\pi)^3$ dimension in the Lattice.

An implied step in [Moschandreou, 2021], in particular Inequality Eq. (25) on pg. 390 leading to Lemma 2 found in [19], the division by δ^5 for arbitrary small δ is taken with a division by the norm $\|u_z\|$ following. In starred variables this is consistent with the infinity norm being of strong growth as supported by Tran et al. [42] and summed up well in their conclusion. In fact $\|u_z^*\|\delta$ in [42] becomes finite, note: $\|u_z^*\delta\| = |\delta|$ $\|u_z^*\|$ is of a finite form $0 \cdot \infty$. Furthermore, there exists a high correlation between low pressure and high velocity for the flows considered which implies that the L^q norm is of strong growth and this was observed for u_z and P in the present work. Integrating P_{zz} in Eq. (6.10) twice with respect to z led to the result $\{\exists z_0 \mid P''(z_0) > 0, u''(z_0) < 0\}$ as $t \to \infty$. It is evident that a concave up pressure graph exists for the geometry studied in this paper together with a minimum of pressure located near the maximum of velocity. The corroborative finding in [42] brings into light the importance of the quantitative size of the infinity norm as used in [19].

6.2.4 Norm Analysis of Eq. (6.7)

From [19] with $q \to \infty$, $\theta = 1/2$ and $\delta \to 0$ the following inequalities were proven with an instant of a strict inequality,

$$C \mid \Omega \mid \|u_z\|_q^{1-\theta} \sup_{x \in \Omega} \mid \frac{\partial u_z}{\partial t} \mid \geq C \int_\Omega \left(\|u_z\|_q^{1-\theta} \mid \frac{\partial u_z}{\partial t} \mid \right) dv$$

$$\geq C \int_\Omega \left(\|u_z\|_q^{1-\theta} \frac{\partial u_z}{\partial t} \right) dv \qquad (6.29)$$

$$C \int_\Omega \left(\|u_z\|_q^{1-\theta} \frac{\partial u_z}{\partial t} \right) dv = C \int_\Omega \|u_z\|_q^{1-\theta} \left(\nabla^2 u_z + \mid \Phi \mid - \vec{u} \cdot \nabla u_z \right) dv$$

$$C \int_\Omega \left(\|u_z\|_q^{1-\theta} \frac{\partial u_z}{\partial t} \right) dv = C \int_\Omega \|u_z\|_q^{1-\theta} \left(\nabla^2 u_z + \mid \Phi \mid - \vec{b} \cdot \nabla u_z \right) dv$$

$$C \int_\Omega \left(\|u_z\|_q^{1-\theta} \frac{\partial u_z}{\partial t} \right) dv > C \|u_z\|_q^{1-\theta} \int_\Omega \nabla^2 u_z dv$$

$$\geq C \|u_z\|_q^{1-\theta} \left[\mid \Omega \mid^{\frac{1}{2}} \|\nabla^2 u_z\|_r^\theta \right]$$

$$C \int_\Omega \left(\|u_z\|_q^{1-\theta} \frac{\partial u_z}{\partial t} \right) dv \geq \mid \Omega \mid^{\frac{1}{2}} \|\nabla u_z\|_2 \qquad (6.30)$$

Recalling the Energy Equality [52], then $\mid \Omega \mid^{\frac{1}{2}} \|\nabla u_z\|_2 = -\frac{1}{\sqrt{2\nu}} \mid \Omega \mid^{\frac{1}{2}}$
$\left(\frac{d}{dt} \|u_z\|_2^2 \right)^{1/2}$.

From Inequality Eq. (6.30), it follows upon integrating by parts that,

$$-\frac{1}{\sqrt{2\nu}} \left(\frac{d}{dt} \|u_z\|_2^2 \right)^{1/2} \leq C \int_\Omega \|u_z\|_q^{1/2} \frac{\partial u_z}{\partial t} dV$$

$$-\delta^{-1} \frac{1}{\sqrt{2\nu}} \int \left(\frac{d}{dt^*} \|\delta u_z^*\|_2^2 \right)^{1/2} dt^* \leq \frac{C}{\delta^2} \left(\int_\Omega u_z^* dV^* \right) \|u_z\|_q^{1/2}$$

$$-\frac{C}{\delta^2} \int \left(\int_\Omega u_z^* dV^* \right) \frac{d}{dt^*} \|u_z\|_q^{1/2} dt^* \qquad (6.31)$$

For δ sufficiently small, since,

$$-\delta^{-1} \frac{1}{\sqrt{2\nu}} \int \frac{d}{dt^*} \|\delta u_z^*\|_2^2 dt^* \leq -\delta^{-1} \frac{1}{\sqrt{2\nu}} \int \left(\frac{d}{dt^*} \|\delta u_z^*\|_2^2 \right)^{1/2} dt^* \qquad (6.32)$$

then

$$-\delta^{-1} \frac{1}{\sqrt{2\nu}} \int \frac{d}{dt^*} \|\delta u_z^*\|_2^2 dt^* < \frac{C}{\delta^2} \left(\int_\Omega u_z^* dV^* \right) \|u_z\|_q^{1/2}$$

$$-\frac{C}{\delta^2} \int \left(\int_\Omega u_z^* dV^* \right) \frac{d}{dt^*} \|u_z\|_q^{1/2} dt^* \qquad (6.33)$$

and in terms of the Energy Integral,

$$\mid \delta \mid \frac{1}{\sqrt{2\nu}} \frac{d}{dt^*} E(t^*) < \frac{C}{\delta^2}(\int_\Omega u_z^* dV^*)\|u_z\|_q^{1/2} - \frac{C}{\delta^2} \int (\int_\Omega u_z^* dV^*)\frac{d}{dt^*}\|u_z\|_q^{1/2} dt^* \tag{6.34}$$

where

$$E(t^*) = \tfrac{1}{2} \int \|u_z^*\|_2^2 dt^* \tag{6.35}$$

Here the derivative of the norm is taken to be greater than the norm for an inflated norm, that is, one that is increasing quickly with respect to time. Transforming u_z to δu_z^* and obtaining $\mid \delta \mid^q$ powers for δ arbitrarily small, since $q \to \infty$, then the right side of previous strict inequality is zero. Here it is seen that the derivative of Energy is less than zero and therefore there is a loss of energy in the system of the viscous incompressible Navier–Stokes equations.

6.3 NOVEL VARIATIONAL FORMULATION OF CAVITATION DYNAMICS

In Section 6.2, the existence of eddies and vortices in incompressible fluids was demonstrated. As discussed earlier, in most hydrodynamic events, cavitation dynamics bubble formation can occur where the kinetic energy and pressure variations in the body of fluid give rise to separated phases of gas known as bubbles [45]. Normally, this is governed by the Rayleigh Plesset Equation derived phenomenologically from the NSEs shown in Eq. (6.2). Here, an alternative statistical mechanical derivation will be provided where a Lagrangian Density assuming an energy dissipation that results in the Rayleigh Plesset Equation will be presented, and an application to cavitation dynamics in eddies and vortices will be proposed.

6.3.1 Membrane Statistical Dynamics as a Variational Technique

In previous studies, a technique for identifying minimal energy configurations for moving membranes that minimizes a dynamic Lagrangian density was obtained [49]. This utilized both statistical mechanics techniques and a modern classical differential geometry approach to moving surfaces termed the Calculus of Moving Surfaces (CMS) [53]. In the previous studies, it was established that for a surface Σ embedded in $\mathbb{R}^3$ by $\mathbf{R} : \Sigma \to \mathbb{R}^3$, if its Surface Action may be expressed by:

$$\mathcal{S} = \int_t \int_\Sigma \mathcal{L}(\mathbf{R}, S_{\alpha\beta}, B_{\alpha\beta}, C) \, d\Sigma dt \qquad (6.36)$$

where $S_{\alpha\beta}$ and $B_{\alpha\beta}$ are its first and second fundamental forms respectively, and $C = \mathbf{V} \cdot \mathbf{N}$ is its normal velocity, assuming it possesses a velocity of $\mathbf{V} = \partial_t \mathbf{R}$ and normal of $\mathbf{N}$. Under the following definitions, and using the conventions from earlier usages [49], a dynamic membrane's CMS-Invariant evolution may be given by the following allowing the operator $\tilde{\delta}_t = \left(\dot{\nabla} - CB_\alpha^\alpha \right)$ to be defined:

$$\delta\mathcal{S} = 0 \rightarrow \mathbf{N} \cdot \left(\frac{\delta\mathcal{L}}{\delta\mathbf{R}} + \nabla_\alpha \mathbf{F}^\alpha - \tilde{\delta}_t \left(\frac{\delta\mathcal{L}}{\delta C} \mathbf{N} \right) \right) = 0 \qquad (6.37)$$

where ∇_α is Σ's co-variant derivative, $\dot{\nabla}$ is Σ's CMS-invariant time derivative [53], and the surface's stress tensor $\mathbf{F}^\alpha$ is given by:

$$\mathbf{F}^\alpha = \mathbf{N}\nabla_\beta \left(\frac{\delta\mathcal{L}}{\delta B_{\alpha\beta}} \right) - \left(\mathcal{L}S^{\alpha\beta} + 2\frac{\delta\mathcal{L}}{\delta S_{\alpha\beta}} + B_\gamma^\alpha \frac{\delta\mathcal{L}}{\delta B_{\beta\gamma}} \right) \mathbf{S}_\beta \qquad (6.38)$$

6.3.2 Identifying a Lagrangian Density for the Rayleigh-Plesset Equations

One of the first observations of Eqs. (6.36), (6.37), and (6.38) is that all terms are linear with respect to the Lagrangian density $\mathcal{L}$. This means that if $\mathcal{L} = \mathcal{L}_{(1)} + \mathcal{L}_{(2)}$, then Eq. (6.37) can be decomposed as the following:

$$\mathbf{N} \cdot \Bigg(\overbrace{\frac{\delta\mathcal{L}_{(1)}}{\delta\mathbf{R}} + \nabla_\alpha \mathbf{F}_{(1)}^\alpha - \tilde{\delta}_t \left(\frac{\delta\mathcal{L}_{(1)}}{\delta C} \mathbf{N} \right)}^{\text{Dynamic Equilibrium of } \mathcal{L}_{(1)}} + \overbrace{\frac{\delta\mathcal{L}_{(2)}}{\delta\mathbf{R}} + \nabla_\alpha \mathbf{F}_{(2)}^\alpha - \tilde{\delta}_t \left(\frac{\delta\mathcal{L}_{(2)}}{\delta C} \mathbf{N} \right)}^{\text{Dynamic Equilibrium of } \mathcal{L}_{(2)}} \Bigg) = 0$$

$$(6.39)$$

where $\mathbf{F}_{(1)}^\alpha$ and $\mathbf{F}_{(2)}^\alpha$ are the stress tensor contributions from $\mathcal{L}_{(1)}$ and $\mathcal{L}_{(2)}$, respectively. In searching for a Lagrangian density for the Rayleigh-Plesset equation, if particular terms are known to be accounted for by various Lagrangian densities, then the objective of finding a total Lagrangian density which results in the Rayleigh-Plesset equation is simplified. Conversely, if this is unfeasible for a term in the Rayleigh-Plesset equation, then it is known that this component either cannot fit in a single conserved Lagrangian density or represents a dissipation of energy. The component of Eq. (6.2) which corresponds to the Young-Laplace-Gauss equation is known by adding a surface and volume constraint onto the Lagrangian density[46,49,53]:

$$\mathcal{L}_{(\text{Young})} = \int_{\Omega} \Delta P \, d\Omega + \int_{\Sigma} \gamma \, d\Sigma \qquad (6.40)$$

where Ω is a volume element, indicating an integral over the bubble's volume, identified by the fact that since the bubble is closed, $\partial\Omega = \Sigma$. The rest of the equation presents difficulty in identifying a single homogeneous Lagrangian density. One option is to assume a simple ansatz for the Lagrangian density and proceed to attempt to match terms. The first step in this process would be identifying differential geometry terms in the CMS-invariant variational framework with their manifestation on a sphere with radius $R(t)$ that the Rayleigh-Plesset equation is formulated on.

6.3.3 Spherical Decomposition of Lagrangian Density

For a sphere, using spherical coordinates with orthonormal basis $\{\hat{\mathbf{r}}, \hat{\boldsymbol{\theta}}, \hat{\boldsymbol{\phi}}\}$ and allowing $R(t)$ to be the sphere's radius, these differential geometry parameters take the following form:

$$\mathbf{R} = R\hat{\mathbf{r}} \,,\; \mathbf{S}_{\alpha} = \begin{pmatrix} R\hat{\boldsymbol{\theta}} & R\sin\theta\hat{\boldsymbol{\phi}} \end{pmatrix} \,,\; \mathbf{N} = \hat{\mathbf{r}} \,,\; \mathbf{V} = (\partial_t R)\hat{\mathbf{r}} \,,\; C = \partial_t R$$

$$S_{\alpha\beta} = \begin{pmatrix} 1 & 0 \\ 0 & \sin^2\theta \end{pmatrix} \,,\; B_{\alpha}^{\beta} = -\frac{1}{R}\begin{pmatrix} 1 & 0 \\ 0 & 1 \end{pmatrix}$$

This is the geometrical configuration in order to apply the Rayleigh Plesset Equation to a geometrical sphere. This schematic of a sphere controlled by a parameter of $R(t)$ can be seen below in the schematic of Figure 6.3:

Observing the Rayleigh-Plesset equation, it can be seen that it is possible a Lagrangian density $\mathcal{L}_{(\text{RP})}$ of the form $\mathcal{L}_{(\text{RP})}(|\mathbf{R}|, C) = \mathcal{L}_{(\text{RP})}(R, \partial_t R)$ may account for the terms not accounted for by the Gauss-Young-Laplace Lagrangian density $\mathcal{L}_{(\text{Young})}$. For such a Lagrangian density, it can be seen that:

$$\mathbf{F}_{(\text{RP})}^{\alpha} = -\mathcal{L}_{(\text{RP})}\mathbf{S}^{\alpha} \qquad (6.41)$$

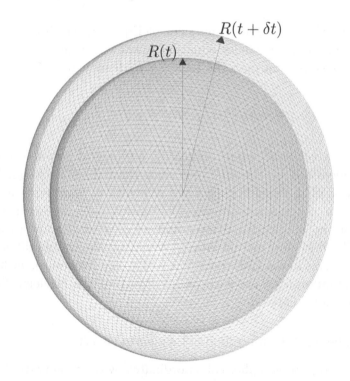

Figure 6.3 Schematic depicting a sphere with a time-varying radius $R(t)$. This is measured from the center of the sphere outward and is the primary variable of interest in a bubble process bubble process such as a Rayleigh-Plesset Process.

And thus it can be seen that the equilibrium equation for $\mathcal{L}_{(\text{RP})}$ after taking the dot product with the normal is given by:

$$\left(\frac{\mathbf{N}\cdot\mathbf{R}}{|\mathbf{R}|}\right)\frac{\delta\mathcal{L}_{(\text{RP})}}{\delta|\mathbf{R}|} - B_\alpha^\alpha \mathcal{L}_{(\text{RP})} - \tilde{\delta}_t\left(\frac{\delta\mathcal{L}_{(\text{RP})}}{\delta C}\right) = 0 \qquad (6.42)$$

If a simple product is assumed as $\mathcal{L}_{(\text{RP})} = \mathcal{A}|\mathbf{R}|^m C^n$ is assumed, it can be seen that this equation can be simplified in CMS terms to:

$$(1-m)|\mathbf{R}|^n C^{m-2}\left(m\dot{\nabla}C + \left(\frac{n(\mathbf{N}\cdot\mathbf{R})}{|\mathbf{R}|^2} - B_\alpha^\alpha\right)C^2\right) = 0 \qquad (6.43)$$

After substituting the sphere representations of C, $\mathbf{R}$, and B_α^α, the following ODE arises:

$$m(1 - m)R^{n-1}(\partial_t R)^{m-2}\left(R\partial_t^2 R + \left(\frac{n+2}{m}(\partial_t R)^2\right)\right) = 0 \quad (6.44)$$

This can be seen to replicate the Kinetic terms in Eq. (6.2) if $m = 2$ and $n = 1$. By this substitution, it can be seen that the action $\mathcal{S}_{(RP)} = \int_t \int_\Sigma \mathcal{L}_{(RP)}\, d\Sigma\, dt$ where $\mathcal{L}_{(RP)} = -\frac{1}{2}\rho_L|\mathbf{R}|C^2$ will produce the following CMS-invariant variational contribution:

$$\delta\mathcal{S}_{(RP)} = 0 \;\rightarrow\; \rho_L\left(R\partial_t^2 R + \frac{3}{2}(\partial_t R)^2\right) = 0 \quad (6.45)$$

The Kinetic Lagrangian density, $\mathcal{L}_{(RP)} = -\frac{1}{2}\rho_L|\mathbf{R}|C^2$ not only has units of J/m^2 as a Lagrangian density on a manifold in $\mathbb{R}^2$ should, but also contains the term $\rho_L|\mathbf{R}|$ which implies in some capacity that the surface mass distribution of the bubble ρ_Σ is related to the density distribution of the external liquid by the relation $\rho_\Sigma = \rho_L|\mathbf{R}|$. Here, the Kinetic Lagrangian density is so named as a general method of representing the kinetic energy of **any** closed surface, which is not only of proper dimensional analysis, but is also CMS-invariant, and can be formulated for any closed surface that is homeomorphic to a sphere.

6.3.4 Accounting for Energy Dissipation in a Rayleigh-Plesset Process

In summary, the functional for the Rayleigh-Plesset Equation thus far assuming a conservative energy process is provided by:

$$\mathcal{S} = \mathcal{L}_{(RP)} + \mathcal{L}_{(Young)} = \int_t\left(\int_{\partial\Omega}\gamma - \frac{1}{2}\rho_L|\mathbf{R}|C^2\,d\Sigma + \int_\Omega \Delta P d\Omega\right)dt \quad (6.46)$$

It is seen that the vanishing of the variation in the previous equation accounts for all the terms in the Rayleigh-Plesset equation with the exception of the viscosity term. This is not a failure, but a representation of the principle that the NSEs contain in their terms to account for the dissipation of energy, in contrast to Lagrangian Mechanics, which is ensured by Noether's theorem that quantities are conserved [54]. This being said, it can be proven that assuming a certain energy dissipation rate ψ in the system, sufficient ground on the basis of energy loss in a damped system and dimensional analysis, may be made to imply that the missing term in the equation must be linear in speed. From Eq. (6.46), we can see that the action is an integral in time of the Lagrangian density integrated over a surface. The expression for the Lagrangian density may have its conjugate taken, to re-write it as an expression for the total energy $\Sigma\mathcal{E}$ in the system:

$$\Sigma\mathcal{E} = \int_{\partial\Omega} \gamma + \frac{1}{2}\rho_L |\mathbf{R}| C^2 \, d\Sigma + \int_{\Omega} \Delta P \, d\Omega \tag{6.47}$$

Here it is seen for a moving surface, $\Sigma\mathcal{E} = \Sigma\mathcal{E}(t)$. Its rate of change in time can be seen to be:

$$\tfrac{d\Sigma\mathcal{E}}{dt} = \int_{\Sigma} \left(\rho_L |\mathbf{R}| \dot{\nabla} C + \rho_L \left(\tfrac{\mathbf{R}\cdot\mathbf{N}}{|\mathbf{R}|} - |\mathbf{R}| B_\alpha^\alpha \right) C^2 + \Delta P - \gamma B_\alpha^\alpha \right) C \, d\Sigma \tag{6.48}$$

This expression in brackets is none other than the Rayleigh-Plesset expression assuming no loss of energy due to viscosity. Therefore, the Lagrangian density produces a conservative system as in the absence of viscosity, $\left(\text{ie. } \frac{d}{dt}\Sigma\mathcal{E} = 0\right)$. In a damped oscillator it can be shown that energy is lost with a rate proportional to its kinetic energy. If we allow this to be identified with $\mathcal{L}_{(\mathrm{RP})}$, then the change of the total energy in time may be satisfied if the term in brackets equals $-\frac{1}{2}\mathcal{B}\rho_L|\mathbf{R}|C$, which would yield an energy loss of:

$$\frac{d\Sigma\mathcal{E}}{dt} = -\int_{\Sigma} \mathcal{B}|\mathcal{L}_{(\mathrm{RP})}| \, d\Sigma \tag{6.49}$$

In this case, this would provide an energy loss similar to a damped oscillator, and would also imply the full Rayleigh-Plesset equation must be:

$$\rho_L |\mathbf{R}| \dot{\nabla} C + \rho_L \left(\frac{\mathbf{R}\cdot\mathbf{N}}{|\mathbf{R}|} - |\mathbf{R}| B_\alpha^\alpha \right) C^2 + \frac{1}{2}\mathcal{B}\rho_L|\mathbf{R}|C + \Delta P - \gamma B_\alpha^\alpha = 0 \tag{6.50}$$

Allowing the proportionality constant in the energy dissipation to be equal to

$$\mathcal{B} = \frac{8\nu_L}{|\mathbf{R}|^2} \tag{6.51}$$

Allows for full identification with the non-conservative Rayleigh-Plesset equation in Spherical Coordinates. For a particular case, a bubble evolving in water for a two-step Heaviside Pressure profile can be observed in Figure 6.4:

Using the above variational formulation of the Rayleigh-Plesset Equation, in addition to the identification of the $\mathcal{B}$ constant, it can be concluded that the energy dissipation in a Rayleigh-Plesset process is given by:

$$\frac{d\Sigma\mathcal{E}}{dt} = -\int_{\Sigma} \frac{4\nu_L}{|\mathbf{R}|} C^2 \, d\Sigma \tag{6.52}$$

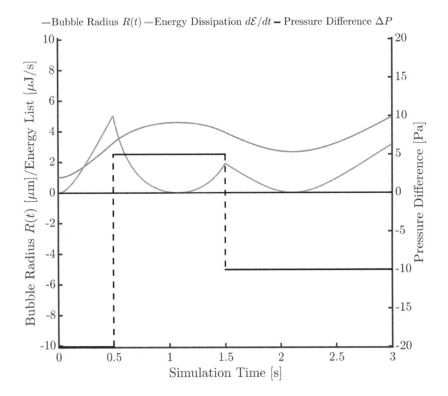

Figure 6.4 Numerical Solution of the Rayleigh-Plesset Equation displaying the bubble radius [cm] evolving in time $R(t)$, with units on the left axis. Physical parameters were chosen to model an immersion in water; thus a kinematic viscosity $\nu_L = (10^{-6})$ m^2/s, density $\rho_L = (10^3)$ kg/m^3, surface tension $\gamma = (7.2 \cdot 10^{-2})$ N/m. Also displayed is a test pressure difference which begins at -20 Pa, then evolves to 5 Pa, and then decreases to -10 Pa displayed against the right axis.

It can be seen on a sphere, recalling $\{C = \partial_t R, |\mathbf{R}| = R, \int_\Sigma d\Sigma = 4\pi R^2\}$ that this can reduce to:

$$\frac{d\Sigma\mathcal{E}}{dt} = -4\pi\nu_L R(\partial_t R)^2 \tag{6.53}$$

Using this novel formulation, we can see the energy loss associated with the Rayleigh-Plesset Process given in Figure 6.4 in $[\mu J/s]$:

It is worth noticing that the rate of Energy Loss changes with discontinuities that align with the discontinuities in the pressure difference ΔP. In addition, comparison of Figure 6.5 against Figure 6.4

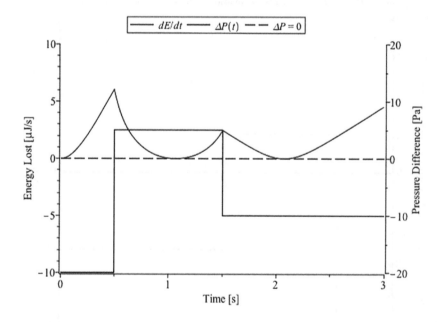

Figure 6.5 Theorized energy loss obtained through the CMS-Lagrangian formulation of the Rayleigh-Plesset Equation. Here, the energy loss is assumed to be negative, and is plotted on the y-axis as per the numerical solution for the bubble radius, and in accordance with Eq. (6.53) with units on the left axis. Physical parameters were chosen to model an immersion in water; thus, a kinematic viscosity $\nu_L = (10^{-6})$ m²/s, density $\rho_L = (10^3)$ kg/m³, surface tension $\gamma = (7.2 \cdot 10^{-2})$ N/m. Also displayed is a test pressure difference which begins at -20 Pa, then evolves to 5 Pa, and then decreases to -10 Pa displayed against the right axis.

indicates that instances where there exists extrema in the bubble radius coincide with instances where the rate of energy loss vanishes.

6.4 CONCLUSION: INCOMPRESSIBLE EDDIES/VORTICES & CAVITATION DYNAMICS

In Section 6.2, velocity and pressure profiles were demonstrated in eddies and vortices from a previous decomposition of the NSEs, while in Section 6.3 a variational formulation of the Rayleigh-Plesset Equations was obtained.

The results in Section 6.2 on fluid mechanics in particular the viscous incompressible 3D-Navier–Stokes equations brought to light an existence proof for a no finite time blowup. These results are an extension by application of the proofs found in [19]. The results show that there exist vortices and eddies in each cell of the periodic Lattice. Here it was found that orbits of an associated saddle surface exist. The different cases of the constant c_3 in the solution whether it be low or high were plotted. Also it was verified that the solution has saddle points which indicates there exist instabilities leading to turbulence.

As mentioned in Section 6.3, it is known in eddy processes for bubbles to form on account of the variation in kinetic energy and pressure of the fluid. This process can be observed using the variational formulation presented in Section 6.3. As was mentioned, to analyze a general energy density on a membrane, the general Lagrangian form may be taken:

$$\mathcal{S} = \int_t \int_\Sigma \mathcal{L}(\mathbf{R}, S_{\alpha\beta}, B_{\alpha\beta}, C)\, d\Sigma dt \tag{6.54}$$

The Lagrangian density which replicated the conservative components of the Rayleigh-Plesset Equation was identified as the following action:

$$\mathcal{S}_0 = \mathcal{L}_{(\mathrm{RP})} + \mathcal{L}_{(\mathrm{Young})} = \int_t \left(\int_{\partial\Omega} \gamma - \frac{1}{2}\rho_L |\mathbf{R}| C^2 \, d\Sigma + \int_\Omega \Delta P d\Omega \right) dt \tag{6.55}$$

And the Rayleigh-Plesset Equation was obtained assuming an energy loss which can be stated as:

$$\frac{d\Sigma\mathcal{E}}{dt} = -\int_\Sigma \frac{4\nu_L}{|\mathbf{R}|} C^2 \, d\Sigma \tag{6.56}$$

In addition, pressure profiles in an eddy process may be given by the Poisson equation:

$$\frac{1}{\rho}\Delta P = -\Delta(\frac{1}{2}u_z^2) + (\frac{\partial u_z}{\partial x})^2 + (\frac{\partial u_z}{\partial y})^2 - u_z\nabla_{xy}^2 u_z \qquad (6.57)$$

What can be seen is that in the case of an eddy process, the pressure P in the Eddy decomposition of the NSEs above contributes to the pressure difference between the inside and outside of the cavitation bubble ΔP in the Variational Formulation of the Rayleigh-Plesset Equation. Accounting for this and sampling the pressure at discrete points within the Cubical Lattice considered in the Eddy NSE region of interest can give insight into the cavitation dynamics within a vortex process. In addition, the versatility of the Variational Framework may permit for the addition of an NSE-related velocity Action which can couple cavitation dynamics to the formation of vortices in an incompressible fluid as such:

$$\mathcal{S}_{(\text{total})} = \mathcal{S}_0 + \delta\mathcal{S}_{(\text{NSE})},$$

$$\text{where } \mathcal{S}_0 = \int_t \left(\int_{\partial\Omega} \gamma - \frac{1}{2}\rho_L|\mathbf{R}|C^2 \, d\Sigma + \int_\Omega \Delta P d\Omega \right) dt \qquad (6.58)$$

In this fashion, the perturbative Action $\delta\mathcal{S}_{(\text{NSE})}$ can depend on u_z, P_{zz} or other terms to replicate the coupling between Cavitation Dynamics and the Navier–Stokes Equations.

6.5   APPENDIX

$$\left(\delta^{-1}-1\right)\left(\frac{\partial u_z}{\partial t}\right)^2 - \frac{(B+u_z)\mu\left(\frac{\partial u_z}{\partial z}\right)\left(\frac{\partial^2 u_z}{\partial x^2}+\frac{\partial^2 u_z}{\partial y^2}+\frac{\partial^2 u_z}{\partial z^2}\right)}{\rho}\left(1-\delta^{-1}\right)$$

$$+\frac{(\delta^{-1}-1)\left(\frac{\partial u_z}{\partial t}\right)K}{\rho} + \delta^2\left(2\,F_{T_1}\,u_z\frac{\partial u_z}{\partial x} + 2\,F_{T_2}\,u_z\frac{\partial u_z}{\partial y}\right)$$

$$+2\,\frac{uz\left(\frac{\partial u_z}{\partial x}\right)(4\,\cos(4\,t)x - 4\,\sin(4\,t)y)}{\delta} + 2\,\frac{uz\left(\frac{\partial u_z}{\partial y}\right)(-4\,\sin(4\,t)x - 4\,\cos(4\,t)y)}{\delta}$$

$$-\left(\frac{(\sin(4\,t)x+(\cos(4\,t)+2)y)\frac{\partial u_z}{\partial x}}{\delta} + \frac{((\cos(4\,t)-2)x-\sin(4\,t)y)\frac{\partial u_z}{\partial y}}{\delta}\right)A$$

$$+\frac{(A-K)A\mu}{\rho}\left(1-\delta^{-1}\right) + u_z^2\frac{\partial^2 u_z}{\partial t\partial z} + \delta^3 u_z\frac{\partial u_z}{\partial t}\frac{\partial u_z}{\partial z}F_z$$

$$+\delta^3\vec{b}\cdot\nabla(u_zF_z) + (B+u_z)\frac{\partial u_z}{\partial t}\frac{\partial u_z}{\partial z} = 0$$

$$(6.59)$$

Setting $K = A$, that is,

$$K - A = \frac{\partial P}{\partial z} - \frac{\partial^2 u_z}{\partial z^2} = 0 \tag{6.60}$$

with solution,

$$u_z = \int_a^z P(x, y, s, t)ds + F_1(x, y, t)(z + a) + F_2(x, y, t) \tag{6.61}$$

For $a = -\pi$ and $z = \pi$, we have,

$$\begin{aligned} u_z &= \int_{-\pi}^{\pi} P(x, y, s, t)ds + F_1(x, y, t)(z + a) + F_2(x, y, t) \\ &= F_2(x, y, t) \\ &= 0 \end{aligned} \tag{6.62}$$

due to periodicity of P on the interval $[-\pi, \pi]$ and zero wall boundary condition at $z = \pi$. In the above $K = \frac{\partial P}{\partial z}$ and $A = \frac{\partial^2 u_z}{\partial z^2}$ are indeterminate forms $(0/0)$ as $\delta \to 0$. In Eq. (6.7) the term $\frac{\mu}{\rho} \frac{1}{u_z^2}(1 - \frac{1}{\delta})\frac{\partial u_z}{\partial t}\nabla^2 u_z$ is expressed in terms of remaining terms in the z-component Navier–Stokes equation. For $\delta \to 0$, there is a non-homogeneous term above which can be absorbed into other terms when a lift of the velocity in the form $u_z \to B + u_z$ is used, where B is a sufficiently large constant. It is immediate that a homogenous equation results since $\frac{\partial u_z}{\partial y}$, and $\frac{\partial u_z}{\partial x}$ are constant due to assumption that vorticity is constant. The force terms are taken to be proportional to $\frac{1}{\delta^2}$ for $\vec{F_T}$ and $\frac{1}{\delta^3}$ for F_z. The solution of the homogeneous equation leads to Eq. (6.13) when Poisson's Equation is used.

## CHAPTER 7

Introduction to Flows and Dynamical Systems

Terry E. Moschandreou

TVDSB

7.1 TANGENT VECTORS

Consider an arbitrary fixed but open set U as a subset of $\mathbb{R}^n$. It could be all of $\mathbb{R}^n$. Fixing $p \in U$ we now describe the tangent space $T_p(U)$ of U at p.

The vector $\mathbf{a} = \begin{pmatrix} a^1 \\ a^2 \\ \vdots \\ a^n \end{pmatrix}$ defines a *directional derivative* $D_{\mathbf{a}}$ at p by the following definition,

$$D_{\mathbf{a}}(f) = \lim_{h \to 0} \frac{f(p + h\mathbf{a} - f(p)}{h} = \sum_{i=1}^{n} a^i \frac{\partial f}{\partial x^i}(p) \tag{7.1}$$

where f is an arbitrary smooth function which is defined on an open neighborhood of p. Note that $D_{\mathbf{a}}(x^i) = a^i$. Alternatively we consider a curve,

$$\tau : (-\eta, \epsilon) \to U \tag{7.2}$$

where ϵ, $\eta > 0$, and

$$\tau(t) = (x^1(t), \ldots, x^n(t)), \tag{7.3}$$

DOI: 10.1201/9781003452256-7

where each $x^i \in C^1$ and we have that,

$$
\begin{aligned}
\tau(0) &= p \\
\dot{\tau}(0) &= \mathbf{a} \\
x^i(0) &= p^i, \\
\dot{x}^i(0) &= a^i,
\end{aligned}
\tag{7.4}
$$

for $1 \leq i \leq n$. The definition of derivative is,

$$
D_{\mathbf{a}}(f) = \lim_{h \to 0} \frac{f(\tau(h)) - f(p)}{h}
\tag{7.5}
$$

Note that straight lines do not have meaning on general manifolds, the notion of a C^1 curve does.

Definition 7.1 *Given a point $p \in U$, $C^\infty(U,p)$ denotes the set of smooth, real valued functions f with domain(f) an open subset of U and $p \in dom(f)$.*

Definition 7.2 *The set of all C^1 curves $\tau\colon (-\eta, \epsilon) \to U$, $\eta > 0$ and $\epsilon > 0$ depend on τ, such that $\tau(0) = p$ is denoted as $S(U,p)$.*

Definition 7.3 *If $\tau_1, \tau_2 \in S(U,p)$, then τ_1 and τ_2 are infinitesimally equivalent at p and we write $\tau_1 \approx_p \tau_2$ iff,*

$$
\frac{d}{dt} f(\tau_1(t)) \big|_{t=0} = \frac{d}{dt} f(\tau_2(t)) \big|_{t=0} .
$$

for all $f \in C^\infty(U,p)$, where $\approx_p$ is an equivalence relation on $S(U,p)$.

Definition 7.4 *The equivalence class of τ in $S(U,p)$ is represented by $\langle \tau \rangle_p$ and is named an infinitesimal curve at p. This is also called a tangent vector to U at p and the set $T_p(U) = S(U,p)$ (wrt to equivalence relation $\approx_p$) of all tangent vectors at p is called the tangent space to U at p.*

Due to infinitesimal equivalence, we have that,

Lemma 7.1 *For each $\langle \tau \rangle_p \in T_p(U)$, the operator,*

$$
D_{\langle \tau \rangle_p} : C^\infty(U,p) \to \mathbb{R}
$$

is a well defined operation when we choose any representative $\tau \in \langle\tau\rangle_p$ and we set,

$$D_{\langle\tau\rangle_p}(f) = \frac{d}{dt}f(s(t))\mid_{t=0},$$

for all $f \in C^\infty(U,p)$. Also, $\langle\tau\rangle_p$ is uniquely determined by $D_{\langle\tau\rangle_p}$.

As is usual we define a natural vector space structure on $T_p(U)$ that is coordinate free, we have,

Lemma 7.2 *Begin with $\langle\tau_1\rangle_p$, $\langle\tau_2\rangle_p \in T_p(U)$ and $a,b \in \mathbb{R}$. Then there is a unique infinitesimal curve $\langle\tau_1\rangle_p$ such that the associated derivatives on $C^\infty(U,p)$ satisfy,*

$$D_{\langle\tau\rangle_p} = aD_{\langle\tau_1\rangle_p} + bD_{\langle\tau_2\rangle_p}.$$

Proof 7.1 *We can write the following,*

$$\tau(t) = a\tau_1(t) + b\tau_2(t) - (a+b-1)p$$

which is defined for all values of t sufficiently near to 0, and is a C^1 curve in U with $s(0) = p$ and represents the required infinitesimal curve $\langle\tau\rangle_p \in T_p(U)$ Here this is completely determined by $D_{\langle\tau\rangle_p}$.

Definition 7.5 *Two elements $f,g \in C^\infty(U,p)$ are germinally equivalent at p and write $f \approx_p g$ if there is an open neighborhood W of p in U such that $W \subseteq dom(f) \cap dom(g)$ and $f \mid W = g \mid W$.*

Definition 7.6 *The germinal equivalence class $[f]_p$ of $f \in C^\infty(U,p)$ is called the germ of f at p. The set $C^\infty(U,p)/ \approx_p$ of germs at p is denoted by $\mathcal{G}_p$.*

Definition 7.7 *For each $\tau \in S(U,p)$ the operator*

$$D_{\langle\tau\rangle_p} : \mathcal{G}_p \to \mathbb{R}$$

is defined by,

$$D_{\langle\tau\rangle_p}[f]_p = \frac{d}{dt}f(s(t))\mid_{t=0}.$$

Definition 7.8 *A derivative operator on the space* $\mathcal{G}_p$ *is an* $\mathbb{R}$-*linear map* $D : \mathcal{G}_p \to \mathbb{R}$ *s.t.,*

$$D(ab) = D(a)e_p(b) + e_p(a)D(b),$$

for all $a, b \in \mathcal{G}_p$. *The derivative on* $\mathcal{G}_p$ *is a tangent vector to* U *at* p. *Finally, the set of all derivatives on* $\mathcal{G}_p$ *is denoted as* $T_p(U)$ *and is the tangent space to* U *at* p.

7.2 LOCAL FLOWS

Let $X \in \mathcal{X}(U)$ the set of all vector fields on U. Now a vector field X can be seen as an autonomous system of ODEs on U.

Definition 7.9 *If* $U \subseteq \mathbb{R}^n$ *and* $V \subseteq \mathbb{R}^m$ *are open, a map* $\Phi : U \to V$ *is a diffeomorphism if it is smooth and bijective and if* $\Phi^{-1} : V \to U$ *is also smooth.*

Definition 7.10 *If* $z := U \to V$ *is a smooth map between open subsets of Euclidean spaces, then* $z^* : C^\infty(V) \to C^\infty(U)$ *is defined by setting,*

$$z^*(f) = f \circ z.$$

for all $f \in C^\infty(V)$.

Lemma 7.3 *If* $z : U \to V$ *is a diffeomorphism between open subsets of* $\mathbb{R}^n$, *then* $z^* : C^\infty(V) \to C^\infty(U)$ *is an isomorphism of algebras.*

Definition 7.11 *If* $z : U \to V$ *is a diffeomorphism between open subsets of* $\mathbb{R}^n$, *then* $z_* : \mathcal{X} \to \mathcal{X}(V)$ *is defined by setting,*

$$z_*(X)(f) = (z^{-1})^*(X(z^*(f))) = X(f \circ z) \circ z^{-1},$$

for all $X \in \mathcal{X}(U)$ *for all* $f \in C^\infty(V)$.

Definition 7.12 *Let* $\tau : (a, b) \to U$ *be smooth. The velocity vector of* τ *at* $t_0 \in (a, b)$ *is,*

$$\dot{\tau}(t_0) = \tau_{*t_0}\left(\frac{d}{dt}\,|_{t_0}\right). \in T_{\tau(t_0)}(U).$$

Note that $T = t + t_0$, $a - t_0 < t < b - t_0$, then $\sigma(t) = \tau(T)$ has velocity $\dot{\sigma}(t) = \dot{\tau}(T) = \dot{\tau}(t + t_0)$. In particular, $\dot{\tau}(t_0) = \dot{\sigma}(0) = \langle \sigma \rangle_{\tau(t_0)}$. We write this infinitesimal curve as $\langle \tau \rangle_{\tau(t_0)}$.

Definition 7.13 *The map* $\dot{\tau} : (a, b) \to T(U)$ *is the velocity field of the smooth curve* $\tau : (a, b) \to U$.

Definition 7.14 *Let* $X \in \mathcal{X}(U)$ *and* $x_0 \in U$. *An integral curve to* X *through* x_0 *is a smooth curve* $\tau : (-\eta, \epsilon) \to U$, *defined for suitable* $\eta, \epsilon > 0$ *such that* $\tau(0) = x_0$ *and* $\dot{\tau}(t) = X_{\tau(t)}$, $-\eta < t < \epsilon$.

Suppose that τ is an integral curve to $X \in \mathcal{X}(U)$ through $x_0 \in U$. We write,

$$X = \sum_{i=1}^{n} f^i D_i$$

and

$$\tau(t) = (x^1(t), \dots, x^n(t)).$$

At each $t \in (-\eta, \epsilon)$, the Jacobian matrix of τ is, $J\tau(t) = \begin{pmatrix} \frac{dx^1}{dt}(t) \\ \frac{dx^2}{dt}(t) \\ \vdots \\ \frac{dx^n}{dt}(t) \end{pmatrix} \in \mathbb{R}^n$

$$= \sum_{i=1}^{n} \frac{dx^i}{dt}(t) D_{i, \tau(t)} \in T_{\tau(t)}(U).$$

Now recall that,

$$X_{\tau(t)} = \sum_{i=1}^{n} f^i(x^1(t), \dots, x^n(t)) D_{i, \tau(t)}.$$

Here we have that τ is an integral curve to X if and only if

$$\frac{dx^i}{dt}(t) = f^i(x^1(t), \dots, x^n(t)),$$

$-\eta < t < \epsilon, 1 \leq i \leq n$. Here we have an autonomous system of ODEs with solution τ subject to the initial condition $\tau(0) = x_0$. We have the following theorem,

Theorem 7.1 *Let $V \subseteq \mathbb{R}^r$ and $U \subseteq \mathbb{R}^n$ be open subsets, let $c > 0$, let $f^i \in C^\infty((-c, c) \times V \times U), 1 \leq i \leq n$. Now consider the nonautonomous system with parameters $b = (b^1, \ldots, b^r) \in V$ of ODE:*

$$\frac{dx^i}{dt} = f^i(t, b, x^1(t, b), \ldots, x^n(t, b)), \quad 1 \leq i \leq n.$$

Let $(a^1, \ldots, a^n) \in U$. Then there are smooth functions $x^i(t, b), 1 \leq i \leq n$, which is defined on some nondegenerate interval $[-\eta, \epsilon]$ about 0, which satisfies the system above and the initial condition,

$$x^i(0, b) = a^i, \quad 1 \leq i \leq n.$$

Next if the functions $\tilde{x}_i(t, b)$ is another solution, defined on $[-\tilde{\eta}, \tilde{\epsilon}]$ and satisfy the same initial condition, then the two solutions will be the same on $[-\eta, \epsilon] \cap [-\tilde{\eta}, \tilde{\epsilon}]$. Finally writing the solutions as $x^i = x^i(t, b, a)$, there is a neighborhood A of $a \in U$, a neighborhood B of b in V, and an $\epsilon > 0$ such that the solutions $x^i(t, y, x)$ are defined and smooth on the open set $(-\epsilon, \epsilon) \times B \times A \subseteq \mathbb{R}^{n+r+1}$.

Definition 7.15 *A local flow Φ around $x_0 \in U$ is a smooth map,*

$$\Phi : (-\epsilon, \epsilon) \times A \to U$$

written $\Phi(t, x) = \Phi_t(x)$, where A is a carefully chosen open neighborhood of $x_0 \in U$, such that,

1. *$\Phi_0 : A \to U$ is the inclusion $A \hookrightarrow U$,*

2. *$\Phi_{t_1+t_2}(x) = \Phi_{t_1}(\Phi_{t_2}(x))$ whenever both sides are defined.*

If $z \in U$, the flow line through z is the curve $\sigma(t) = \Phi_t(z), \ -\epsilon < t < \epsilon$.

7.3   APPLIED DYNAMICAL SYSTEMS AND BIFURCATION

One Dimensional Systems are basically those in Eq. (7.6) with $n = 1$ for $\vec{x} = (x_1, x_2, \ldots, x_n)$. That is

$$\dot{x} = f(x) \tag{7.6}$$

The simplest dynamical system is,

$$\dot{x} = 0 \quad \to \quad x = c \tag{7.7}$$

This equation, Eq. (7.7) is quite useful to gain insight into the dynamical structure of non-linear systems, as it is the condition for **steady states** or **fixed points**.

Next is an equation of exponential growth, where change $\dot{x}$ is proportional to x.

$$\dot{x} = \lambda x \quad \rightarrow \quad \int \frac{dx}{x} = \int \lambda \, dt \quad \rightarrow \quad x(t) = ce^{\lambda t} \qquad (7.8)$$

As seen in Figure 7.1, the linear differential equation of exponential growth has solutions that are exponentially increasing, exponentially decaying or simply constant depending on the parameter λ.

If we plot $\dot{x}$ as a function of x, we get a **phase space** plot. The graphs are straight lines given by $\dot{x} = \lambda x$ with a negative, vanishing and positive slope, respectively. We now introduce phase plots (Figure 7.2).

7.3.1 Flows on the Line

We use a geometric approach to study difficult and even unsolvable nonlinear systems. Suppose that we have a fluid flowing along the real line with a local velocity $f(x)$. This imaginary fluid is called the **phase fluid**, and the real line is the **phase space**.

The flow is to the right where $f(x) > 0$ and to the left where $f(x) < 0$. (see chapter one for how a fluid can flow due to pressure gradient) *Fixed points* have $f(x) = 0$ (Figure 7.3).

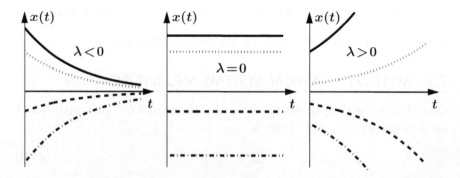

Figure 7.1 Solutions $x(t)$ for the Eq. (7.8) for different values of c (solid, dashed, dotted and dash-dotted).

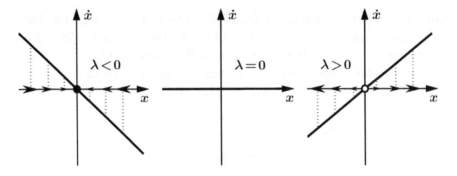

Figure 7.2 Phase space plots, $\dot{x}$ as a function of x, for the Eq. (7.8).

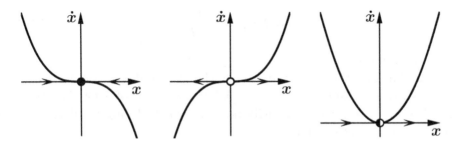

Figure 7.3 Phase portrait, solid black dot(left) is a **stable** fixed point (the local flow is toward it) and the open dot(middle) is an **unstable** fixed point (the flow is away from it) and the right one is **half-stable** (the flow is towards it on left and away on right).

7.3.2 Linear Stability Analysis

Consider a dynamical system of the form $\dot{x} = f(x)$ and the Taylor expansion around one of its fixed points $\tilde{x}$ and let $\epsilon = x - \tilde{x}$, then $\dot{\epsilon} = \dot{x}$

$$\dot{x} = f(x) \approx \underbrace{f(\tilde{x})}_{=0} + \underbrace{f'(\tilde{x})(x - \tilde{x})}_{=\epsilon} + \underbrace{\frac{1}{2!}f''(\tilde{x})(x - \tilde{x})^2}_{=\epsilon^2} + \cdots \qquad (7.9)$$

$f(\tilde{x})$ vanishes as it is a fixed point, $|\epsilon|$ is small as we assume x to be in the vicinity of $\tilde{x}$, and therefore ϵ^2 is negligible. Taken together, we obtain an approximation of (6.11) around the fixed point $\tilde{x}$.

$$\dot{\epsilon} = f'(\tilde{x})\,\epsilon = \lambda\,\epsilon \qquad (7.10)$$

Now, Eq. (7.10) is similar to Eq. (7.8). The slope $f'(\tilde{x})$ at the fixed point determines its stability,[1] if $f'(\tilde{x}) > 0$ then $\tilde{x}$ is a *unstable fixed point* and if $f'(\tilde{x}) < 0$ then $\tilde{x}$ is a *stable fixed point*. $1/f'(\tilde{x})$ is a **characteristic time scale**, it determines the time required for $x(t)$ to vary significantly in the neighborhood of $\tilde{x}$.

7.3.3 Potential Functions

A second way of representing the dynamics of such systems graphically is the landscape defined by its potential function.

The potential of a dynamical system is a function $V(x)$ where

$$\dot{x} = f(x) = -\frac{dV(x)}{dx} \tag{7.11}$$

All one-dimensional systems have a potential function.

As time evolves, either the value of the potential function decreases, or it stays constant if a local or global minimum of $V(x)$ has been reached. This behavior can easily be seen by calculating the derivative of V with respect to time

$$\dot{V} = \frac{dV(x)}{dt} = \underbrace{\frac{dV}{dx}}_{-\dot{x}} \underbrace{\frac{dx}{dt}}_{\dot{x}} = -\left\{\frac{dV}{dx}\right\}^2 = -\dot{x}^2 \le 0 \tag{7.12}$$

Local *minima* of V correspond to **stable** fixed points, local *maxima* correspond to **unstable** fixed points, **half stable** fixed points are found at locations where the tangent at an *inflection point* is horizontal.

7.3.3.1 Oscillations Never Occur for First-Order Systems

Fixed points dominate the dynamics of first-order systems. Trajectories either approach a fixed point, or diverge to $\pm\infty$. Trajectories either increase or decrease monotonically, or remain constant. If a fixed point is regarded as an equilibrium solution, the approach to equilibrium is always monotonic-overshoot and damped oscillations can never be when we consider a first-order system. It follows that there are no periodic solutions to $\dot{x} = f(x)$.

---

[1]Our analysis fails for some $f(x)$ in case where solution is not unique or does not exist.

7.3.4 Flows on the Circle

Periodic systems can be realized in one dimension with dynamics that occur on a circle. The state variable is θ with $0 \leq \theta < 2\pi$ or $-\pi \leq \theta < \pi$. The general form here is

$$\dot{\theta} = f(\theta) \text{ with } f(\theta + 2\pi) = f(\theta) \tag{7.13}$$

The function $f(\pi)$ has to be 2π−periodic to make sure that it has the same value at $\theta = 0$ and $\theta = 2\pi$.

7.3.5 Nonuniform Oscillator

The non-uniform oscillator has the equation,

$$\dot{\theta} = \omega - a\sin\theta \tag{7.14}$$

The parameter a introduces a non-uniformity in the flow around the circle, the flow is fastest at $\theta = -\pi/2$ and slowest at $\theta = \pi/2$.

- For $a < \omega$, the period of the oscillation ($T = \int dt = \int_0^{2\pi} \frac{dt}{d\theta} d\theta = \int_0^{2\pi} \frac{d\theta}{\omega - a\sin\theta}$, $T = \dfrac{2\pi}{\sqrt{\omega^2 - a^2}}$.

- When a is slightly less than ω, the oscillation is very extreme: the phase point $\theta(t)$ takes a long time to pass through a *bottleneck* near $\theta = \pi/2$, after which it moves around the rest of the circle in a quicker time scale (Figure 7.4a).

- When $a = \omega$, the system stops oscillating altogether: a half-stable emerges in a saddle-node bifurcation at $\theta = \pi/2$ (Figure 7.4b).

- Finally, when $a > \omega$, the half-stable fixed point breaks apart into a stable and unstable fixed point (Figure 7.4c). All trajectories are attracted to the stable fixed point as $t \to \infty$.

The same information can be shown by plotting the vector fields on the circle (Figure 7.4).

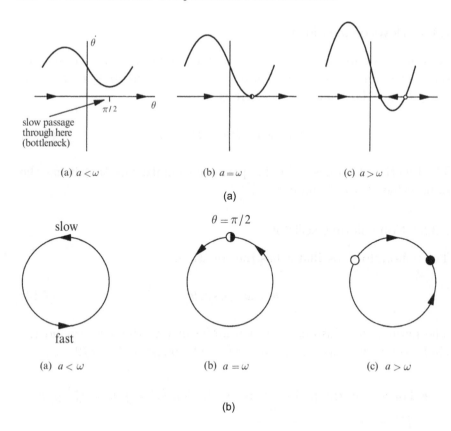

Figure 7.4

EXERCISES

7.1 Set up the operations of scalar multiplication and addition $(\cdot, +)$ on the space $T_p(U)$ and in turn prove that the tangent space $T_p(U)$ is a vector space over $\mathbb{R}$ with respect to these operations.

7.2 Define $D_{i,p} : \mathcal{G}_p \to \mathbb{R}$ by the formula, $D_{i,p}[f]_p = \frac{\partial f}{\partial x^i}(p)$. Show that this is a well-defined $\mathbb{R}-$ linear map and that $D_{i,p}$ is a derivative for all i s.t $1 \le i \le n$.

7.3 If $\langle \tau \rangle_p$ is an infinitesimal curve then prove that $D_{\langle \tau \rangle} : \mathcal{G}_p \to \mathbb{R}$ is indeed a derivative.

7.4 Prove the following, if c is a constant function on U and $D \in T_p(U)$, then $D[c]_p = 0$.

7.5 Prove that the set of tangent vectors $\{D_{1,p}, \ldots, D_{n,p}\}$ is a basis of the vector space $T_p(U)$. Hint: show that this set is linearly independent and is a spanning set.

7.6 Let $U = \mathbb{R}^2$ and $X = x^1 D_1 + x^2 D_2$. The integral curves $\tau(t) = (x^1(t), x^2(t))$ satisfies,

$$\frac{dx^1}{dt} = x^1$$

$$\frac{dx^2}{dt} = x^2.$$

Determine the solution curve with initial condition $\tau(0) = (a^1, a^2)$, $a^i \in \mathbb{R}$.

Prove that there is a local flow Φ around $\tau_0 \in U$. Show that the speed of the trajectory increases proportionally with the distance from the origin.

7.7 Let $U = \mathbb{R}^2$ and $X = x^1 D_1 - x^2 D_2$. The integral curves $\tau(t) = (x^1(t), x^2(t))$ satisfy,

$$\frac{dx^1}{dt} = x^1$$

$$\frac{dx^2}{dt} = -x^2.$$

Determine the solution curve with initial condition $\tau(0) = (a^1, a^2)$, $a^i \in \mathbb{R}$. Prove that there is a local flow Φ around $\tau_0 \in U$. When is the curve stationary? Describe the trajectories of points on the x^1 and x^2 axis, on the curves. What kind of paths do these trajectories follow relative to the coordinate axes?

7.8 Repeat the proof as in exercise 7 for the vector field,

$$X = x^2 D_1 - x^1 D_2$$

on $\mathbb{R}^2$. Sketch the flow lines and show that they are defined for $-\infty < t < \infty$.

7.9 On $U = \mathbb{R}$, consider the vector field $X = e^x \frac{d}{dx}$. Let $a \in \mathbb{R}$ and calculate the integral curve τ_a to X through a. Also find the largest open interval (δ_1, δ_2) on which $\tau_a(t)$ is defined.

7.10 Let $X \in (U)$ and $x_0 \in U$. Then prove that there is a local flow around x_0 such that the flow lines are integral curves to X. Prove that two such local flows agree on their common domain.

## CHAPTER 8

Numerical Analysis of 3D Periodic Navier–Stokes Equations and the Maple Environment

Terry E. Moschandreou

TVDSB

8.1 INTRODUCTION TO THE PERIODIC NAVIER–STOKES EQUATIONS

The following two paragraphs are related to the work in Liu et al. [18]. In 1845, Stokes derived the equation of motion of a viscous flow by adding Newtonian viscous terms and finalized the Navier–Stokes equations, which have been used for almost two centuries. However, there are few studies to find how to understand the physical meaning of the viscous terms in NS equations. As is well known, Stokes had three assumptions: (1) The force on fluids is the stationary pressure when the flow is stationary. (2) Fluid viscosity is isotropic. (3) Fluid flow follows Newton's law that fluid stress and strain have linear relations. These assumptions finalized the NSE. In [18], the vorticity part is considered the only source of fluid stress for the purpose of computation

DOI: 10.1201/9781003452256-8

cost reduction. In fact, fluid shear stress is contributed by both strain and vorticity. In mathematics, the computation of stress can be performed by strain only, vorticity only, or both. The computational results are exactly the same. The NSE equation adopts strain, which is symmetric and stress based on Stokes's assumption. In [18], a new governing equation adopts a new assumption that accepts that fluid stress has a linear relation with vorticity, which is anti-symmetric. They further modify the linear relation by considering the second-order terms in future research on turbulence, but a linear relation is given in [18]. According to the mathematical analysis, the new governing equation is identical to NS equations in mathematics, but in a physical sense, the new governing equation is just the opposite to NSEs as it assumes that fluid stress is proportional to vorticity, where both are anti-symmetric, but not strain, against Stokes's assumption and against the current NSE.

Although both NSEs and the new governing equation in [18] lead to the same computational results for laminar flow, the new governing equation has several advantages: (1) The vorticity tensor is anti-symmetric, which has three elements, but NSEs use the strain tensor, which has six elements. The computational cost is reduced to half for the viscous term. (2) The anti-symmetric matrix is independent of the coordinate system change or Galilean invariant, but the symmetric matrix that NSE uses is not. (3) The physical meaning is clear that the viscous term is generated by vorticity, not by strain only. (4) The viscosity is obtained by experiments, which are based on vorticity but not strain, since both strain and stress are hard to measure by experiments. (5) Vorticity can be further decomposed to rigid rotation and pure anti-symmetric shear, which is very useful for further study of turbulent flow. However, the NS equation has no vorticity term, which is a hurdle for further turbulence research. Ref. [27] in [18], studied the mechanism of turbulence generation and concluded that shear instability and transformation from shear to rotation are the paths of flow transition from laminar flow to turbulent flow. Using Liutex and the third generation of vortex identification methods, a lot of new physics has been found (see Dong et al., Liu et al., and Xu et al. references 24–26 in [18]) In Ref. 28 in [18], Zhou et al. elaborated the hydrodynamic instability induced turbulent mixing in wide areas, including inertial confinement fusion, supernovae, and their transition criteria. Since the new governing equation has a vorticity term, which can be further decomposed to shear and rigid rotation, the new governing equation would be helpful in studying flow instability and transition to turbulence.

The purpose of this chapter is to show that the periodic NS equations with high energy assumption can breakdown in finite time but with sufficient low energy scaling the equations simply do not exhibit finite time blowup. This chapter gives a general model using specific periodic special functions, that is elliptic Weierstrass P functions . The definition of vorticity in [18], is that vorticity is a rotational part added to the sum of antisymmetric shear and compression and stretching. Satisfying a divergence free vector field and periodic boundary conditions respectively with a general spatiotemporal forcing term $f(\vec{x}, t)$ which is smooth and periodic, the existence of solutions which blowup in finite time can occur. On the other hand, if u_0 is not smooth, then there exist global in-time solutions on $t \in [0, \infty)$ with a possible blowup at $t = \infty$. The control of turbulence is possible to maintain when the initial conditions and boundary conditions are posed properly for (PNS) [27]. This leads to the following two questions for (PNS),

1. *Is there a decaying of turbulence in the 3D case for the z component of velocity when non-smooth initial conditions are considered for x, y components of velocity?*

   and

2. *Are the vorticity sheets in the small scales of 3D turbulence dominating the flow to the extent that non-smooth temporal solutions exist for the z velocity for smooth initial data for the x, y components of velocity?*

A positive answer exists for both of the above questions [27,28]. In this chapter it is shown explicitly that for smooth forcing that is both spatial and temporal and product functions in space for velocity u_x and u_y that the equivalent form of the Navier–Stokes equations derived in [19], [29] and [30], has as one of the possible solutions for u_z a separable product of spatial functions in the three space variables together multiplied by a general function of t which is a blowup at infinity. On the other hand if $\mathbf{f}(\vec{x}, t)$ is a smooth reciprocal of a general Weierstrass P function defined on the 3-Torus, then when u_x and u_y in 3D Navier–Stokes equations are both in the smooth reciprocal form of the Weierstrass P function then this implies that u_z is not smooth in time. In [19],[29], and [30] the z component of vorticity was chosen to be constant. Extending the vorticity definition, in particular in this chapter, u_x, u_y satisfy a non-constant spatial or time-dependent vorticity for 3D vorticity $\vec{\omega}$. Finally new equations [18] are conjectured to possess smooth solutions appearing to not have finite time singularities using the correct definition of vorticity in this study. For (PNS) it is shown in [57] that there exists a vortex in each cell of the lattice associated with using the decomposition of pure rotation(Liutex), antisymmetric shear and compression and stretching. Furthermore, it is observed in [57] that a singular cusp bifurcation occurs along a principle main axis for the case of smooth and non-smooth initial inputs of velocity. In this chapter, it is shown that no finite time blowup occurs for sufficiently low energy scaling of the PNS equations. Here the pressure is a decreasing form of a general Cantor function $(1\text{-}C_F)$ where the C_F term is an arbitrary in length product of Cantor functions. For high energy scaling associated with the continuum model of the NS equations, it is shown that there are finite time singularities.

Note: All Appendices in this chapter are in the supplementary section of the book on Taylor and Francis website.

8.2 EQUIVALENT FORM OF 3D PERIODIC NAVIER–STOKES EQUATIONS

Recall the Gagliardo-Nirenberg and Prékopa-Leindler inequalities introduced in Chapter 5.

$$||\nabla^j u||_p \leq C ||\nabla^k u||_r^\theta ||u||_q^{1-\theta}, \quad \text{for} \quad \text{all} \quad u \in L^q(\mathbb{R}^n) \cap W^{k,r}(\mathbb{R}^n) \quad (8.1)$$

$$\int_{\mathbb{R}^n} h(x)dx \geq \left(\int_{\mathbb{R}^n} f(x)dx \right)^{1-\lambda} \left(\int_{\mathbb{R}^n} g(x)dx \right)^\lambda, \quad x,y \in \mathbb{R}^n \quad (8.2)$$

The 3D incompressible unsteady Navier–Stokes Equations (NSEs) in Cartesian coordinates may be expressed [19,29,30], as the coupled system Eqs. (8.3–8.8) below, for the velocity field $\mathbf{u}^* = u^{*i}\vec{e}_i, u^{*i} = \{u_x^*, u_y^*, u_z^*\}$ from the original NSE's,

$$\mathcal{G} = \mathcal{G}_{\delta 1} + \mathcal{G}_{\delta 2} + \mathcal{G}_{\delta 3} + \mathcal{G}_{\delta 4} = 0 \qquad (8.3)$$

where

$$u_x^* \rightarrow \frac{u_x}{\delta}, \quad u_y^* \rightarrow \frac{u_y}{\delta}, \quad u_z^* \rightarrow \frac{u_z}{\delta}$$

$$x^* \rightarrow \delta x, \quad y^* \rightarrow \delta y, \quad z^* \rightarrow \delta z, \quad t^* \rightarrow \delta^2 t$$

$$\frac{\partial}{\partial x^*} \rightarrow \delta^{-1} \frac{\partial}{\partial x}, \quad \frac{\partial}{\partial y^*} \rightarrow \delta^{-1} \frac{\partial}{\partial y}, \quad \frac{\partial}{\partial z^*} \rightarrow \delta^{-1} \frac{\partial}{\partial z}, \quad \frac{\partial}{\partial t^*} \rightarrow \delta^{-2} \frac{\partial}{\partial t}$$

and where $\delta = \epsilon + 1$ with ϵ being arbitrarily small when $\delta \approx 1$. In [19] $\mathcal{G}$ there is defined without the tensor product term $\mathbf{Q}$. Calculating the tensor product term $\mathbf{Q}$ in Eq. (8.13) see [19] and using Eq. (21) in [19] shows its volume integral to approach zero due to $\|\nabla u_z\|_2^2$ approaching zero. See Theorem 1 and 2, where the Prekópa-Leindler and Gagliardo-Nirenberg inequality is used to show this when $r = 1$ and $\lambda = \theta = \frac{1}{2}$. In this paper it is shown that the problem in [19] can be extended to all three velocity components u_i. The G.A. decomposition used there shows that there is a missing term (intended to be present) when multiplying Eq. 4(there) by u_z and adding to the product of $\vec{b}$ and the z momentum equation. This sum is precisely $\|\nabla u_z\|^2$ which is bounded by $\|\nabla u_z\|_2^2$ which is shown below to approach zero as the volume Ω approaches infinity. Also the pressure due to conservation of force s theorem is a regular function in t. As a result, it is assumed that it can be written as $P = \tilde{P}(x, y, z) P_{t_0}(t - t_0)$ where $P_{t_0}(t - t_0)$ approaches zero as $t \rightarrow t_0$ and $\nabla \cdot \left(\frac{1}{\delta \rho} u_z^2 \nabla_{xy} P + \vec{b} \frac{1}{\delta} u_z \frac{\partial P}{\partial z} \right) = \Phi(t)$ where $\vec{r} = x\vec{i} + y\vec{j} + z\vec{k}$. In Eq. (6) of [19], for the vector $\vec{B} = u_z \nabla \cdot (u_z \vec{b}) \vec{b}$, $L\vec{u} = \vec{B}$. Furthermore, in Eq. (10) [19] there, the term Ω_4 is the divergence of the vector $\vec{B}$. Using Ostrogradsky's formula in terms of the vorticity $\vec{\omega}$ and velocity $\vec{b}$, $\int_\Omega u_z \vec{b} \nabla \cdot (u_z \vec{\omega} \mid \vec{r} \mid) \, d\vec{x} = - \int_\Omega \mid \vec{r} \mid u_z \vec{\omega} \cdot \nabla (\vec{b} u_z) \, d\vec{x}$. Now for a specific pressure P on an R-sphere, $\int_\Omega \mathcal{G}_{\delta 1} + \mathcal{G}_{\delta 2} + \mathcal{G}_{\delta 4} \, d\vec{x} = \Phi(t)$, where $\Phi(t)$ assumed to be bounded and contain the pressure terms in Eq. (8.6). The sphere is $\mid \vec{r} \mid = R$. Since the 3-Torus is compact there are m closed sets covering it. The outer measure is used where the infimum is taken over all finite subcollections $\mathcal{M}$ of closed spheres $\{E_j\}_{j=1}^n$ covering a specific space associated with $\mathbb{T}^3$. This space is $S_{\mathbb{T}^3}$ and is within ϵ measure of the 3-Torus and is obtained by minimally smoothening the vertices of $[-L, L]^3$ and slightly puffing out its facets. Also inner measure is used where the supremum is taken over all finite subcollections $\mathcal{N}$ of closed spheres $\{F_j\}_{j=1}^p$ inside $S_{\mathbb{T}^3}$. Generally by Hölder's inequality and for sufficiently small $c > 0$,

$$\sup_{\mathcal{N}} \mid c \left[\int_{F=\bigcup_{j=1}^p F_j} \vec{r} \times \nabla_D^{-1} \left[\sum_{i=1,i\neq 3}^4 \mathcal{G}_{\delta i} \right] d\vec{x} \right]$$

$$\mid \leq \inf_{\mathcal{M}} \mid c \int_{\Omega_R=\bigcup_{j=1}^m E_j} \vec{r} \times \nabla_D^{-1} \left[\sum_{i=1,i\neq 3}^4 \mathcal{G}_{\delta i} \right] \mid d\vec{x}$$

$$= \mid c \int_\Omega \mid \vec{r} \mid\mid \nabla_D^{-1} \left[\sum_{i=1,i\neq 3}^4 \mathcal{G}_{\delta i} \right] \mid \sin(\theta)\vec{n} \, d\vec{x} \mid \leq \mid c$$

$$\int_\Omega \mid \vec{r} \mid \nabla_D^{-1} \left[\sum_{i=1,i\neq 3}^4 \mathcal{G}_{\delta i} \right] \mid d\vec{x} \mid$$

$$= \mid -c \int_\Omega \mid \vec{r} \mid^2 \vec{\omega_1} \cdot \nabla(u_z \vec{b}) \, d\vec{x} \mid$$

$$= c \mid \int_\Omega \mid \vec{r} \mid^2 \vec{\omega_1} \cdot \nabla(u_z \vec{b}) \, d\vec{x} \mid \leq c \int_\Omega \mid\mid \vec{r} \mid^2 \vec{\omega_1} \cdot \nabla(u_z \vec{b}) \mid$$

$$d\vec{x} \leq c \left(\int_\Omega \mid \vec{r} \mid^4 d\vec{x} \right)^{\frac{1}{4}} \left[\left(\int_\Omega \mid \vec{\omega_1} \cdot \nabla(u_z \vec{b}) \mid^2 \right)^{\frac{3}{4}} \right]$$

$$\leq c \left(\int_\Omega \mid \vec{r} \mid^4 d\vec{x} \right)^{\frac{1}{4}} \left[\left[\left(\int_\Omega \mid \vec{\omega_1} \cdot (\nabla u_z \otimes \vec{b}) \mid^2 d\vec{x} \right)^{1/2} \right]^{\frac{3}{2}} + \right.$$

$$\left. \left[\left(\int_\Omega \mid u_z \nabla(\vec{b}) \cdot \vec{\omega_1} \mid^2 d\vec{x} \right)^{1/2} \right]^{\frac{3}{2}} \right]$$

$$\leq c \left(\int_\Omega \mid \vec{r} \mid^4 d\vec{x} \right)^{\frac{1}{4}} \left[\left[\int_\Omega \mid \vec{\omega_1} \mid^2 \mid \vec{u} \mid^2 \mid \nabla u_z \mid^2 d\vec{x} \right]^{\frac{3}{4}} \right.$$

$$\left. + \left[\int_\Omega \mid u_z \mid^2 \mid \nabla \vec{b} \mid^2 \mid \vec{\omega_1} \mid^2 d\vec{x} \right]^{\frac{3}{4}} \right]$$

$$\leq c \left(\int_\Omega \mid \vec{r} \mid^4 d\vec{x} \right)^{\frac{1}{4}} \left[\left[\|\vec{\omega_1}\|_4^2 \|\vec{u}\|_8^2 \|\nabla u_z\|_{16}^2 \right]^{\frac{3}{4}} \right.$$

$$\left. + \left[\int_\Omega \mid u_z \mid^2 \mid \nabla \vec{b} \mid^2 \vec{\omega_1} \mid^2 d\vec{x} \right]^{\frac{3}{4}} \right]$$

$$\leq \frac{152}{15} c L^{7/4} \left[\frac{1}{c} \|\vec{\omega_1}\|_4^{\frac{3}{2}} \|\vec{u}\|_8^{\frac{3}{2}} \|\nabla u_z\|_2^3 + \left[\frac{1}{c} \|u_z\|_8^{\frac{3}{2}} \|\nabla \vec{b}\|_2^3 \|\vec{\omega_1}\|_4^{\frac{3}{2}} \right] \right]$$

$$\boxed{\leq \kappa}$$

The last inequality in chain of inequalities above is true if and only if all the norms in Inequality above it are presumed to be finite. The goal of this chapter is to show non-smoothness for higher derivatives in time for the continuum case associated with high energy. The operator on the left most side of the above sequence of inequalities is essentially bounded and its essential supremum is zero,

$$||f||_\infty \equiv \inf\{\kappa \geq 0 : |f| \leq \kappa \text{ for almost every } \mathbf{x}$$

$$||f||_\infty \equiv 0$$

Considering inscribed spheres in each puffed cell of the perturbed lattice $S_{\mathbb{T}^3}$ and integrating over the union of all such spheres where each cell is of arbitrary small measure containing spheres results in a finite well defined integral bounded by a now finite supremum. The supremum on the left most side of above chain of inequalities may exist for certain Elliptic functions defined on the 3-Torus with limiting parameters as the present study aims to show. Then integration over $\mathbb{T}^3$ is well defined and we can replace $S_{\mathbb{T}^3}$ by $\mathbb{T}^3$. Now $\vec{\omega_1}$ is vorticity $\vec{\omega}$ multiplied with u_z and to begin with ∇_D^{-1} is the inverse of the divergence operator integrated over an arbitrary cell of the 3-Torus lattice. Here the identity $A \times B = | A \,||\, B \,| \sin(\theta)\vec{n}$ was used and in the previous two integrals over a periodic lattice, the first one approaches zero as R becomes arbitrarily large, where the 3-Torus being a compact manifold is bounded by an R-Ball with the norm of the gradient of u_z approaching zero and the second integral consisting of $| \nabla \vec{b} |$ approaches zero for the finite but large volume 3-Torus. Now any topological space that is homeomorphic to $S^1(\mathbf{x}, \alpha) \times S^1(\mathbf{x}, \alpha) \times S^1(\mathbf{x}, \alpha)$ for radius $\alpha = 1$ is defined as the 3-Torus and since increasing concentric circles($\alpha > 1$) are homeomorphic to the unit circle S^1, we can extend the 3-Torus to larger homeomorphic sets as $\alpha \to \infty$. The Preköpa-Leindler and Gagliardo-Nirenberg Inequalities have been used for each component equation indexed by j associated with the Laplacian of the Navier–Stokes Equations,

$$| \Omega |^{\frac{1}{2}} \, \|\nabla u_j\|_2 \leq C\|u_j\|_q^{1-\theta} \left[| \Omega |^{\frac{1}{2}} \, \|\nabla^2 u_j\|_r^\theta \right] \leq C\|u_j\|_q^{1-\theta} \int_\Omega \nabla^2 u_j d\vec{x}$$

Here $\sum_{j=1}^3 u_j = \Psi * w$, where $*$ is convolution and $\int_{\mathbb{T}^3} w \, dx = \mathbf{s} \in \mathbb{R}$. If Ψ is the fundamental solution of the scalar Laplacian on the 3-Torus

$\mathbb{T}^3 = S^1 \times S^1 \times S^1$ noting that $\triangle(\Psi * w) = w$ then the integral of the Laplacian is in general non zero [33]. We must rely on the dimension of the Lattice to ensure the limit value is zero upon dividing by large enough $|\Omega|$. Off of the associated compact set, the velocity is zero or the velocity has compact support . The chain of inequalities at the top of this page imply that, in general $\nabla_D^{-1}(\mathcal{G}_{\delta 1} + \mathcal{G}_{\delta 2} + \mathcal{G}_{\delta 4}) = \vec{r}\Phi(t)$ since the two vectors $\vec{r}$ and ∇_D^{-1} can be in the same direction. A term $\Phi(t)$ is also multiplied by $\vec{r}$. A group G of transformations of $\vec{u}(\vec{r}, t)$ is a symmetry group of NS if for all $g \in G$, $\vec{u}$ a NS solution implies $g\vec{u}$ is a NS solution. The group is $\mathbb{Z}$. The Navier–Stokes equations are invariant under the dilation group as shown after Eq. (3). Next there will be an application of the group transformations seen in Eqs. (8.4–8.7):

$$\mathcal{G}_{\delta 1} = -\left(1 - \frac{1}{\delta}\right)\left(\frac{\partial u_z}{\partial t}\right)\left(\frac{\partial u_z}{\partial t} - \frac{\mu}{\rho}\nabla^2 u_z + \frac{1}{\rho}\frac{\partial P}{\partial z}\right) \tag{8.4}$$

$$\mathcal{G}_{\delta 2} = u_z \frac{\partial u_z}{\partial z}\frac{\partial u_z}{\partial t} + u_z^2 \frac{\partial^2 u_z}{\partial z \partial t} + \frac{2u_z}{\delta}\left(\frac{\partial \mathbf{u}}{\partial t} \cdot \nabla u_z\right) \tag{8.5}$$

$$\mathcal{G}_{\delta 3} = \iint_{\partial \Omega} \frac{u_z}{\rho}\left(\frac{1}{\delta}u_z \nabla_{xy} P + \frac{\partial P}{\partial z}\vec{b}\right) \cdot \vec{n} dS$$
$$- \int_{\Omega} \left\|\left(\frac{\vec{b}}{\|\vec{b}\|} \cdot \left(\vec{b} \otimes \nabla u_z\right)\right)\frac{\partial u_z}{\partial t}\right\| dV \tag{8.6}$$

$$\mathcal{G}_{\delta 4} = \delta^2 \vec{F_T} \cdot \nabla u_z^2 - \delta^3 u_z \frac{\partial u_z}{\partial z} F_z + \delta^3 \vec{b} \cdot \nabla(u_z F_z) + \delta^3\left(1 - \frac{1}{\delta}\right)\frac{F_z}{u_z}\frac{\partial u_z}{\partial t} \tag{8.7}$$

where $\vec{b} = \frac{1}{\delta}\left(u_x\vec{i} + u_y\vec{j} + u_z\vec{k}\right)$, and $\vec{i}$, $\vec{j}$ and $\vec{k}$ are the standard unit vectors. For Poisson's Equation seen in Eq. (8.8) (see [19,30]), the second derivative P_{zz} is set equal to the second derivative obtained in the $\mathcal{G}_{\delta 1}$ expression as part of $\mathcal{G}$,

$$P_{zz} = -2u_z\nabla^2 u_z - \left(\frac{\partial u_z}{\partial z}\right)^2 + \frac{1}{\eta}\frac{\partial}{\partial z}\left(\frac{\partial u_z}{\partial x} + \frac{\partial u_z}{\partial y}\right) - \delta\, u_x \frac{\partial^2 u_z}{\partial z \partial x}$$
$$- \delta\, u_y \frac{\partial^2 u_z}{\partial z \partial y} + \left(\frac{\partial u_x}{\partial x}\right)^2 + 2\frac{\partial u_x}{\partial y}\frac{\partial u_y}{\partial x} + \left(\frac{\partial u_y}{\partial y}\right)^2 \tag{8.8}$$

where the last three terms on rhs of Eq. (8.8) can be shown to be equal to $-(P_{xx} + P_{yy})$ [16]. Along with Eqs. (8.3–8.8), the continuity equation in Cartesian co-ordinates, is $\nabla^i u_i = 0$. Furthermore, the right-hand side of the one parameter group of transformations are next mapped to η variable terms,

$$u_i = \frac{1}{\eta}v_i, P = \frac{1}{\eta^2}Q, x_i = \eta y_i, t = \eta^2 s, \quad i = 1, 2, 3. \tag{8.9}$$

and Eq. (8.3) becomes,

$$\mathcal{G}(\eta) = \mathcal{G}(\eta)_{\delta 1} + \mathcal{G}(\eta)_{\delta 2} + \mathcal{G}(\eta)_{\delta 3} + \mathcal{G}(\eta)_{\delta 4} = 0 \tag{8.10}$$

where

$$\mathcal{G}(\eta)_{\delta 1} = \frac{1}{\eta^6}\left[\left(\delta^{-1} - 1\right)\left(\frac{\partial v_3}{\partial s}\right)^2\right.$$
$$+ \frac{\mu\left(\frac{\partial v_3}{\partial s}\right)\left(\frac{\partial^2 v_3}{\partial y_1^2} + \frac{\partial^2 v_3}{\partial y_2^2} + \frac{\partial^2 v_3}{\partial y_3^2}\right)}{\rho}\left(1 - \delta^{-1}\right)$$
$$\left. + \frac{\left(\delta^{-1} - 1\right)\left(\frac{\partial v_3}{\partial s}\right)\frac{\partial Q}{\partial y_3}}{\rho}\right] \tag{8.11}$$

$$\mathcal{G}(\eta)_{\delta 2} = \frac{v_3}{\eta^6}\left(\frac{\partial v_3}{\partial y_3}\right)\frac{\partial v_3}{\partial s} + \frac{(v_3)^2}{\eta^6}\frac{\partial^2 v_3}{\partial y_3 \partial s}$$
$$+ \frac{2\left(\frac{\partial v_1}{\partial s}\right)v_3\frac{\partial v_3}{\partial y_1} + 2\left(\frac{\partial v_2}{\partial s}\right)v_3\frac{\partial v_3}{\partial y_2} + 2\left(\frac{\partial v_3}{\partial s}\right)v_3\frac{\partial v_3}{\partial y_3}}{\delta\eta^6} \tag{8.12}$$

$$\mathcal{G}(\eta)_{\delta 3} = \frac{1}{\eta^3} \times \left[\iint_S \left(\frac{1}{\delta\rho}v_3^2 \nabla_{y_1 y_2} Q + \frac{1}{\delta}\frac{1}{\rho}\vec{v} v_3 \frac{\partial Q}{\partial y_3}\right) \cdot \vec{n}\, dS\right.$$
$$\left. - \int_\Omega \frac{\left\|\frac{\partial v_3}{\partial s}\vec{b} \cdot \left(\vec{b} \otimes \nabla v_3\right)\right\|}{\|\vec{b}\|}\, dV\right] \tag{8.13}$$

$$\mathcal{G}(\eta)_{\delta 4} = \frac{1}{\eta^3}\left[\delta^2 \vec{F_T} \cdot \nabla_{y_1 y_2} v_3^2 - \delta^3 v_3 \frac{\partial v_3}{\partial y_3} F_z + \delta^2 \vec{v} \cdot \nabla(v_3 F_z)\right] \tag{8.14}$$

where $\vec{v} = (v_1, v_2, v_3)$ and $\vec{F_T} = F_{T_1}\vec{i} + F_{T_2}\vec{j}$. The body force $\mathbf{F}^* = F^{*i}\mathbf{e_i}$, with $\vec{F_T} = (F_{T_1}(y_1, y_2, y_3, s), F_{T_2}(y_1, y_2, y_3, s))$ and F_z is the z−component of the force vector. P depends on η as $P = \frac{1}{\eta^2}Q$. Thus $\frac{\partial P}{\partial z} = \frac{1}{\eta^3}\frac{\partial Q}{\partial y_3}$. We solve for $\frac{\partial P}{\partial z}$ and using Poisson's equation Eq. (8.8), set second derivatives of P w.r.t. z equal to each other, and then set

$\delta \approx 1$ after multiplying a factor of $\delta - 1$ out of the equation. This makes $\epsilon = \delta - 1$, a (small) perturbation parameter. The Kinematic Viscosity is $\nu = \mu/\rho$.

$$\mathcal{L} = \mathcal{L}_1 + \mathcal{L}_2 + \eta^3 \mathcal{L}_3 + \eta^3 \mathcal{L}_4 = 0 \tag{8.15}$$

where the four components are defined as:

$$\mathcal{L}_1 = (\delta^{-1} - 1)\left(\frac{\partial v_3}{\partial s}\right)^2 + (1 - \delta^{-1})\left(\frac{\partial v_3}{\partial s}\right)\left(\frac{\partial^2 v_3}{\partial y_1{}^2} + \frac{\partial^2 v_3}{\partial y_2{}^2} + \frac{\partial^2 v_3}{\partial y_3{}^2}\right)$$
$$+ (\delta^{-1} - 1)\frac{\left(\frac{\partial v_3}{\partial s}\right)\frac{\partial Q}{\partial y_3}}{\rho} \tag{8.16}$$

$$\mathcal{L}_2 = v_3\left(\frac{\partial v_3}{\partial y_3}\right)\frac{\partial v_3}{\partial s} + v_3^2\frac{\partial^2 v_3}{\partial y_3 \partial s} + \frac{1}{\delta}\left[2\left(\frac{\partial v_1}{\partial s}\right)v_3\frac{\partial v_3}{\partial y_1}\right.$$
$$\left. + 2\left(\frac{\partial v_2}{\partial s}\right)v_3\frac{\partial v_3}{\partial y_2} + 2\left(\frac{\partial v_3}{\partial s}\right)v_3\frac{\partial v_3}{\partial y_3}\right] \tag{8.17}$$

$$\mathcal{L}_3 = \iint_S \left(\frac{1}{\delta\rho}v_3^2\nabla_{y_1 y_2}Q + \frac{1}{\delta}\vec{v}\frac{1}{\rho}v_3\frac{\partial Q}{\partial y_3}\right) \cdot \vec{n}\, dS$$
$$- \int_\Omega \frac{\left\|\frac{\partial v_3}{\partial s}\vec{b}\cdot\left(\vec{b}\otimes\nabla v_3\right)\right\|}{\|\vec{b}\|}dV \tag{8.18}$$

$$\mathcal{L}_4 = \delta^2 \vec{F_T}\cdot\nabla_{y_1 y_2}v_3^2 - \delta^3 v_3\frac{\partial v_3}{\partial y_3}F_z + \delta^2\vec{v}\cdot\nabla(v_3 F_z) \tag{8.19}$$

As for Eq. (8.18) see the Appendix at the end of this chapter. It can be seen that the expressions in square brackets in Eqs. (8.13–8.14) are $\mathcal{L}_3$ & $\mathcal{L}_4$ in Eqs. (8.18–8.19), respectively. Finally, in subsequent sections, the Weierstrass degenerate elliptic P function will be used. Letting the Weierstrass P function be denoted by $\wp(z, g2, g3)$, the degenerate case can be denoted as $P_m(z) = \wp(z, 3m^2, m^3)$.

8.2.1 Decomposition of NSEs

For Eqs. (8.3)–(8.8) the Dirichlet condition $\vec{u}^*(\vec{x}^*, 0) = \vec{\xi}(\vec{x}^*)$ such that $\nabla \cdot \vec{\xi} = 0$ describes the NSEs together with an incompressible initial condition. Considering periodic boundary conditions defined on 3-torus with associated Lattice is a periodic BVP for the NSEs. Solutions were found to be in the form,

$$\mathbf{u} = (u_x, u_y, u_z) : \mathbb{R}^+ \times \mathbb{R}^3/\mathbb{Z}^3 \to \mathbb{R}^3 \tag{8.20}$$

where u_x, u_y and u_z satisfy Eqs. (8.3), (8.11–8.15).

8.2.2 Liutex Vector and Respective Governing Equations

Theorem 1 in [19] is used in the above decomposition of (PNS) and is the basis theorem of this chapter. By using the generalized divergence theorem for scalar products, the term $\nabla u_z^2 \cdot \frac{\partial \vec{b}}{\partial t}$ in Eq. (8.5) is extended to 3-components of velocity (see Eq. (12) in [30] which only incorporated a 2-component velocity field). No finite time blowup was obtained for the simplified case there. Solutions are obtained symbolically with Maple 2021 software, and with the use of Poisson's equation, Eq. (8.8), Eqs. (8.3)–(8.8), lead to the following,

$$L = 0 \tag{8.21}$$

with L an operator. See Appendix A2 at the end of this chapter for the entire solution script in Maple for the following part.(Also see Appendix A4) Solving symbolically for $L = 0$ individually for the mixed partial derivatives in the expression $\left(\frac{\partial^2 v_2}{\partial y_3 \partial s} - \frac{\partial^2 v_1}{\partial y_3 \partial s} \right)$, using the following definition of κ in terms of v_1, v_2 and v_3, [see Eq. (32) in [18]. The new equation there has a viscous term reduced to half and one only needs to calculate three elements.]

$$-\left(\frac{\partial^2 v_2}{\partial y_3 \partial s} - \frac{\partial^2 v_1}{\partial y_3 \partial s} \right) = \frac{\partial}{\partial s} \kappa \left(y_1, y_2, y_3, s \right) - \frac{\partial^2 v_3}{\partial y_2 \partial s} + \frac{\partial^2 v_3}{\partial y_1 \partial s} \tag{8.22}$$

Both,

$$M_1 = \frac{\partial^2 v_1}{\partial y_3 \partial s} \tag{8.23}$$

and

$$M_2 = \frac{\partial^2 v_2}{\partial y_3 \partial s} \tag{8.24}$$

are nonlinear partial differential equations. The v_1 and v_2 velocities are chosen respectively as the following general spatial-temporal functions, which are assumed to fulfill compatibility conditions in [35] and [36],

$$v_1(y_1, y_2, y_3, s) = (U_1(y_1, y_2, y_3) + A) \times f_1(s) \tag{8.25}$$

and

$$v_2(y_1, y_2, y_3, s) = (U_2(y_1, y_2, y_3) + A) \times f_2(s) \tag{8.26}$$

where $A \ll 0$. Note that the magnitude of Liutex (scalar form) is obtained in the plane perpendicular to the local axis, which is twice the angular speed of local fluid rotation ,

$$\vec{\omega_L} = \frac{2}{r^2}(\vec{r} \times \vec{v}) \tag{8.27}$$

where ω_L is associated with the Liutex vector part of vorticity, $r = y_1\vec{i} + y_2\vec{j} + y_3\vec{k}$ and where the Liutex magnitude difference is calculated as follows,

$$\kappa(y_1, y_2, y_3, s) = 2\frac{y_2 v_3 - y_3 v_2 - y_1 v_3 + y_3 v_1}{y_1^2 + y_2^2 + y_3^2} \tag{8.28}$$

Substituting $M_1 - M_2$ expressions in Eqs. (8.23–8.24) and κ from Eq. (8.28) into Eq. (8.22) gives a new PDE,

$$\mathcal{F} = \frac{\mathcal{F}_1}{H} + 2\frac{\rho}{H}\frac{\partial v_3}{\partial s}\left[\mathcal{F}_2 + \mathcal{F}_3 + \mathcal{F}_4\right] = \mathcal{F}_5, \tag{8.29}$$

$$H = 2\rho\delta\left(\frac{\partial v_3}{\partial s}\right)v_3\left(\frac{\partial v_3}{\partial y_1}\right)\frac{\partial v_3}{\partial y_2}$$

$$\mathcal{F}_1 = \left(\frac{\partial v_3}{\partial s}\right)^2 \mu(-1+\delta)\left(\frac{\partial v_3}{\partial y_1} - \frac{\partial v_3}{\partial y_2}\right)\frac{\partial^3 v_3}{\partial y_3 \partial y_1^2}$$

$$+ \left(\frac{\partial v_3}{\partial s}\right)^2 \mu(-1+\delta)\left(\frac{\partial v_3}{\partial y_1} - \frac{\partial v_3}{\partial y_2}\right)\frac{\partial^3 v_3}{\partial y_3 \partial y_2^2}$$

$$+ \left(\frac{\partial v_3}{\partial s}\right)^2 \mu(-1+\delta)\left(\frac{\partial v_3}{\partial y_1} - \frac{\partial v_3}{\partial y_2}\right)\frac{\partial^3 v_3}{\partial y_3^3}$$

$$+ (v_3)^2 \left(\frac{\partial v_3}{\partial s}\right)\rho\delta\left(\frac{\partial v_3}{\partial y_1} - \frac{\partial v_3}{\partial y_2}\right)\frac{\partial^3 v_3}{\partial y_3^2 \partial t}$$

$$- (v_3)^2 \rho\delta\left(\frac{\partial v_3}{\partial y_1} - \frac{\partial v_3}{\partial y_2}\right)\left(\frac{\partial^2 v_3}{\partial y_3 \partial t}\right)^2$$

$$- 2\rho\left(\frac{\partial v_3}{\partial y_1} - \frac{\partial v_3}{\partial y_2}\right)\left((\delta/2 - 1/2)\left(\frac{\partial v_3}{\partial s}\right)^2 - v_3\left(\frac{\partial v_3}{\partial s}\right)\left(\frac{\partial v_3}{\partial y_3}\right)\delta\right.$$

$$+ \delta \left(v_3 \left(F_{T_1} + (A + U_{y_1}) \frac{\mathrm{d}}{\mathrm{d}t} f_1(s) \right) \frac{\partial v_3}{\partial y_1} \right.$$

$$+ v_3 \left(F_{T_2} + (A + U_{y_2}(y_1, y_2, y_3)) \frac{\mathrm{d}}{\mathrm{d}s} f_1(s) \right) \frac{\partial v_3}{\partial y_2}$$

$$\left. + 1/2 \, \Lambda(y_1, y_2, y_3, s) + 1/2 \, \Phi(s) \right) \right) \rho \frac{\partial^2 v_3}{\partial y_3 \partial s} \left(\frac{\partial v_3}{\partial y_1} - \frac{\partial v_3}{\partial y_2} \right)$$

$$(8.30)$$

$$\mathcal{F}_2 = \left(\frac{\partial v_3}{\partial y_1} - \frac{\partial v_3}{\partial y_2} \right) \left(1/2 \, (-1 + \delta) \left(A \delta \, f_1(s) \right. \right.$$

$$+ \, U_{y_1}(y_1, y_2, y_3) \, \delta \, f_1(s) - 1 \right) \frac{\partial v_3}{\partial s} + v_3 \rho \, \delta \left(F_{T_1}(x, y, z, s) \right.$$

$$+ (A + U_{y_1}(y_1, y_2, y_3)) \frac{\mathrm{d}}{\mathrm{d}s} f_1(s) \right) \right) \frac{\partial^2 v_3}{\partial y_3 \partial y_1}$$

$$+ \left(1/2 \, (-1 + \delta) (A \delta \, f_1(s) + U_{y_2}(y_1, y_2, y_3) \, \delta \, f_1(s) - 1) \frac{\partial v_3}{\partial s} \right.$$

$$+ \, v_3 \rho \, \delta \left(F_{T_2}(y_1, y_2, y_3, s) \right.$$

$$+ (A + U_{y_2}(y_1, y_2, y_3)) \frac{\mathrm{d}}{\mathrm{d}s} f_1(s) \right) \right) \frac{\partial^2 v_3}{\partial y_3 \partial y_2} \qquad (8.31)$$

$$\mathcal{F}_3 = \frac{3}{2} \left(\frac{\partial v_3}{\partial s} \right) \left(-\frac{2}{3} + \left(\rho + \frac{2}{3} \right) \delta \right) \left(\frac{\partial v_3}{\partial y_1} - \frac{\partial v_3}{\partial y_2} \right) v_3 \frac{\partial^2 v_3}{\partial y_3{}^2}$$

$$+ v_3 \left(\frac{\partial v_3}{\partial s} \right) (-1 + \delta) \left(\frac{\partial v_3}{\partial y_1} - \frac{\partial v_3}{\partial y_2} \right) \frac{\partial^2 v_3}{\partial y_1{}^2}$$

$$+ v_3 \left(\frac{\partial v_3}{\partial s} \right) (-1 + \delta) \left(\frac{\partial v_3}{\partial y_1} - \frac{\partial v_3}{\partial y_2} \right) \frac{\partial^2 v_3}{\partial y_2{}^2}$$

$$+ 1/2 \left((-1 + (3\rho + 1) \delta) \left(\frac{\partial v_3}{\partial y_3} \right)^2 \right.$$

$$+ (f_1(s))^2 (-1 + \delta) \times \left(\left(\frac{\partial U_{y_1}}{\partial y_1}(y_1, y_2, y_3) \right)^2 \right.$$

$$+ 2 \left(\frac{\partial U_{y_1}}{\partial y_2}(y_1, y_2, y_3) \right) \frac{\partial U_{y_2}}{\partial y_1}(y_1, y_2, y_3)$$

$$+ \left(\frac{\partial U_{y_2}}{\partial y_2}(y_1, y_2, y_3) \right)^2 \right) \right) \times \left(\frac{\partial v_3}{\partial y_1} - \frac{\partial v_3}{\partial y_2} \right) \frac{\partial v_3}{\partial s} \qquad (8.32)$$

$$\mathcal{F}_4 = \delta\rho\Bigg(\Bigg(\Bigg(F_{T_1}(y_1, y_2, y_3, s) + (A + U_{y_1}(y_1, y_2, y_3))\frac{\mathrm{d}f_1}{\mathrm{d}s}(s)\Bigg)\frac{\partial v_3}{\partial y_3}$$

$$+ v_3\Bigg(\Bigg(\frac{\mathrm{d}}{\mathrm{d}s}f_1(s)\Bigg)\frac{\partial U_{y_1}(y_1, y_2, y_3)}{\partial y_3} + \frac{\partial F_{T_1}(y_1, y_2, y_3, s)}{\partial y_3}\Bigg)\Bigg)\Bigg(\frac{\partial v_3}{\partial y_1}\Bigg)^2$$

$$+ \Bigg(\Bigg(\Bigg(-F_{T_1}(y_1, y_2, y_3, s) + F_{T_2}(y_1, y_2, y_3, s)$$

$$+ (-U_{y_1}(y_1, y_2, y_3) + U_{y_2}(y_1, y_2, y_3))\frac{\mathrm{d}}{\mathrm{d}s}f_1(s)\Bigg)\frac{\partial v_3}{\partial y_3}$$

$$- v_3\Bigg(\frac{\partial F_{T_1}(y_1, y_2, y_3, s)}{\partial y_3} - \frac{\partial F_{T_2}(y_1, y_2, y_3, s)}{\partial y_3}\Bigg)\Bigg)\frac{\partial v_3}{\partial y_2}$$

$$+ 1/2\frac{\partial\Lambda(y_1, y_2, y_3, s)}{\partial y_3}\Bigg)\frac{\partial v_3}{\partial y_1} - \Bigg(\frac{\partial v_3}{\partial y_2}\Bigg)\Bigg(\Bigg(\Bigg(F_{T_2}(y_1, y_2, y_3, s)$$

$$+ (A + U_{y_2}(y_1, y_2, y_3))\frac{\mathrm{d}}{\mathrm{d}s}f_1(s)\Bigg)\frac{\partial v_3}{\partial y_3}$$

$$+ v_3\Bigg(\Bigg(\frac{\mathrm{d}}{\mathrm{d}s}f_1(s)\Bigg)\frac{\partial U_{y_2}(y_1, y_2, y_3)}{\partial y_3} + \frac{\partial F_{T_2}(y_1, y_2, y_3, s)}{\partial y_3}\Bigg)\Bigg)\frac{\partial v_3}{\partial y_2}$$

$$+ 1/2\frac{\partial\Lambda(y_1, y_2, y_3, s)}{\partial y_3}\Bigg)\frac{\partial v_3}{\partial y_2}\Bigg) \tag{8.33}$$

$$\mathcal{F}_5 = \frac{2\,y_2\frac{\partial v_3}{\partial s} - 2\,y_3\big(A + U_{y_2}(y_1, y_2, y_3)\big)\frac{\mathrm{d}}{\mathrm{d}s}f_1(s) - 2\,y_1\frac{\partial v_3}{\partial s} + 2\,z\big(A + U_{y_1}(y_1, y_2, y_3)\big)\frac{\mathrm{d}}{\mathrm{d}s}f_1(s)}{y_1{}^2 + y_2{}^2 + y_3{}^2} - \frac{}{\frac{\partial^2 v_3}{\partial y_2 \partial s} + \frac{\partial^2 v_3}{\partial y_1 \partial s}}$$

$$\tag{8.34}$$

$$F_{T_1}(y_1, y_2, y_3, s) = f_0(s)\,(F(y_1, y_2, y_3) + A) \tag{8.35}$$

$$F_{T_2}(y_1, y_2, y_3, s) = f_0(s)\,(G(y_1, y_2, y_3) + A)$$

Also $\Lambda = \mathcal{L}_4$ which contains the y_3 external forcing term and it's derivatives. Equation (8.21) becomes,

$$L = \Bigg(\frac{\partial v_3}{\partial s}\Bigg)^2\mu(\delta - 1)\frac{\partial^3 v_3}{\partial y_3\partial y_1^2} + \Bigg(\frac{\partial v_3}{\partial s}\Bigg)^2\mu(\delta - 1)\frac{\partial^3 v_3}{\partial y_3\partial y_2^2}$$

$$+ \Bigg(\frac{\partial v_3}{\partial s}\Bigg)^2\mu(\delta - 1)\frac{\partial^3 v_3}{\partial y_3^3} + \Bigg(\frac{\partial v_3}{\partial s}\Bigg)(v_3)^2\Bigg(\frac{\partial^3 v_3}{\partial y_3^2\partial s}\Bigg)\delta\rho$$

$$- (v_3)^2\Bigg(\frac{\partial^2 v_3}{\partial y_3\partial s}\Bigg)^2\delta\rho - 2\rho\Bigg(\Bigg(\frac{\delta}{2} - \frac{1}{2}\Bigg)\Bigg(\frac{\partial v_3}{\partial s}\Bigg)^2$$

$$- v_3\Bigg(\frac{\partial v_3}{\partial s}\Bigg)\Bigg(\frac{\partial v_3}{\partial y_3}\Bigg)\delta + \Bigg(v_3\Bigg(F_{T_1}(y_1, y_2, y_3, s) + \frac{\partial v_1}{\partial s}\Bigg)\frac{\partial v_3}{\partial y_1}\Bigg)$$

$$+ v_3 \left(F_{T_2} \left(y_1, y_2, y_3, s \right) + \frac{\partial v_2}{\partial s} \right) \frac{\partial v_3}{\partial y_2} + \frac{\Lambda \left(y_1, y_2, y_3, s \right)}{2} + \frac{\Phi \left(s \right)}{2} \right) \delta \right) \frac{\partial^2 v_3}{\partial y_3 \partial s}$$

$$+ \left(\left(\left(\delta - 1 \right) \left(\delta v_1 \left(y_1, y_2, y_3, s \right) - 1 \right) \frac{\partial v_3}{\partial s} + 2 v_3 \rho \delta \left(F_{T_1} \left(y_1, y_2, y_3, s \right) \right. \right.$$

$$+ \frac{\partial v_1}{\partial s} \right) \right) \frac{\partial^2 v_3}{\partial y_3 \partial y_1} + \left(\left(\delta - 1 \right) \left(v_2 \left(y_1, y_2, y_3, s \right) \delta - 1 \right) \frac{\partial v_3}{\partial s}$$

$$+ 2 v_3 \rho \delta \left(F_{T_2} \left(y_1, y_2, y_3, s \right) + \frac{\partial v_2}{\partial s} \right) \right) \frac{\partial^2 v_3}{\partial y_3 \partial y_2}$$

$$+ 3 v_3 \left(-\frac{2}{3} + \left(\rho + \frac{2}{3} \right) \delta \right) \left(\frac{\partial v_3}{\partial s} \right) \frac{\partial^2 v_3}{\partial y_3^2} + 2 v_3 \left(\frac{\partial v_3}{\partial s} \right) \left(\delta - 1 \right) \frac{\partial^2 v_3}{\partial y_1^2}$$

$$+ 2 v_3 \left(\frac{\partial v_3}{\partial s} \right) \left(\delta - 1 \right) \frac{\partial^2 v_3}{\partial y_2^2} + 2 \left(\frac{\partial^2 v_1}{\partial y_3 \partial s} \right) v_3 \left(\frac{\partial v_3}{\partial y_1} \right) \rho \delta$$

$$+ 2 \left(\frac{\partial^2 v_2}{\partial y_3 \partial s} \right) v_3 \left(\frac{\partial v_3}{\partial y_2} \right) \rho \delta + \left(\left(-1 + \left(3 \rho + 1 \right) \delta \right) \left(\frac{\partial v_3}{\partial y_3} \right)^2 \right.$$

$$+ \left(\delta - 1 \right) \left(\left(\frac{\partial v_1}{\partial y_1} \right)^2 + 2 \left(\frac{\partial v_1}{\partial y_2} \right) \frac{\partial v_2}{\partial y_1} + \left(\frac{\partial v_2}{\partial y_2} \right)^2 \right) \right) \frac{\partial v_3}{\partial s}$$

$$+ 2 \rho \left(\left(\left(F_{T_1} \left(y_1, y_2, y_3, s \right) + \frac{\partial v_1}{\partial s} \right) \frac{\partial v_3}{\partial y_1} + \left(\frac{\partial v_3}{\partial y_2} \right) \left(F_{T_2} \left(y_1, y_2, y_3, s \right) \right. \right. \right.$$

$$+ \frac{\partial v_2}{\partial s} \right) \right) \frac{\partial v_3}{\partial y_3} + v_3 \left(\frac{\partial v_3}{\partial y_1} \right) \frac{\partial F_{T_1}}{\partial y_3} + v_3 \left(\frac{\partial v_3}{\partial y_2} \right) \frac{\partial F_{T_2}}{\partial y_3}$$

$$+ \frac{1}{2} \frac{\partial \Lambda \left(y_1, y_2, y_3, s \right)}{\partial y_3} \right) \delta \right) \frac{\partial v_3}{\partial s} = 0$$

and $\Lambda \left(y_1, y_2, y_3, s \right)$ is given as,

$$\Lambda \left(y_1, y_2, y_3, s \right) = 2 \frac{f_0(s) F \left(y_1, y_2, y_3 \right) v_3 \left(y_1, y_2, y_3, s \right) \frac{\partial v_3}{\partial y_1}}{\delta}$$

$$+ 2 \frac{f_0(s) G \left(y_1, y_2, y_3 \right) v_3 \left(y_1, y_2, y_3, s \right) \frac{\partial v_3}{\partial y_2}}{\delta} - \delta^3 v_3 \left(\frac{\partial v_3}{\partial y_3} \right) F_{sz} \left(y_1, y_2, y_3, s \right)$$

$$+ \delta^2 \left(\left(\frac{\partial v_3}{\partial y_3} \right) F_{sz} \left(y_1, y_2, y_3, s \right) + v_3 \frac{\partial F_{sz}}{\partial y_3} \right)$$

where $\vec{f} = \left(F_{T_1,}, F_{T_2}, F_{sz} \right)$ is the forcing vector and $\vec{v} = \left(v_1, v_2, v_3 \right)$ is the velocity in each cell of the 3-Torus.

For the three forcing terms, set them equal to products of reciprocals of degenerate WeierstrassP functions shifted in spatial coordinates from the center $\left(a_i, b_i, c_i \right)$, $i = 1..N$.

Here the $\left(a_i, b_i, c_i \right)$ is the center of each cell of the lattice belonging to the flat torus. Upon substituting the WeierstrassP functions and their reciprocals(unity divided by P-function) into Eq.(8.21) together with the forcing terms given by Λ, it can be observed that in the equation that terms in it are multiplied by reciprocal WeierstrassP functions which touch the centers of the cells of the lattice, thus simplifying Eq. (8.21). The initial condition in v_3 at $t = 0$ is instead of a product of reciprocal degenerate WeierstrassP functions for forcing, is a sum of these functions. The parameter m in the degenerate WeierstrassP function, if chosen to be small gives a ball,

$$B_r = \{y \in \mathbb{R}^3 : \|y\|_2 = (|y_1|^2 + |y_2|^2 + |y_3|^2)^{\frac{1}{2}} \le r\}$$

Here we are in Cartesian space $\mathbb{R}^3$ with 2-norm L_2. Since the terms are squared in length in the initial condition for v_3 we require to multiply by dynamic viscosity μ to obtain units of velocity. In the above, the forcing is taken to be different than the gradient of pressure.

Introducing the space $\mathfrak{I}(y_3, s) = \{s \in \mathbb{R}^+, y_3 \in B(y_{3c_i}; \varepsilon) : 2y_1 v_1 + v_2 = 0 \& Ay_1 + By_2 + C = 0, \forall y_1, y_2 \in I \times I (I \subset \mathbb{R}) \& y_2 = y_1^2 \& v_3(y_1, y_2, y_3, s) \in C^0(\mathbb{T}^3)\}$,

where $B(y_{3c_i}; \varepsilon)$ is the 1-dimensional ε- ball centered at y_{3c_i}, i=1,2,...N, ranging through the expanding lattice generated by the flat torus.. The point y_{3c_i} coincides with the center point (a_i, b_i, c_i), where $\vec{r} = (y_1 - a_i, y_2 - b_i, y_3 - c_i), i = 1, 2, \ldots N$.

The y_3 points are along segments parallel to the y_3-axis, throughout the lattice. For points belonging to the space $\mathfrak{I}(y_3, s)$, the following part of Eq.(1) is exactly zero:

$$
\begin{aligned}
X = & \left((\delta - 1)v_1 \frac{\partial v_3}{\partial s} + 2\rho v_3 \frac{\partial v_1}{\partial s}\right) \frac{\partial^2 v_3}{\partial y_3 \partial y_1} + \left((\delta - 1)v_2 \frac{\partial v_3}{\partial s}\right. \\
& \left. + 2\rho v_3 \frac{\partial v_2}{\partial s}\right) \frac{\partial^2 v_3}{\partial y_3 \partial y_2} - \frac{\partial v_3}{\partial s}\left[v_3 \frac{\partial v_3}{\partial y_1}\frac{\partial^2 v_1}{\partial y_3 \partial s} + v_3 \frac{\partial v_3}{\partial y_2}\frac{\partial^2 v_2}{\partial y_3 \partial s}\right. \\
& \left. - \frac{\partial v_3}{\partial y_1}\frac{\partial v_3}{\partial y_3}\frac{\partial v_1}{\partial s} - \frac{\partial v_3}{\partial y_2}\frac{\partial v_3}{\partial y_3}\frac{\partial v_2}{\partial s}\right] + v_3 \frac{\partial v_3}{\partial y_1}\frac{\partial v_1}{\partial s}\frac{\partial^2 v_3}{\partial s \partial y_3} \\
& + v_3 \frac{\partial v_3}{\partial y_2}\frac{\partial v_2}{\partial s}\frac{\partial^2 v_3}{\partial s \partial y_3}
\end{aligned}
$$

Also $F_{T_2}(y_1, y_2, y_3, s) = -2y_1 F_{T_1}(y_1, y_2, y_3, s)$, so that the individual forces are not zero, (but cancel in Eq.(8.21) including $F_{sz}(y_1, y_2, y_3, s)$ in an arbitrary small ball where the WeierstrassP functions touch the center of given epsilon-ball. In Eq. (8.21) by writing as $\delta \to 1$ the sum of the three viscous terms as

$$M_\mu = \mu \frac{\partial^3 v_3}{\partial y_3 \partial y_1^2} + \mu \frac{\partial^3 v_3}{\partial y_3 \partial y_2^2} + \mu \frac{\partial^3 v_3}{\partial y_3^3}$$

and inserting this in place of original, in Eq.(8.21), results in the following form of solution:

$$v_3 = F_4(s) F_5(y_1, y_2, y_3)$$

where, $\Phi(s) = -\dfrac{F_4(s)\left(F_4(s)\left(\frac{d}{ds}F_4(s)\right)^2 - \mu\lambda^2\right)}{\rho\lambda^2 \frac{d}{ds}F_4(s)}$ (I)

$\left(\left(\frac{\partial v_3}{\partial s}\right)^2 = \frac{1}{\delta-1}\right.$ near the blowup where as $\delta \to 1$, $v_3 \to \infty$ from the right of a potential blowup point $s = s_0$. We will show that $v_3 \to -\infty$ from the left of $s = s_0$.)

Now in section 8.9 of this chapter, an analysis is completed for the PNS equation with $\delta = 1$. There in Eq. (8.77) and Eq. (8.78) $\Phi(s)$ is replaced by Eq.(I) to include viscosity effects.

The term λ is an eigenvalue for the separation problem associated with the separable solution. Proceeding as in section 8.9.1 with this new "viscous" $\Phi(s)$ we obtain a solution with an oscillating pressure gradient term there, given as $C_L(s) = \exp(is)/\sqrt{sR}$. See Figures A and B below for solution $F_4(t)$.

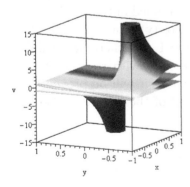

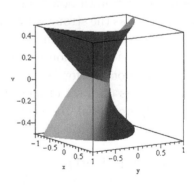

Figure A Riemann Surface for solution v_3. Here singularities appear off centered from the center of each cell of the Lattice of the 3-Torus

Figure B Riemann Surface $z = w^2$ for solutions v_1 and v_2 when $X = 0$. Here the functions both have spatial derivatives at the center which exhibit blowup.

The idea here is to set the R term(see section 8.9.1) multiplied by λ^2 term to be unity. So for infinite $\lambda^2 = \left(\frac{dF_4}{ds}\right)^2$(First derivative blows up in time), $R \to 0$.

8.2.3 Case 1

In this section we introduce case 1 of the PNS problem. It has been shown that the 1/3 law of Onsager for the Euler equations holds as is proven by [32]. There it is shown that for $\alpha < 1/3$ there exists a weak solution that fails to conserve energy and $\alpha = 1/3$ is the threshold of kinetic energy with $\alpha > 1/3$ occurring for weak solutions that conserve energy. Actually what Onsager conjectured is that there may occur turbulent or anomalous dissipation for $\alpha < 1/3$. Now for PNS it can be seen that a strong solution exists and hence a weak one at the threshold $\alpha = 1/3$ in Holder space. Now Cheskidov et al. have proven that energy is conserved for velocities in the Besov space of tempered distributions $B_{3,p}^{1/3}$ as weak solutions to Euler equations and also prove the result in the slightly larger space $B_{3,c(N)}^{1/3}$ (see Section 3 in [31]), thus it is proposed that there may be a weak solution of PNS at zero viscosity such that the system conserves energy at the above fractional threshold and therefore the 3-Torus possesses a definite fractional differentiability. Solutions of PNS may possibly conserve energy but it is not known until now that this may occur for PNS at $\alpha = 1/3$ for any viscosity as the strong solution further obtained indicates. At $\delta = 1$ for zero viscosity (including zero) ν, all viscosity terms are associated with a $1 - 1/\delta$ expression and vanish at $\delta = 1$. Furthermore, a finite time blowup occurs for the first derivative of v_3 with respect to s iff blowup occurs on a set of measure zero, that is it will be proven that the blowup occurs at the center of each cell of the 3-Torus. Also the following theorem proven in [32],

Theorem 8.1 (Isett) *For any $\alpha < 1/3$, there is a nonzero weak solution to incompressible Euler in the class $f \in C_{s,y}^{\alpha}$ where $\mathbf{v} \in C_{s,y}^{\alpha}(\mathbb{R} \times (\mathbb{R}/\mathbb{Z})^3), p \in C_{s,y}^{2\alpha}(\mathbb{R} \times (\mathbb{R}/\mathbb{Z})^3$ such that $\mathbf{v}$ is identically 0 outside a finite time interval. In particular, the solution $\mathbf{v}$ fails to conserve energy.*

Now f is in the class $C_{s,y}^{\alpha}$ if there exists $C \geq 0$ such that $| f(s + \triangle s, y + \triangle y) - f(s,y) | \leq C(| \triangle s | + | \triangle y |)^{\alpha}$ uniformly in $s, y, \triangle s, \triangle y$. It is noteworthy to observe in this chapter that if $\triangle s = s - s_0$ and $\triangle y = y - y_0$, then on a set of measure zero about y_0 that is at the point y_0, $| \triangle y | = 0$, reducing to $| f(s + \triangle s) - f(s) | \leq C | \triangle s |^{\alpha}$. For the following, choose $\Phi(s)$ such that $F_4(s) = C(s - s_0)^{\frac{1}{3}}$,

$$F_4(s) = \sqrt[3]{3 c_4 \int \Phi(s) \, ds + C_1}, \quad \text{smooth } f_0(s) \tag{8.36}$$

note that a strong solution v_3 is produced in this chapter which is in the class $C_{s,y}^{\alpha}$, for $\alpha = 1/3$.

Next we set,

$$F(y_1, y_2, y_3) = \mathcal{P}(y_1, 3m^2, m^3)^{-1} \times \mathcal{P}(y_2, 3m^2, m^3)^{-1}$$
$$\times \mathcal{P}(y_3, 3m^2, m^3)^{-1}$$
$$G(y_1, y_2, y_3) = \mathcal{P}(y_1, 3m^2, m^3)^{-1} \times \mathcal{P}(y_2, 3m^2, m^3)^{-1}$$
$$\times \mathcal{P}(y_3, 3m^2, m^3)^{-1} \tag{8.37}$$

Setting v_1 and v_2 , as

$$v_1 = f_1(s)\big(\mathcal{P}(y_1, 3m^2, m^3)^{-1} \times \mathcal{P}(y_2, 3m^2, m^3)^{-1}$$
$$\times \mathcal{P}(y_3, 3m^2, m^3)^{-1} + A\big)$$
$$v_2 = f_1(s)\big(\mathcal{P}(y_1, 3m^2, m^3)^{-1} \times \mathcal{P}(y_2, 3m^2, m^3)^{-1}$$
$$\times \mathcal{P}(y_3, 3m^2, m^3)^{-1} + A\big) \tag{8.38}$$

Here the relationship between forcing f_0 and each of v_i is $\frac{d}{ds} f_1 = -f_0(s) + \frac{\text{Conv}(y_1, y_2, y_3, s)}{U_{y_i} + A}$, $i = 1, 2$ where the latter expression is the nonlinear convective term of PNS equation which is divided by the negatively shifted product of the reciprocal of the WeierstrassP functions. Here the oscillations do not intersect the y_i axes. This is bounded in the L_2 norm by the norm of the Laplacian that approaches zero as the lattice gets larger as observed in a series of inequalities using the Preköpa-Leindler and Gagliardo-Nirenberg Inequalities. This becomes,

$$0 \geq \|\frac{\text{Conv}}{U_{y_i} + A}\|_2^2 = \|L_1\|_2^2 \|L_2\|_2^2 \geq 0$$

where L_1, L_2 are equations so that $L = L_1 L_2 = \frac{L_1}{L_D}$ where $L_D \neq 0$ and this occurs when $y_3 \neq G(y_1, y_2)$ for any function G of two space variables. Necessarily $L_1 = 0$ upon multiplication by L_D. Finally $\mathcal{P}^{-1}$ is the reciprocal of the degenerate Weierstrass $\mathcal{P}$ function with parameter m. The definition of degenerate function is,

$$\mathcal{P}(z, 3m^2, m^3) = -\frac{m}{2} + \frac{3}{2}m \csc\left(\frac{z\sqrt{6}\sqrt{m}}{2}\right)^2 \tag{8.39}$$

For sufficiently negative $A < 0$ substituting, $\frac{d}{ds}f_1 = -f_0(s) + \frac{\text{Conv}}{U_{y_i}+A}$, v_1, v_2 with corresponding terms U_{y_1} and U_{y_2} from Eq. (8.38) and F, G from Eq. (8.37) reduces Eq. (8.29) to L_1,

$$L_1 = Y_1 + Y_2 = 0 \tag{8.40}$$

where

$$
Y_1 = \left(y_1{}^2 + y_2{}^2 + y_3{}^2\right)\left(\frac{\partial v_3}{\partial y_1} - \frac{\partial v_3}{\partial y_2}\right) \times \left(2\left(\frac{\partial v_3}{\partial s}\right)\left(\frac{\partial^2 v_3}{\partial y_3 \partial s}\right) v_3 \frac{\partial v_3}{\partial y_3}\right.
$$
$$
\left. + v_3^2\left(\frac{\partial v_3}{\partial s}\right)\frac{\partial^3 v_3}{\partial y_3{}^2 \partial s} - v_3^2\left(\frac{\partial^2 v_3}{\partial y_3 \partial s}\right)^2 - \left(\frac{\partial^2 v_3}{\partial y_3 \partial s}\right)\Phi\left(t\right)\right) \tag{8.41}
$$

$$
Y_2 = -\left(\frac{\partial v_3}{\partial s}\right)\left[-3/2\left(\frac{\partial v_3}{\partial s}\right)uz\left(y_1{}^2 + y_2{}^2 + y_3{}^2\right)\left(\frac{\partial v_3}{\partial y_1} - \frac{\partial v_3}{\partial y_2}\right)\right.
$$
$$
\times \frac{\partial^2 v_3}{\partial y_3{}^2} + v_3\left(\frac{\partial v_3}{\partial y_1}\right)\left(\frac{\partial v_3}{\partial y_2}\right)\left(y_1{}^2 + y_2{}^2 + y_3{}^2\right)\left[\frac{\partial^2 v_3}{\partial y_1 \partial s} - \frac{\partial^2 v_3}{\partial y_2 \partial s}\right]
$$
$$
- 3/2\left[\left(4/3\, v_3\left(y_1 - y_2\right)\frac{\partial v_3}{\partial y_2} + \left(\frac{\partial v_3}{\partial y_3}\right)^2\left(y_1{}^2 + y_2{}^2 + y_3{}^2\right)\right)\frac{\partial v_3}{\partial y_1}\right.
$$
$$
\left.\left. - \left(\frac{\partial v_3}{\partial y_3}\right)^2\left(\frac{\partial v_3}{\partial y_2}\right)\left(y_1{}^2 + y_2{}^2 + y_3{}^2\right)\right]\frac{\partial v_3}{\partial s}\right] \tag{8.42}
$$

where Λ and its partial derivative with respect to y_3 have been isolated and whose norm approaches zero using Eq. (9) of the Millennium problem hypothesis on the decay of the forcing terms with respect to $s \to \infty$ [44]. The solution of Eq. (8.40) is in the form $v_3 = F_5(y_1, y_2, y_3)F_4(s)$, where F_4 is in Eq. (8.36) and F_5 solves the following

PDE,

$$\left(y_1{}^2 + y_2{}^2 + y_3{}^2\right)\left(\frac{\partial F_5}{\partial y_1} - \frac{\partial F_5}{\partial y_2}\right) \times \left[2\left(\frac{\partial F_5}{\partial y_1}\right)\left(\frac{\partial F_5}{\partial y_2}\right)\left(F_5\left(y_1, y_2, y_3\right)\right)^2 c_4\right.$$

$$\left. -4\left(F_5\left(y_1, y_2, y_3\right)\right)^3\left(\frac{\partial^2 F_5}{\partial z^2}\right)c_4 - 4\left(F_5\left(y_1, y_2, y_3\right)\right)^2\left(\frac{\partial F_5}{\partial y_3}\right)^2 c_4 + \frac{\partial F_5}{\partial y_3}\right] = 0$$

$$(8.43)$$

Here the solution for the problem and Maple program with Λ and $\frac{\partial \Lambda}{\partial z}$ terms included in the analysis is worked through. The idea here is that $\Phi(t)$ blows up in finite time and so we can let $\Phi(t) \to \infty$ for some finite time $t = T_c$. Here we have divided by Φ and let the limit approach infinity in the PDE $L_1 = 0$ given by Eq. (8.40). Here the result is that the solution is $u_z = F_4(t)F_5(x, y, z) = \frac{1}{f_1(t)}F_5(x, y, z)$. So there exists finite time singularity when $f_1(t)$ is a polynomial in t for example.

8.2.4 JacobiSN Solution

The logarithm of a sum can be expanded as follows,

$$\log(a + b) = \log\left(a\left(1 + \tfrac{b}{a}\right)\right) \tag{8.44}$$

Let a be expressed as,

$$a = -4\,F_5^3\left(\frac{\partial^2 F_5}{\partial z^2}\right)c_4 - 4F_5^2\left(\frac{\partial F_5}{\partial y_3}\right)^2 c_4 \tag{8.45}$$

$$b = 2\left(\frac{\partial F_5}{\partial y_1}\right)\left(\frac{\partial F_5}{\partial y_2}\right)F_5^2 c_4 + \frac{\partial F_5}{\partial y_3} \tag{8.46}$$

Since $\frac{b}{a} \to 0$ if $F_5 = \alpha + G$ as $\alpha \to \infty$ $(O(1/\alpha) \to 0)$,

$$\log(a + b) = \log(a) = -4\,F_5^3\left(\frac{\partial^2 F_5}{\partial z^2}\right)c_4 - 4\,F_5^2\left(\frac{\partial F_5}{\partial y_3}\right)^2 c_4 \tag{8.47}$$

Next it is true that,

$$\log(a) = \log(c + d) = \log(c) + \log\left(1 + \frac{d}{c}\right) \tag{8.48}$$

where

$$c = -4\,F_5^3\left(\frac{\partial^2 F_5}{\partial z^2}\right)c_4 \tag{8.49}$$

$$d = -4\, F_5^2 \left(\frac{\partial F_5}{\partial y_3}\right)^2 c_4 \tag{8.50}$$

but since $d/c \to 0$ if $F_5 = \alpha + G$ as $\alpha \to \infty$ and $\log(E) = \log(a + b)$
then,

$$\log E + \log\left(-4F_5^3 \frac{\partial^2 F_5}{\partial y_3^2}\right) + \log(|\, c_4\,|) = 0$$
$$\log\left(-F_5^3 \frac{\partial^2 F_5}{\partial y_3^2}\right) = \log(E) - \log(\tfrac{E}{4c_4}) \tag{8.51}$$
$$\log(-F_5^3) + \log\left(\frac{\partial^2 F_5}{\partial y_3^2}\right) = \log(-\,(4c_4))$$

Taking the exponential of both sides where the log function takes on
negative values, gives,

$$F_5^{-3}\left(\frac{\partial^2 F_5}{\partial y_3^2}\right) = 1 \tag{8.52}$$

$$\frac{\partial^2 F_5}{\partial y_3^2} = F_5^3 \tag{8.53}$$

In the above equalities, $E > 1$. The solution for F_5 is,

$$F_5 = F_2\,(y_1, y_2)\, \text{JacobiSN}\left[\left(i/2\sqrt{2}y_3 + F_1\,(y_1, y_2)\right) F_2\,(y_1, y_2), i\right] \tag{8.54}$$

8.2.4.1 Continuum of z-poles on Imaginary Axis

The problem of the instability of a part of branch of viscous
incompressible fluid flows induced by a shrinking sheet has been studied
in [55]. To show the instability of a viscous flow due to arbitrarily
small perturbations, it is required to show the existence of an unstable
eigenvalue corresponding to the linearization around the flow. For
the case of a flow of an incompressible fluid over a shrinking plate
with suction, both eigenfunctions corresponding to 0 eigenvalue of the
linearization are shown in [55], and there it is proven that the eigenvalue
crosses an imaginary axis transversally.

In the present work it is shown that a similar type of result is
obtained. It is shown for the spatial z direction of viscous flow of 3D
PNS equations that there exists a continuum of z values that lie on
the imaginary axis of the complex Z plane. Here there are no special
restrictions or types of flow conditions except external smooth periodic

forcing and divergence-free condition as stated in the hypothesis of the Millennium problem [44]. From the continuity equation, $\nabla^i u_i = 0$, setting $u_{y_1} = u_{y_2} = (P(y_1, 3m^2, m^3)^{-1} \times P(y_2, 3m^2, m^3)^{-1} \times P(y_3, 3m^2, m^3)^{-1})$ and $u_{y_3} = F(y_1, y_2, y_3, s)$ and from the continuity equation, $\nabla^i u_i = 0$, set $v_1 = v_2 = f_1(s)(P(y_1, 3m^2, m^3)^{-1} \times P(y_2, 3m^2, m^3)^{-1} \times P(y_3, 3m^2, m^3)^{-1})$ and v_3 is,

$$v_3 = F_4(s) F_2(y_1, y_2) \text{ JacobiSN} \left[\left(i/2\sqrt{2} y_3 + F_1(y_1, y_2) \right) F_2(y_1, y_2), i \right] \tag{8.55}$$

It is noticed here that $F_4(s)$ and, $f_1(s)$ appearing twice, are all common to all three velocities, so that we can factor it out and obtain the following spatial part of the continuity equation,

$$-\frac{P'\left(y_1, 3\,m^2, m^3\right)}{\left(P\left(y_1, 3\,m^2, m^3\right)\right)^2 P\left(y_2, 3\,m^2, m^3\right) P\left(y_3, 3\,m^2, m^3\right)}$$
$$-\frac{P'\left(y_2, 3\,m^2, m^3\right)}{P\left(y_1, 3\,m^2, m^3\right) \left(P\left(y_2, 3\,m^2, m^3\right)\right)^2 P\left(y_3, 3\,m^2, m^3\right)}$$
$$+ i/2 \left(F_2(y_1, y_2)\right)^2 \sqrt{2} \text{JacobiCN} \left(\left(i/2\sqrt{2} y_3 \right. \right.$$
$$\left. + F_1(y_1, y_2) \right) F_2(y_1, y_2), i \right)$$
$$\times \text{JacobiDN} \left(\left(i/2\sqrt{2} y_3 + F_1(y_1, y_2) \right) F_2(y_1, y_2), i \right) = 0 \tag{8.56}$$

Now $F_2^2(y_1, y_2)$. is,

$$F_2^2 = -\frac{P'\left(y_2, 3\,m^2, m^3\right) P\left(y_1, 3\,m^2, m^3\right) + P'\left(y_1, 3\,m^2, m^3\right) P\left(y_2, 3\,m^2, m^3\right)}{\left(P\left(y_1, 3\,m^2, m^3\right)\right)^2 \left(P\left(y_2, 3\,m^2, m^3\right)\right)^2} \tag{8.57}$$

If y_1 and y_2 are extended to complex variables, upon taking the real and imaginary parts of F_2, for example there exists specific y_1 and y_2 such that the following is true,

$$\text{JacobiCN}(i\beta, i) + \frac{1}{P(y_3, 3m^2, m^3)} \text{JacobiDN}(i\beta, i)^{-1} = 0 \tag{8.58}$$

where $F_2 = \beta = c \left(\frac{\sqrt{2}}{2} + i\frac{\sqrt{2}}{2} \right)$, and

$$y_3 = \sqrt{2} i F_1 + i\beta \tag{8.59}$$

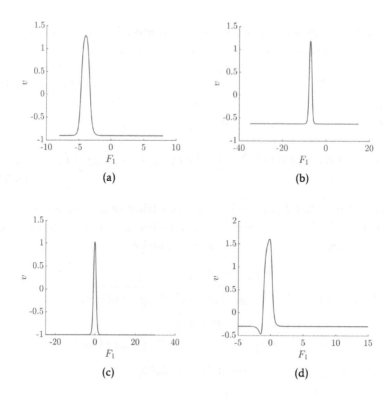

Figure 8.1 Imaginary axis crossovers in Z-plane. (a) Case of $c = 8$. (b) Case of $c = 14$. (c) Case of $c = 0.25$. (d) Case of $c = 1$.

When c is 8 for example, the plot of $V = \text{JacobiCN}(i\beta, i) + \frac{1}{P(y_3, 3m^2, m^3)}\text{JacobiDN}(i\beta, i)^{-1}$, shows that there exists zeros of F_1, thereby proving there exists poles of the P functions on the imaginary axis in the Z-plane.

The existence of poles s on the imaginary axis is due to cancellation of two equal in magnitude but alternate signs leaving β real and thus y_3 is on the imaginary axis. This is true for c independent of sign. See Figure 8.1

8.2.5 The Comparison of Blowup for Each of $\frac{dF_4^{**}}{ds}$ and $\Phi(s)$.

The blowup of $\frac{dF_4^{**}}{ds}$ (F_4^{**} is the occurring separable part of the New PDE obtained by solving for Φ in Eq. (8.40) for the present case 1 and multiplying by $y_1^2 + y_2^2 + y_3^2 = \epsilon$) occurs at the same time as the blowup of Φ on the time line, which means that we can prove blowup on an ϵ ball about the center of each cell of the Lattice of the 3-Torus. Here ϵ is the radius of the sphere about the center of each cell of the lattice .(The nonhomogeneous PDE with $\epsilon\Phi(s)$ on RHS appears.) Note $\epsilon\Phi$ will approach zero as $y_1^2 + y_2^2 + y_3^2 = \epsilon \to 0$. The resulting PDE has a blowup for $\frac{dF_4^{**}}{ds}$ at the same point than the blowup of Φ. This is true since $\Phi(s)$ is not part of the resulting PDE and s_0 can be shifted. The blowup occurs at the center of each cell of the lattice throughout the 3-Torus. The PDE obtained by solving algebraically for $\epsilon\Phi(t)$ has the following separable solution, upon taking the limit as $\epsilon \to 0$, $v_3 = F_4^{**}(s)F_5^{**}(y_1, y_2, y_3)$, where

$$\left(\frac{d}{ds}F_4^{**}(s)\right)^2 = \frac{c_4}{\left(F_4^{**}(s)\right)^3} \tag{8.60}$$

$$F_5^{**} = F_2(y_1, y_2) \, \text{JacobiSN}\left[\left(i/2\sqrt{2}y_3 + F_1(y_1, y_2)\right)F_2(y_1, y_2), i\right] \tag{8.61}$$

8.3 ANALYSIS OF $F_4^{**}(S)$ AND $\Phi(S)$

Using the solution of Eq. (8.60) which gives two real valued functions we retain the important one here. This is,

$$F_4^{**} = \frac{(80C_1c_4^3 - 80c_4^3s)^{2/5}}{4c_4} \tag{8.62}$$

Next we set both the solutions F_4^{**} and F_4 equal to each other,

$$\left(C_1 + 3c_4\int\Phi(s)ds\right)^{1/3} = \frac{(80C_1c_4^3 - 80c_4^3s)^{2/5}}{4c_4} \tag{8.63}$$

Next solving the integral equation for $\Phi(s)$ we obtain,

$$\Phi(s) = \frac{3\left(-80C_1c_4^3 + 80c_4^3s\right)^{1/5}}{6c_4} \tag{8.64}$$

Substituting Φ into $\left(C_1 + 3c_4\int\Phi(s)ds\right)^{1/3}$ and substituting $c_4 = -1$ and $C_1 = 1$, for example we obtain the results in Figure below. Here it

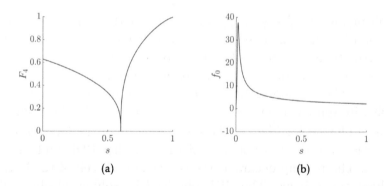

Figure 8.2 Plot of forcing Schwarzian term $f_0(s)$ versus s. (a) Plot of $F_4(s)$ versus s. (b) Case of $c = 8$.

is evident that the blowup of F_4 occurs at a value ($s = 0.6$) sooner in time than that of the Φ^s blowup which occurs at $s = 1$. Also for these values $f_0(s)$ will be a Schwarzian function that decreases as $s \to \infty$ (Figure 8.2).

$$\sqrt[3]{2 + 3 \int \left(\ln\left(C + \Phi_1\left(s\right)\right)\right)^{-1} ds} = C_1 + \sqrt[3]{s - s_0} \qquad (8.65)$$

8.3.1   Setting the Time Derivative of $F_4(s)$ Equal to $-f_0(s)$

Here in this section the time derivative of $F_4(s) = (C_1 c_4 + 3 \int \Phi(s)ds \; c_4^2)^{1/3}/c_4$ is set to the negative of the temporal part $-f_0(s)$ of the external force term f_{y_3}. Thus,

$$\frac{c_4 \Phi\left(s\right)}{\left(\left(C_1 \, c_4 + 3 \int \Phi\left(s\right) ds\right) c_4^2\right)^{2/3}} = -f_0\left(s\right) \qquad (8.66)$$

It necessarily follows that C_1 is positive and c_4 is negative. Solving the integral equation for Φ gives,

$$\Phi(s) = -c_4 f_0\left(s\right) \left(C_1 + \int f_0\left(s\right) ds\right)^2 \qquad (8.67)$$

Setting Φ to a logarithmic blowup at the time value of $s = \frac{s_0 + C}{2}$ which is at the blowup of F_4^{**} when $C > s_0$ gives,

$$c_4 f_0\left(s\right)\left(C_1 + \int f_0\left(s\right)\mathrm{d}s\right)^2 = \ln\left(s_0/2 + C/2 - s\right) \qquad (8.68)$$

Solving this equation for $\int f_0\left(s\right)\mathrm{d}s$ and in turn for $f_0(s)$ gives 3 roots, one is real and two are complex. The real root is,

$$f_0(s) = -\frac{18^{\frac{2}{3}}}{9}\frac{\begin{pmatrix} c_4\left(\ln\left(s_0 + C - 2\,s\right)C + 2\,\ln\left(2\right)s \\ -2\,\ln\left(s_0 + C - 2\,s\right)s \\ +\ln\left(s_0 + C - 2\,s\right)s_0 - C + C_1 + 2\,s - s_0\right)^2\end{pmatrix}^{2/3}}{\begin{pmatrix}\left(\ln\left(2\right) - \ln\left(s_0 + C - 2\,s\right)\right) \\ c_4\left(\ln\left(s_0 + C - 2\,s\right)C \\ +2\,\ln\left(2\right)s - 2\,\ln\left(s_0 + C - 2\,s\right)s \\ +\ln\left(s_0 + C - 2\,s\right)s_0 - C + C_1 + 2\,s - s_0\right)^2\end{pmatrix}} \qquad (8.69)$$

Next solving algebraically for $s_0 + C - 2s$ with symbolic software package gives an expression in terms of $f_0(s)$ and s_0 with arbitrary constants. Taking the limit of this expression as $s \to \frac{s_0 + C}{2}$ gives,

$$\left(\left(\left(\sqrt[3]{\begin{array}{c} 4\sqrt{((C + s_0)\ln\left(2\right) + C_1)^4\, c_4{}^2} \\ +\left(4C^2 + 8Cs_0 + 4\,s_0{}^2\right)c_4\left(\ln\left(2\right)\right)^2\ \sqrt[3]{18 f_0} \\ +\left(8C + 8\,s_0\right)C_1\, c_4 \ln\left(2\right) + 4C_1{}^2 c_4 \end{array}}\right)^6\right)^{1/24} = 0 \qquad (8.70)$$

Now the only way this equation can be valid at $\Phi's$ blowup point is if $f_0 \to -\infty$ there. Now if the parameter C is approaching positive infinity then the midpoint between the fixed point s_0 and large and positive C, goes to infinity. Hence $f_0(s)$ is continuous on $[0,\infty)$. Substituting $\Phi(s)$ in Eq. (8.67) into Eq. (8.68) one can obtain $f_0(s)$. Here it can be shown that all derivatives of $f_0(s)$ are differentiable and hence continuous at $s = s_0$.

8.3.2 Case 2

An example of a smooth force $f_0(s)$ at $s = s_0$ is considered with it's accompanying solution of Eq. (8.29),

Setting v_1 and v_2, as

$$\begin{aligned} v_1 &= f_a(s)\mathcal{P}(y_1, 3m^2, m^3) \times \mathcal{P}(y_2, 3m^2, m^3) \times \mathcal{P}(y_3, 3m^2, m^3) \\ v_2 &= f_b(s)\mathcal{P}(y_1, 3m^2, m^3) \times \mathcal{P}(y_2, 3m^2, m^3) \times \mathcal{P}(y_3, 3m^2, m^3) \end{aligned} \qquad (8.71)$$

$$f_a(s) = -f_b(s) \tag{8.72}$$

where $\mathcal{P}$ is the degenerate Weierstrass P function with parameter m. Also for case 2, $\Phi(s) = 0$ and the relationship between f_0 and each of v_i is $\frac{d}{ds} f_i = -f_0(s) + \frac{\text{Conv}}{U_{y_i} + A}$, $i = a, b$. The general reducing solution in Section 8.4 will also be considered for the two cases $\Phi(s) = 0$ and $\Phi(s) \neq 0$. It follows that $\Phi(s) = 0$ iff $v_1 = -v_2$, from which it follows that $\iint_S v_3^2 (n_3 \frac{\partial P}{\partial y_3} + (\nabla_{y_1 y_2} P \cdot \vec{n})) dS = 0$.

For case 2, for spatially non-smooth v_1 and v_2, the solution of Eq. (8.29) is in the form $v_3 = F_4(s) F_5(y_1, y_2, y_3)$ which satisfies,

$$\frac{d}{dt} F_4(s) = c_4 f_0(s) \tag{8.73}$$

$$
\begin{aligned}
- \Bigg[&(P(y_2, 3m^2, m^3))^2 \left((y_1{}^2 + y_2{}^2 + y_3{}^2) \left(\frac{\partial F_5}{\partial x} - \frac{\partial F_5}{\partial y_2} \right) \frac{\partial^2 F_5}{\partial y_3 \partial y_1} \right. \\
&- (y_1{}^2 + y_2{}^2 + y_3{}^2) \left(\frac{\partial F_5}{\partial y_1} - \frac{\partial F_5}{\partial y_2} \right) \frac{\partial^2 F_5}{\partial y_3 \partial y_2} - 4 y_3 \left(\frac{\partial F_5}{\partial y_1} \right) \frac{\partial F_5}{\partial y_2} \Big) \\
&\times (P(y_1, 3m^2, m^3))^2 (P(y_3, 3m^2, m^3))^3 + P(y_2, 3m^2, m^3) \\
&\times \left(P(y_2, 3m^2, m^3) P'(y_3, 3m^2, m^3) \left(\left(\frac{\partial F_5}{\partial y_1} \right)^2 + \left(\frac{\partial F_5}{\partial y_2} \right)^2 \right) \right. \\
&\times (y_1{}^2 + y_2{}^2 + y_3{}^2) P(y_1, 3m^2, m^3) - 2 c_4 \Big(F_5(y_1, y_2, y_3) \\
&\times (y_1{}^2 + y_2{}^2 + y_3{}^2) \left(\frac{\partial F_5}{\partial y_1} - \frac{\partial F_5}{\partial y_2} \right) \frac{\partial F_5^2}{\partial y_3{}^2} - \frac{1}{2} \left(\frac{\partial F_5}{\partial y_2} \right) \\
&\times (y_1{}^2 + y_2{}^2 + y_3{}^2) \left(\frac{\partial F_5}{\partial y_1} \right)^2 + \frac{1}{2} \Big((y_1{}^2 + y_2{}^2 + y_3{}^2) \left(\frac{\partial F_5}{\partial y_2} \right)^2 \\
&+ F_5(y_1, y_2, y_3) (y_1 - y_2) \frac{\partial F_5}{\partial y_2} + \left(\frac{\partial F_5}{\partial y_3} \right)^2 (y_1{}^2 + y_2{}^2 + y_3{}^2) \Big) \frac{\partial F_5}{\partial y_1} \\
&- \left(\frac{\partial F_5}{\partial y_2} \right) \left(\frac{\partial F_5}{\partial y_3} \right)^2 (y_1{}^2 + y_2{}^2 + y_3{}^2) \Big) \Big) P(y_1, 3m^2, m^3) \\
&\times (P(y_3, 3m^2, m^3))^2 - (y_1{}^2 + y_2{}^2 + y_3{}^2) \left(\frac{\partial F_5}{\partial y_1} - \frac{\partial F_5}{\partial y_2} \right) \\
&\times \left(\frac{\partial^2 F_5}{\partial y_3 \partial y_1} + \frac{\partial^2 F_5}{\partial y_3 \partial y_2} \right) P(y_3, 3m^2, m^3) + P'(y_3, 3m^2, m^3)
\end{aligned}
$$

$$\times \left(\frac{\partial F_5}{\partial y_1} - \frac{\partial F_5}{\partial y_2}\right)\left(\frac{\partial F_5}{\partial y_1} + \frac{\partial F_5}{\partial y_2}\right)\left(y_1{}^2 + y_2{}^2 + y_3{}^2\right)\Bigg] F_5^2 = 0 \quad (8.74)$$

The solution of Eq. (8.73) for $F_4(s)$ is,

$$F_4 = c_4 \int f_0(s)\ ds + C_1 \qquad (8.75)$$

where it is observed that it is smooth at $s = s_0$ and the limit of F_4 as $s \to \infty$ should be bounded. (In general $f_0(s)$ is assumed to be bounded and the same for both cases 1 and 2.) Solving algebraically for $\|\nabla F_5\|_2^2$ in Eq. (8.74) and setting the resulting normed equation to zero since the norm approaches zero as lattice becomes large and then taking the limit as $c_4 \to \infty$ of the resulting PDE upon division by c_4 gives,

$$\frac{\partial^2 F_5}{\partial y_3{}^2} F_5\left(y_1, y_2, y_3\right) - \frac{1}{2}\frac{\partial F_5}{\partial y_1}\frac{\partial F_5}{\partial y_2} + \left(\frac{\partial F_5}{\partial y_3}\right)^2 = 0$$

The solution of this PDE is in terms of sinusoidal functions in the spatial variables

8.4   GENERAL CASE WHEN Λ IS NOT EXCLUDED

In Appendix A4, the general program for Eq. (8.29) including all terms with Λ and its partial with respect to z is solved in Maple . The idea is to solve for $V = \frac{df_1(s)}{ds}$ symbolically and write out the expression $L_a = AV - 2\frac{df_1(s)}{ds}A$. Moving ahead we next solve symbolically for $F = f_0(s)$ and write out the expression $L_b = (1 - A)F - f_0(s)(1 - A)$. Here if $A = 1$ we obtain $2\frac{df_1(s)}{ds}$ otherwise if $A = 0$ we obtain $f_0(s)$. Adding the two equations $L_a + L_b$ and setting to zero we have the general equation to solve for both $\Lambda(y_1, y_2, y_3; s)$ and $v_3(y_1, y_2, y_3; s)$. We obtain Λ from Eq. (8.14) and substitute into $L_a + L_b = 0$. The computer program in the Appendix solves the problem for $A = 0$, which proves to be a finite time blowup the same as in the previous section for the case where Λ is excluded. The problem of $A \neq 0$ is more difficult to solve for the general case however one solution is also the same as for the case $A = 0$. A general solution of both cases includes the solution $v_3 = F_4(s)F_6(y_3, y_2 + y_1)$, where $F_6(y_3, y_2 + y_1)$ is an arbitrary function of this form and $F_4(s)$ solves the non-linear differential equation: $\frac{dF_4(s)}{ds} = \frac{c_4\Phi(s)}{F_4^2(s)}$.

8.5   NO FINITE TIME BLOWUP WHEN PRESSURE IS DECREASING CANTOR-LIKE FUNCTION

8.5.1   The Cantor Set

The Cantor set is formed by repeatedly diving a line into thirds. See [58] where it is shown that the sum of sufficiently many copies of the Cantor set covers a whole interval of the line. This idea of adding fractals is reintroduced in Section 8.11 for the Cantor function . The Cantor set is a fractal .

1. We take the line from 0 to 1, and divide it into thirds.

2. Omit the middle, open set $A = (1/3, 2/3)$

3. We repeat this procedure to remaining sets ad infinitum

After infinitely many iterations, there is no measure left on the line:

- We started by deleting middle $1/3$ measure set.

- Then we remove two times the remaining middle sets of measure $1/9$.

- Then we remove four times the remaining middle sets of measure $1/27$.

The measure of sets deleted is the geometric series $= \frac{1}{3} + \frac{2}{3^2} + \frac{2^2}{3^3} + \dots$
$$= \frac{1}{3}(1 + \frac{2}{3} + \frac{2^2}{3^2} + \dots)$$
$$= \frac{1}{3}(1 + \frac{1}{1-\frac{2}{3}}) = 1$$
There are always end points left after any finite iteration.

8.6   OBSERVATION OF A RESIDUAL SET

The Cantor Set has as many numbers left behind as there were to begin with. The objective is to show that any number left over can be mapped back to a number we started with. This is where uncountability is used.

8.6.1 Base 3 Arithmetic

Let's use ternary arithmetic to explain what gets omitted in the iteration process. In the first iteration. The end point $1/3$ written in ternary is 0.1_3 since it is one times the first negative exponent of three. Also, $2/3$ is twice that: 0.2_3. By removing the middle third in the first iteration, we are deleting the numbers between $1/3$ and $2/3$. In ternary, we are deleting numbers that are bigger than 0.1_3 and less than 0.2_3. In other words, we remove any number that has 1 in its first decimal place in base 3 (except $1/3$, which can also be written without a 1 as $0.0222..._3$). Here are some examples of numbers we remove:

$0.11_3, 0.12_3, 0.121_3, 0.122_3$

We now have two lines remaining. The lines are a third of the length of the original line. This means any removing that happens will now occur one decimal away in ternary. The line from 0 to 0.1_3 is removed at a third of a third, or 0.01_3 in ternary. It is also removed at the point twice that: 0.02_3. Again, the terms with 1, now in the second decimal place, will be removed. We remove anything between 0.01_3 and 0.02_3. For the other line from 0.2_3 to 1, at the second decimal place. Therefore, we remove anything between $0.2 + 0.01_3$ and $0.2 + 0.02_3$. We now have a convenient way of describing any number in our set: it must consist entirely of 0s and 2s in ternary where end points can be represented with never-ending 2s if required. With only these numbers, we can find any number we originally had. Here the code is different but the numbers are the same and the numbers must be uncountable.

The Cantor set has uncountably more points than the countable endpoints we built up. In fact the set of all endpoints is of measure zero. The measure, $\mu(C) = 0$.

8.7 THE CANTOR FUNCTION

The Cantor function is an example of a function that is continuous, but not absolutely continuous. Though it is continuous everywhere and has zero derivative a.e., its value goes from 0 to 1 as $t \to 1$. This function defies many theorems in analysis and yet can be used to explain turbulence.

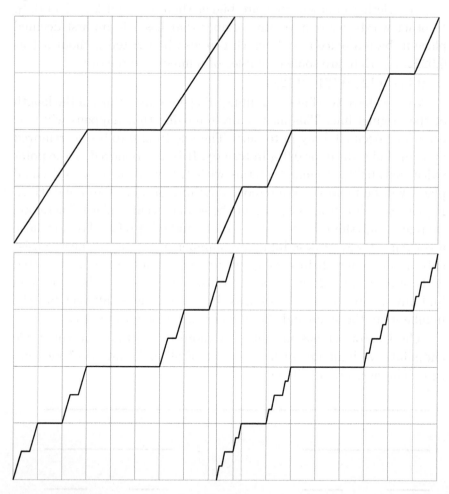

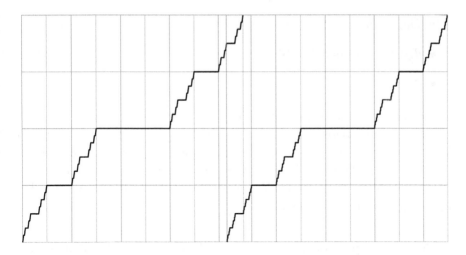

8.8   LAMBERT W FUNCTION

The Lambert W function is defined as the inverse of xe^x. That is: $y = W(x) \iff x = ye^y$ The following is the Taylor series of W, $W(x) = \sum_{k=1}^{\infty} \frac{(-k)^{k-1}}{k!} x^k$ The following interesting derivation is derived and repeated here as found in [59]. This form is appropriate for numerical calculation of the Lambert W function. Taylor's theorem is: $f(x) = \sum_{k=0}^{\infty} f^{(k)}(a) \frac{(x-a)^k}{k!}$.

The theorem could be used directly on $W(x)$, but that involves differentiating $W(x)$ a number of times to find a pattern. This is quite awkward in the given form as can be verified by calculating the Taylor coefficients. A more interesting alternative approach is used. Use the theorems on polynomials instead. This avoids issues of convergence; for polynomials, the Taylor series is really a finite sum, because if k is large enough, then $f^{(k)} = 0$. Polynomials used with Taylor series are easier to handle. Instead of writing f' for the derivative of f, let's write Df. (Here, D is an *operator* – it turns a function into a function.) Additionally, $\frac{x^k}{k!}$ is *such* an important polynomial that has a special name: $d_k(x) := \frac{x^k}{k!}$. Note that:

- $Dd_k = d_{k-1}$

- $d_k(0) = 0$ (when $k \neq 0$)

- $d_0 = 1$

d_k is called the *basic sequence* of D.

The revised Taylor series looks like: $f(x) = \sum_{k=0}^{\infty}(D^k f)(a) \, d_k(x - a)$. We can add together operators. For example, $D + D^2$ is the operator such that $(D + D^2)f = f' + f''$. I is the identity operator, i.e., that $If = f$ for every function f. We have $D^0 = I$. In addition, we can do things such as find e^D – since $e^x = \sum_{k=0}^{\infty} \frac{x^k}{k!}$, we can define e^D to mean $\sum_{k=0}^{\infty} \frac{D^k}{k!}$.

Define the operator E as follows: $(Ef)(x) = f(x + 1)$. That is, E shifts f over one. More generally, $(E^a f)(x) = f(x + a)$. A nice fact is that $DE = ED$ (that is, they commute). It has been proven that $E = e^D$. Start with the Taylor series, and substitute $x \mapsto x + 1$ and $a \mapsto x$:

$$f(x + 1) = \sum_{k=0}^{\infty}(D^k f)(x) \, d_k(x + 1 - x)$$

$$f(x + 1) = \sum_{k=0}^{\infty}(D^k f)(x) \, d_k(1)$$

$$(Ef)(x) = \sum_{k=0}^{\infty} \frac{(D^k f)(x)}{k!}$$

$$E = \sum_{k=0}^{\infty} \frac{D^k}{k!}$$

$$E = e^D$$

Define the *Abel operator* $A := DE$. That is, $(Af)(x) = f'(x + 1)$. By the above theorem, $A = De^D$. Now A is written in terms of D. Can we express D in terms of A? That is, we can find the coefficients c_k of the series: $D = \sum_{k=0}^{\infty} c_k A^k$

Here is the connection to the Lambert W function. Since $W(x)$ is the inverse of xe^x, and $A = De^D$, we have $D = W(A)$. That means the same coefficients c_k will be the coefficients of the series for $W(x)$. This is a variant of Taylor's theorem, and is equally true:

$f(x) = \sum_{k=0}^{\infty}(A^k f)(a)\, a_k(x - a)$ where a_k is the basic sequence for A — that is, $Aa_k = a_{k-1}$, $a_k(0) = 0$ when $k \neq 0$, and $a_0 = 1$. It has been figured out what the a_k are later. Basically, this is the Taylor sequence with all of the Ds replaced by As. Again, f only needs to be a polynomial. The proof of this is similar to how to prove Taylor's theorem for polynomials.

Differentiating: $(Df)(x) = \sum_{k=0}^{\infty}(A^k f)(a)\, a_k'(x - a)$ Set $a = x$:

$$(Df)(x) = \sum_{k=0}^{\infty}(A^k f)(x)\, a_k'(0)$$

$$(Df)(x) = \sum_{k=0}^{\infty} a_k'(0)\, (A^k f)(x)$$

$$D = \sum_{k=0}^{\infty} a_k'(0) A^k$$

This means that the coefficients of the Lambert W function are precisely $a_k'(0)$, where a_k is the basic sequence of A. So, what are the a_k? Let's consider the first few and develop a pattern. Remember that $A = DE$. Also, $Aa_k = a_{k-1}$, $a_k(0) = 0$ when $k \neq 0$, and $a_0 = 1$. So:

- $a_0(x) = 1$

- $a_1(x) = (x - 1) + 1$

  Think of it as doing A backwards, by integrating a_0 and then shifting it. It has been shown that $Aa_1 = a_0$. The 1 is to ensure that $a_1(0) = 0$.

- $a_2(x) = \dfrac{(x - 2)^2}{2} + (x - 2)$

  It's easy to check that $Aa_2 = a_1$. Now, $a_2(0) = \frac{4}{2} - 2 = 0$.

- $a_3(x) = \dfrac{(x - 3)^3}{3!} + \dfrac{(x - 3)^2}{2!}$

  It is easy to verify that $a_3(0) = -\frac{27}{6} + \frac{9}{2} = 0$

Generalizing the pattern, it is noticed that: $a_k(x) = \dfrac{(x - k)^k}{k!} + \dfrac{(x - k)^{k-1}}{(k - 1)!}$ (except for $k = 0$, where $a_k = 1$). The three conditions for a_k are satisfied, as one can check.

Now, compute $a'_k(0)$:

$$a'_k(x) = \frac{(x-k)^{k-1}}{(k-1)!} + \frac{(x-k)^{k-2}}{(k-2)!}$$

$$= \left(\frac{(x-k)^{k-2}}{(k-1)!}\right)((x-k)+(k-1))$$

$$= \left(\frac{(x-k)^{k-2}}{(k-1)!}\right)(x-1)$$

$$a'_k(0) = \frac{(-k)^{k-2}}{(k-1)!}(-1)$$

$$= \frac{(-k)^{k-2}}{(k-1)!}\frac{(-k)}{k}$$

$$= \frac{(-k)^{k-1}}{k!}$$

(except for $k=0$, where $a'_k(0)=0$).

That means that, by the above result:

$$D = \sum_{k=1}^{\infty} \frac{(-k)^{k-1}}{k!}A^k \text{ and, thus: } W(x) = \sum_{k=1}^{\infty} \frac{(-k)^{k-1}}{k!}x^k$$

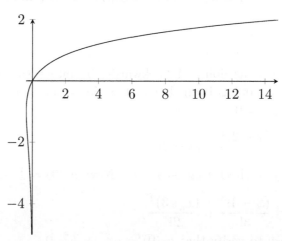

8.9 MATCHING

In this section the solution given in terms of the LambertW function arising from matching the Φ solutions from $\mathcal{G}(\eta)_{\delta4}$ in Eq. (8.14) together with the solution of F_4 given in Section 8.2.2, involves the pressure term P. Comparing Section 10.2.2 for the case of finite time blowup for the continuum-based Periodic Navier–Stokes equations (PNS) with this section's approach which involves top to bottom infinite subdivisions of turbulence energy scales using the Cantor function, lead to velocities that have no blowup in finite time. Here the term Φ containing pressure gradients in all directions is,

$$\Phi(t) = \mathcal{F}(C(t)) \tag{8.76}$$

where $C(t)$ is the Cantor function and $\mathcal{F}$ is a specific function of $C(t)$. The analysis requires that the Φ's are equated to each other. It can be assumed that on a micro or nano scale (Recall Eq. (8.36) where we can solve for $\Phi(t)$ explicitly as the following),

$$\Phi(t) = F_4^2(t)\frac{d}{dt}F_4(t) \tag{8.77}$$

where Φ is in terms of pressure and $F_4(t)$ and is Φ given by Eq. (8.13). Recall that the PNS equations were solved with the remaining equations, Eqs. (8.11, 8.12 and 8.14) combined as a sum. So it is necessary to match the Φ's from each part and solve a new differential equation. Inputing a decreasing function in time such as pressure (using transformation in Eq. (8.9)) $Q = y_3 C_L(s)$ (partial with respect to y_3 is $\frac{\partial Q}{\partial y_3} = C_L(s)$) in one case will be shown to blowup in finite time when using a linear decreasing function in s, however using a decreasing Cantor type function for pressure ($C_L(s)$ is decreasing Cantor function $1 - C_f(s)$) will lead to a no-finite time blowup for velocities defined on a arbitrarily small scale for turbulent "nano" flow. This scale is arbitrarily small due to infinitely small divisions of the Cantor set. Here the PNS equations breakdown but the velocities can be determined by the new differential equation mentioned previously. It seems very plausible that no blowup occurs since the energy at the continuum top scale is much higher than the bottom energy levels. The Equipartition theorem of Statistical Mechanics can be used here. A bottom up approach from a computational perspective is emphasized but in reality a top down approach is needed in an infinite progression of subdividing energy levels associated with the Cantor set for which the pressure is defined. First for the pressure function, that is, $P = \eta^{(-3)}(1 - C_f(t))$, we have defined $C_f(t)$ as the Cantor function. Secondly if $t \in C_c/\{0,1\}$ where C_c is the Cantor set then since the Cantor function acting on the Cantor set is equal to $[0, 1]$, then omitting $0, 1$ leaves us with the open interval $(0, 1)$. Then necessarily the pressure will not be zero and hence the velocity is non zero and $\eta \neq 0$. Next recall the definition of turbulence, turbulent flows contain a wide range of eddy sizes (scales). First large eddies derive their energy from the mean flow and are anisotropic. Second, these eddies are unstable and they break up into smaller eddies, at the scales of the smallest eddies, the turbulent energy is dissipated. The behavior of the small eddies is more universal in nature. In between the large and small eddies, there are some intermediate scales. The turbulent kinetic energy is transferred from the largest eddies to the smallest ones. Finally we say that a function f on $\mathbb{R}^3$ is periodic if $f(x + k) = f(x)$ for all $x \in \mathbb{R}^3$ and $k \in \mathbb{Z}^3$. Every periodic function is thus completely determined by its values on the unit cube, $Q_1 = [-\frac{1}{2}, \frac{1}{2})^n.(n = 3)$ Periodic functions may be regarded as functions on the space $\mathbb{R}^3/\mathbb{Z}^3$ of cosets of $\mathbb{Z}^3$ which we call the 3-dimensional torus and denote by $\mathbb{T}^3$. Now for measure-theoretic purposes, we identify $\mathbb{T}^3$ with the unit cube Q_1, and when we speak of Lebesgue measure on $\mathbb{T}^3$ we mean the measure induced on $\mathbb{T}^3$ by Lebesgue measure on Q_1. In particular, $m(\mathbb{T}^3) = 1$. Functions on $\mathbb{T}^3$ may be considered as periodic functions on $\mathbb{R}^3$ or as functions on Q_1.

8.9.1 Analysis for the Non-Blowup on Turbulent Cantor-Dust

The governing new differential equation mentioned above is obtained by matching and is,

$$- 2F_4^2(t) \int_0^t C_L(s)ds \ \text{EllipticE}(\text{JacobiSN}) \ (1 + C, i), i)R$$

$$- 2RF_4(t) \int_0^t C_L(s)ds(-F_4(t)\text{EllipticE}(\text{JacobiSN}(-1 + C), i), i)$$

$$+ 6C - 2F_4(t) = -F_4^2(t)\frac{d}{dt}F_4(t) \tag{8.78}$$

Solving this differential equation results in,

$$\int^{F_4(t)} \frac{\frac{2}{3}Rt \int_0^t C_L(s)ds - \frac{2R}{3} \int_0^t C_L(s)sds - C_1t -}{a\text{EllipticE}(\text{JacobiSN}(C+1,i),i) - a\text{EllipticE}(\text{JacobiSN}(C-1,i),i) + 6C - 2a}\,da + C_2 = 0 \tag{8.79}$$

The indefinite integral is equal to,

$$I = -\frac{a}{EllipticE(JacobiSN(C+1,i),i) - EllipticE(JacobiSN(C-1,i),i) - 2} +$$

$$6\,\frac{C\ln(a\,EllipticE(JacobiSN(C+1,i),i) - a\,EllipticE(JacobiSN(C-1,i),i) + 6\,C - 2\,a)}{(EllipticE(JacobiSN(C+1,i),i) - EllipticE(JacobiSN(C-1,i),i) - 2)^2}$$

$$(8.80)$$

Solving for a is possible algebraically and is expressed in terms of the LambertW function. WLOG substituting $C = 1$ gives the following for $F_4(t)$,

Taking the derivative of F_4 with respect to t for the Cantor function input and plotting the denominator shows no finite time blowup in the figure below.

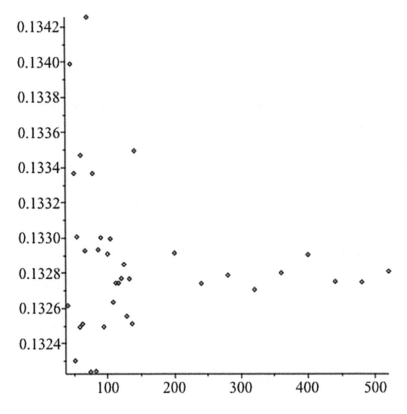

Figure 8.3 Plot of denominator of F_4 versus t when pressure is $1 - P(t)$ in terms of Cantor function.

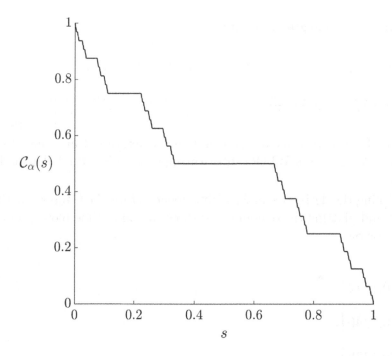

Figure 8.4 Plot of pressure $P = C_L(t) = 1 - C_f(t)$, where $C_f(t)$ is the Cantor function.

$$
\begin{aligned}
F_4 = \bigg[&-6\,\mathrm{W}\bigg[-1/6\exp\bigg[-1/9\left(\mathit{EllipticE}\left(\mathit{JacobiSN}\left(2,i\right),i\right)-2\right)^2 \\
&\times \left[\int_0^t C_L\left(s\right)s\,\mathrm{d}s + Rt\int_0^t C_L\left(s\right)\right) \bigg] + 1/18\left(3\,C_1\,t - 3C_2\right) \\
&\times \left(\mathit{EllipticE}\left(\mathit{JacobiSN}\left(2,i\right),i\right)\right)^2 + 1/18\bigg(-12\,C_1\,t + 12C_2 \bigg) \\
&\times \mathit{EllipticE}\left(\mathit{JacobiSN}\left(2,i\right),i\right) + 2/3\,C_1\,t - 2/3\,C_2 - 1\bigg]\bigg] - 6\bigg] \\
&\times \bigg[\mathit{EllipticE}\left(\mathit{JacobiSN}\left(2,i\right),i\right) - 2\bigg]^{-1}
\end{aligned}
\tag{8.81}
$$

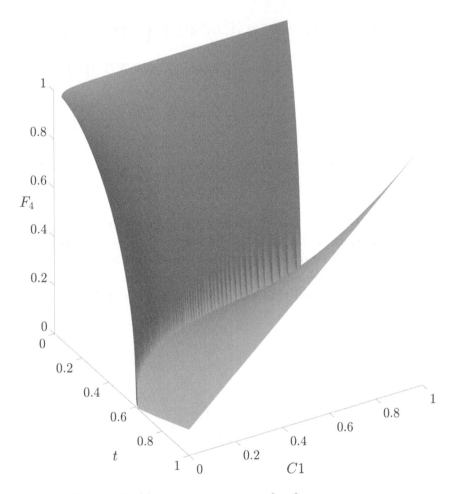

Figure 8.5 Plot of $F4(t)$ versus t and $C_1 \in [0, 1]$.

8.10 CANTOR FUNCTION REPLACED BY LINEAR FORM

The solution F_4 for the Cantor function C_f replaced by $C_L = 1 - s/R$:
Energy conservation on the interval $s \in [0, T_F]$

$$F_4 = \left[-6\,\mathrm{W}\left[-1/6\exp\left[1/9\left(EllipticE\left(JacobiSN\left(2, i\right), i\right) - 2\right)^2 \right.\right.\right.$$
$$\left.\left.\left. \times \int_0^t C_L\left(s\right) s\,\mathrm{d}s + 1/18\left(3\,C_1\,t + t^2 - 2t \int_0^t C_L\left(s\right)\,\mathrm{d}s - 3\,C_2\right) \right.\right.\right.$$

$$\times \left(EllipticE \left(JacobiSN \left(2,i \right), i \right) \right)^2 + 1/18 \left(- 12\, C_1\, t - 4\, t^2 \right.$$

$$+ 8\, t \int_0^t C_L \left(s \right) \, \mathrm{d}s + 12\, C_2 \right) EllipticE \left(JacobiSN \left(2,i \right), i \right)$$

$$+ 2/9\, t^2 + 2/3\, C_1\, t - 4/9\, t \int_0^t C_L \left(s \right) \, \mathrm{d}s - 2/3\, C_2 - 1 \Big] \Big] - 6 \Big]$$

$$\times \left[EllipticE \left(JacobiSN \left(2,i \right), i \right) - 2 \right]^{-1} \tag{8.82}$$

Substituting $C_L = 1 - s/R$ thereby assuming a decreasing linear pressure P, using the initial condition value $C_2 = -6.275$ and letting $C_1 = 0.1e-1$, $R = 1$(both chosen so that the solution is bounded) leads to the graph in Figure 8.6. Above it is a graph showing the change in t and C_1. It is observed that there exists a $t = T_F$ such that the y_3

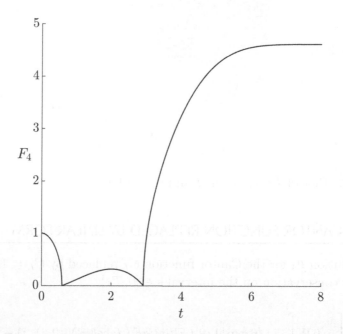

Figure 8.6 Plot of $F4(t)$ versus t for $t \in [0,8]$, and use of condition $W(-1/e) = -1$ for the LambertW function. This is the point where the real solution transitions to complex solution.

velocity F_4 satisfies the following reproductive property,

$$F_4(0) = F_4(T_F) \tag{8.83}$$

Also there is a $t < T_F$ such that the first derivative blows up at this time t. Substituting into the incompressible viscous Navier–Stokes energy balance equation which consists of $F_4(0)$, $F_4(T_F)$, $2\nu \parallel \nabla \vec{u} \parallel_2^2$ and $2 \int_0^t \vec{F} \cdot \vec{u}(s)$, leads to,

$$\parallel \vec{u}(T_F) \parallel_2^2 = \parallel u(0) \parallel_2^2 - 2\nu \parallel \nabla \vec{u} \parallel_2^2 + 2 \int_0^t \vec{F} \cdot \vec{u}(s)ds = 0 \tag{8.84}$$

Here there is an energy balance equation with forcing $\vec{F}$ and the usual viscous term included.

It has been concluded that by Gagliardo/Prekopa Leindler inequalities that for a lattice volume approaching infinity $\parallel \nabla \vec{u} \parallel_2^2 \to 0$ and the Energy balance equation at $t = T_F$ reduces to the following,

$$2 \int_0^t \vec{F} \cdot \vec{u}(s) = 0 \tag{8.85}$$

which is recalling that $\vec{F} = (f_0(s), f_0(s), f_3(s))$,

$$\int_0^{T_F} (f_3(s)F_4(s) + 2f_0(s)f_1(s))ds = 0 \tag{8.86}$$

Recall that $f_1(s)$ is the time component of velocity in the separable solution for v_1 and v_2 solutions.

The following is true,

$$\int_0^{T_F} -\frac{dF_4}{ds}F_4 - 2f_1\frac{df_1}{ds}ds = L \tag{8.87}$$

which becomes,

$$\begin{aligned}
L &= (-F_4^2 - 2f_1^2)|_0^{T_F} \\
&= -F_4^2(T_F) - 2f_1^2(T_F) + 1 + 2f_1^2(0) \\
&= -2f_1^2(T_F) + 2f_1^2(0) \\
&= 2(f_1^2(0) - f_1^2(T_F)) = 0
\end{aligned} \tag{8.88}$$

Thus the following important result follows,

$$f_1(0) = f_1(T_F) \tag{8.89}$$

Here it is shown explicitly that the velocity $\vec{v}$ in $\mathbb{T}^3$ conserves energy on the interval $[0, T_F]$ and that in the interval $[0, T_F]$ there exists a t such that $\frac{\partial v_3}{\partial t}$ blows up in finite time.

8.10.1 Quantitative Bounds for Critically Bounded Solution to Navier–Stokes Equations

Recalling the quantitative bounds[56] for higher regularity norms of solutions to NS equations in terms of the critical L_x^3 bound with a dependence that is triple exponential in nature, we have the following theorem due to Tao. [56]

Theorem 8.2 *Quantitative regularity for critically bounded solutions:*
Let $u : [0, T] \times \mathbb{R}^3 \to \mathbb{R}^3$, $p : [0, T] \times \mathbb{R}^3 \to \mathbb{R}$ be a classical solution to the Navier–Stokes equations with

$$\| u \|_{L_t^\infty L_x^3 ([0,T] \times \mathbb{R}^3)} \leq A \tag{8.90}$$

for some $A \geq 2$. Then we have the derivative bounds

$$| \nabla_x^j u(t, x) | \leq exp\ exp\ exp(A^{\mathcal{O}(1)}) t^{-\frac{i+1}{2}} \tag{8.91}$$

and

$$| \nabla_x^j \omega(t, x) | \leq exp\ exp\ exp(A^{\mathcal{O}(1)}) t^{-\frac{i+2}{2}} \tag{8.92}$$

whenever $0 < t \leq T, x \in \mathbb{R}^3$, and $j = 0, 1$, where $\omega := \nabla \times u$ is the vorticity field.

Using the z solution obtained in terms of the W function gives the following for $\| u \|_{L_t^\infty L_x^3 ([0,T] \times \mathbb{T}^3)}$ in a given cell of dimension $I = [-1, 1]$ (linear dimension z only),

$$v_3 = V_1 V_2 \tag{8.93}$$

$$V_1 = F_2 \left(1/2 \sqrt{2} + i/2\sqrt{2} \right) JSN \left((i/2\sqrt{2} y_3 + F_1) F_2 \left(1/2 \sqrt{2} + i/2\sqrt{2} \right), i \right) \tag{8.94}$$

and

$$V_2 = \frac{-6\,\mathrm{W}\left(-\frac{1}{6}\,e^{\frac{1}{27}\,(E(JSN(2,i),i)-2)^2\xi_1(s)+\xi_2(s)(E(JSN(2,i),i))^2+\xi_3(s)E(JSN(2,i),i)+\xi_4(s)}\right)-6}{E(JSN(2,i),i)-2}$$

(8.95)

where $i = \sqrt{-1}$, JSN is the JacobiSN function and E is the incomplete elliptic integral defined by, $\mathrm{EllipticE}(z,k) = \int_0^z \frac{\sqrt{-k^2 t^2 + 1}}{\sqrt{-t^2 + 1}}$. The ξ_i parameters are defined as,

$$\xi_1(s) = s^3$$

$$\xi_2(s) = \tfrac{1}{18}\left(s^2 - s^3 + 0.003\,s + 18.825\right)$$

$$\xi_3(s) = \tfrac{1}{18}\left(-4\,s^2 + 4\,s^3 - 0.012\,s - 75.300\right)$$

$$\xi_4(s) = \tfrac{2}{9}\,s^2 - \tfrac{2}{9}\,s^3 + 0.00067\,s + 3.183$$

(8.96)

Using the norm estimate in above theorem and integrating first with respect to y_3 alone gives,

$$I(t) = \int_I \left[|\,v_3\,|^3 + 2\,|\,v_i\,|^3\,(1 + \ln\,(1 + t))^3\right] dy_3 \qquad (8.97)$$

Recalling Section 10.4.1, and plotting $\mathfrak{Im}V$ there versus F_1 the zeros F_1 are related to F_2 as $F_2 = -2F_1$. We have the most smooth solution for factor F_5 of v_3 solution if and only if $F_1 = \frac{\sqrt{2}}{2}$.(see Figure where the maximum change is on an interval [-1,1]. Here we have that the relationship of the force terms F_{T_1} and F_{T_2}, equal to each other, are also equal to the velocity in the y_1, y_2 directions respectively at the initial condition. The frictional force is proportional to the velocity. The velocity of a fluid particle has y_3 component with the form,

$$F_{f_{y_3}} = -k_{y_3} v_{y_3} \qquad (8.98)$$

The A value in [56] is assumed for some $A \geq 2$. In the present analysis upon calculation,

$$\|\,u\,\|_{L_t^\infty L_x^3([0,T]\times\mathbb{R}^3)} = 4 \qquad (8.99)$$

We too obtain quantitative bounds for higher regularity norms in terms of the critical L_x^3 bound. This agrees well with the Prekopa-Leindler and Gagliardo-Nirenberg inequality approach followed earlier in this chapter (Figures 8.7 and 8.8.

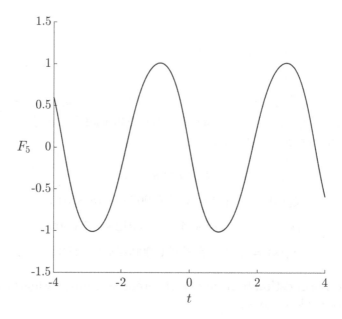

Figure 8.7 Plot of oscillatory function $\Re e F5$ versus y_3. It is seen that the function is periodic and differentiable everywhere along y_3. Also the amplitude is unity agreeing with the initial condition(Note the $F_4(s)$ satisfies the initial condition at $s = 0$, a more general form was used with an unknown constant specified by initial condition), and the value of $F_1 = \frac{\sqrt{2}}{2}$ is unique to obtain an initial condition of unity.

As a body moves forward, frictional force opposes it's motion decreasing the speed of the body of fluid. This is set as an initial condition. We start the motion at the instant the frictional force is applied. The direction is exactly opposite to the direction of the velocity and proportional to its speed. This makes sense in the context of oscillatory flows and as an example simple harmonic motion. The simple harmonic motion SHM would then be a mass which resists acceleration with a term consisting of the mass multiplied by the second derivative of the displacement x with respect to t, a spring which resists displacement with a term $-kx$, and a dashpot (which resists velocity with a term $-c\frac{dv}{dt}$ and surface friction which resists always, any time there is motion with a constant term $-\mu N$) The dashpot and surface friction are just two different types of friction. Referring to the estimate

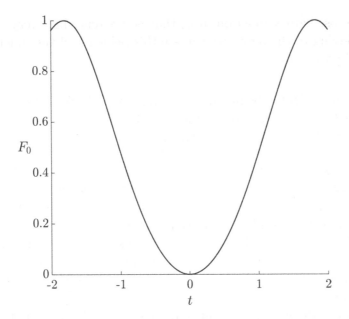

Figure 8.8 Plot of oscillatory function, the reciprocal of the Weierstrass P function(at $\sqrt{2}y_3/2$) , the forcing F_0 versus y_3.It is seen that the function is periodic and differentiable everywhere along y_3. Also the amplitude is unity agreeing with the initial condition(Note the $F_4(s)$ satisfies the initial condition at $s = 0$, and the value of $F_1 = \frac{\sqrt{2}}{2}$ is unique to obtain an initial condition of unity. Here the period is the same as the F_5 function in y_3. This period is the same for any value of m in the P function.

in Eq. (8.97) and raising the result to the exponent 3 since we are working in the normed space L_3 which is a critical space for the Navier–Stokes equations, and finally calculating the time integral in the mixed Lebesgue integral leaves us with the following,

$$I = \left(\int_{[0,T]} I(t)^q \right)^{1/q} \tag{8.100}$$

If the integral I is set to $1/2$ then for the case of the magnitude of velocity equal to the magnitude of the forcing term in the y_i direction, $i = 1, 2,$(as in the present case) one obtains a numerical estimate for complex number F_2. Substituting this back into the expression for v_3

gives an oscillatory function in y_3 that is differentiable everywhere in y_3. Referring to the figure we can see this behavior of the function for the real part of v_3.

8.11 FIGURES CONFIRMING NO BLOWUP FOR THIRD COMPONENT OF VELOCITY OF PNS SOLUTION FOR SUMS OF CANTOR FUNCTIONS

It becomes evident that with the Cantor function multiplied by a non-unity constant α, that is $\alpha \cdot C_f(s)$ (see Appendix 8 for Maple code-Here data was produced at individual points in time for $F_4(t)$), based on decreasing pressure in time s, there is no finite time blowup for $F_4(t)$ Here a Maple procedure was used (not shown) where a numerical trapezoid integration command was introduced to compute the antiderivatives of $C_L(s) = C_F(s) = \frac{1}{2^{2n-2i+2}} \cdot \sum_{i=1}^{n}(1 - C_f(\frac{s}{in}))$ in Eq. (8.82). Comparing to $C_L(s) = \frac{1}{2^{2n-2i+2}} \cdot \sum_{i=1}^{n}(1 - \frac{s}{in})$ it is seen that with the same parameters in F_4 in both cases of stochastic pressure and non-stochastic pressure, that there is a blowup in Figure 8.9a for non-stochastic pressure unlike that in the stochastic case. This is seen in Figure 8.9a where a blowup occurs for non-stochastic P but not for stochastic pressure P. The dotted graph shows that near the blowup of the non-stochastic case the Cantor function incorporated in the F_4 function is a fractal function. A repeating pattern is seen clearly in Figure 8.10 for three different scales zoomed in. Finally a simplified expression $C_L(s) = \sum_{i=1}^{n}(1 - \frac{C_f(s)}{n})$ was considered and substituted into Eq. (8.82). The result shows that a function like $\tanh(t)$ emerges ($n = 420$ is chosen here). If we select η to be small in the transformations between starred and non-starred variables for time, even if η is small then as it approaches zero the solution approaches a constant.

8.12 GENERAL SOLUTION WITH NO RESTRICTIONS ON FORCING AND SPATIAL VELOCITIES

Here there are no assumptions made on forcing and spatial velocities as being separable in space and time. PNS is made of the following component parts C_i and is given by Eq. (8.101) and also Eq. (8.29) written in terms of special terms Φ_1 here [57].

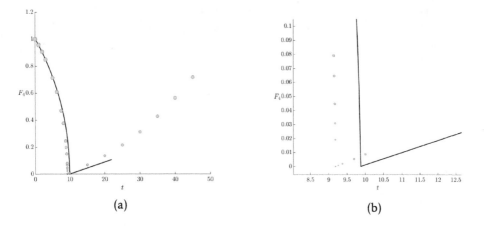

(a)

(b)

Figure 8.9 Plot of non fractal $F_4(t)$ and Fractal $F_4(t)$ versus t. (a) Plot of F_4 with Cantor function $C_F(s)$ substituted in for $C_L(s)$. Solid line is linear case when $C_L(s) = \frac{1}{2^{2n-2i+2}} \cdot \sum_{i=1}^{n}(1-\frac{s}{in})$ $n = 420$. Dotted graph is for Cantor function $C_F(s) = \frac{1}{2^{2n-2i+2}} \cdot \sum_{i=1}^{n}(1 - C_f(\frac{s}{in}))$, $n = 420$. (b) Zoomed in near vertex for fractal $F_4(t)$. Note the repeating fractal on smaller scales.

$$C_1 + C_2 = C_3 \tag{8.101}$$

$$
C_1 = \left(\frac{\partial v_3}{\partial s}\right)(v_3)^2\left(\frac{\partial v_3}{\partial y_1} - \frac{\partial v_3}{\partial y_2}\right)\frac{\partial^3 v_3}{\partial y_3{}^2 \partial s}
$$

$$
- (v_3)^2\left(\frac{\partial v_3}{\partial y_1} - \frac{\partial v_3}{\partial y_2}\right)\left(\frac{\partial^2 v_3}{\partial y_3 \partial s}\right)^2
$$

$$
- \left(-2\left(\frac{\partial v_3}{\partial y_3}\right)v_3\frac{\partial v_3}{\partial s} + \Phi_1 + \Phi\left(s\right)\right)\left(\frac{\partial v_3}{\partial y_1} - \frac{\partial v_3}{\partial y_2}\right)\frac{\partial^2 v_3}{\partial y_3 \partial s}
$$

$$\tag{8.102}$$

$$
C_2 = \left[\left(\frac{\partial v_3}{\partial y_1} - \frac{\partial v_3}{\partial y_2}\right)\frac{\partial \Phi_1}{\partial y_3} + 3\left(\frac{\partial v_3}{\partial s}\right)v_3\left(\frac{\partial v_3}{\partial y_1} - \frac{\partial v_3}{\partial y_2}\right)\frac{\partial^2 v_3}{\partial y_3{}^2}\right.
$$

$$
- \left(2\frac{\partial v_3}{\partial y_1} - 2\frac{\partial v_3}{\partial y_2}\right)v_3\left(\frac{\partial v_3}{\partial y_1}\right)\frac{\partial^2 v_3}{\partial y_1 \partial s} - \left(2\frac{\partial v_3}{\partial x} - 2\frac{\partial v_3}{\partial y_2}\right)
$$

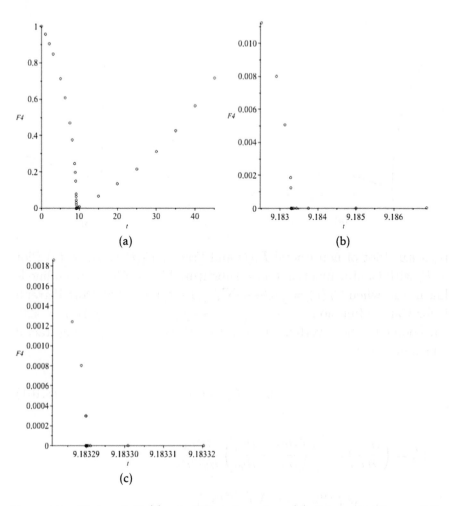

Figure 8.10 Plots of $F_4(t)$ at different scales. (a) Same as Figure 8.9a (Not Zoomed in). (b) Same as before but Zoomed in X2. (c) Zoomed in X3.

$$\times v_3 \left(\frac{\partial v_3}{\partial y_2}\right) \frac{\partial^2 v_3}{\partial y_2 \partial s} + 2\, v_3 \left(\frac{\partial v_3}{\partial y_1}\right)^2 \left(\frac{\partial^2 v_3}{\partial y_1 \partial s} + \frac{y_3 N}{y_1^2 + y_2^2 + y_3^2}\right)$$

$$- 2\, v_3 \left(\frac{\partial v_3}{\partial y_2}\right)^2 \left(\frac{\partial^2 v_3}{\partial y_2 \partial s} - \frac{\partial}{\partial s}\left(\frac{2(y_2 - y_1)v_3}{y_1^2 + y_2^2 + y_3^2}\right)\right)$$

$$+ 3 \left(\frac{\partial v_3}{\partial y_3}\right)^2 \left(\frac{\partial v_3}{\partial s}\right) \left(\frac{\partial v_3}{\partial y_1} - \frac{\partial v_3}{\partial y_2}\right) \Bigg] \frac{\partial v_3}{\partial s} \tag{8.103}$$

$$C_3 = \left[2 \frac{(-2\,y_1 + 2\,y_2)\frac{\partial v_3}{\partial s} + 2\,y_3\left(\frac{\partial v_1}{\partial s} - \frac{\partial v_2}{\partial s}\right)}{y_1{}^2 + y_2{}^2 + y_3{}^2} - 2\frac{\partial^2 v_3}{\partial y_2 \partial s} + 2\frac{\partial^2 v_3}{\partial y_1 \partial s} \right] \times$$
$$v_3 \left(\frac{\partial v_3}{\partial y_1}\right)\left(\frac{\partial v_3}{\partial s}\right)\frac{\partial v_3}{\partial y_2}$$
(8.104)

where Φ_1 is given as,

$$\Phi_1 = -\left(\frac{\partial v_3}{\partial y_3}\right)v_3\frac{\partial v_3}{\partial s} - v_3^2\frac{\partial^2 v_3}{\partial y_3 \partial s} - \frac{1}{2}\Phi\,(s) \tag{8.105}$$

The expression for Φ_1 is determined by solving for $\vec{F}\cdot\nabla v_3^2 + \nabla v_3^2 \cdot \frac{\partial \vec{b}}{\partial s}$ in Eq. (8.3) when $\delta \approx 1$.

Simplifying Eq. (8.101) using Eq. (8.105) and using the definition of the Liutex part of vorticity in Eqs. (8.27–8.28) and solving algebraically for $N = \frac{\partial v_1}{\partial s} - \frac{\partial v_2}{\partial s}$ in Eq. (8.101) the problem is simplified to,

$$-\frac{\left(\frac{\partial v_3}{\partial y_1} - \frac{\partial v_3}{\partial y_2}\right)\left(-1/4\,\Phi\,(s)\left(y_1{}^2 + y_2{}^2 + y_3{}^2\right)\frac{\partial^2 v_3}{\partial y_3 \partial s}\right.}{ }$$
$$\left. + \left(\frac{\partial v_3}{\partial s}\right)^2\left(v_3\left(y_1{}^2 + y_2{}^2 + y_3{}^2\right)\frac{\partial^2 v_3}{\partial y_3{}^2} + 2\,v_3\left(-y_2 + y_1\right)\frac{\partial v_3}{\partial y_2}\right.\right.}{ }$$
$$\frac{\left.\left. + \left(\frac{\partial v_3}{\partial y_3}\right)^2\left(y_1{}^2 + y_2{}^2 + y_3{}^2\right)\right)\right)}{y_3\left(\frac{\partial v_3}{\partial s}\right)\left(\frac{\partial v_3}{\partial y_1}\right)v_3\left(\frac{\partial v_3}{\partial y_1} - 2\frac{\partial v_3}{\partial y_2}\right)} - $$
$$\frac{\partial v_1}{\partial s} + \frac{\partial v_2}{\partial s} = 0$$
(8.106)

This PDE is separable as $v_3 = F_7(y_1, y_2, y_3)F_4(s)$, where

$$4\,F_7^3\left(\frac{\partial^2 F_7}{\partial y_3{}^2}\right)y_1{}^2 + 4\,F_7^3\left(\frac{\partial^2 F_7}{\partial y_3{}^2}\right)y_2{}^2 + 4\,F_7^3\left(\frac{\partial^2 F_7}{\partial y_3{}^2}\right)y_3{}^2$$
$$+ 4\,F_7^2\left(\frac{\partial F_7}{\partial y_3}\right)^2 y_1{}^2 + 4\,F_7^2\left(\frac{\partial F_7}{\partial y_3}\right)^2 y_2{}^2 + 4\,F_7^2\left(\frac{\partial F_7}{\partial y_3}\right)^2 y_3{}^2$$
$$+ 8\,F_7^3\left(\frac{\partial F_7}{\partial y_2}\right)y_1 - 8\,(F_7^3)\left(\frac{\partial F_7}{\partial y_2}\right)y_2 - \left(\frac{\partial F_7}{\partial y_3}\right)y_1{}^2 c_4$$
$$- \left(\frac{\partial F_7}{\partial y_3}\right)y_2{}^2 c_4 - \left(\frac{\partial F_7}{\partial y_3}\right)y_3{}^2 c_4 = 0 \tag{8.107}$$

We have used the following equalities,

$$\frac{\partial^2 v_1}{\partial y_3 \partial s} = \frac{\partial^2 v_3}{\partial y_1 \partial s} + \frac{y_3 N}{y_1^2 + y_2^2 + y_3^2} \tag{8.108}$$

$$\frac{\partial^2 v_2}{\partial y_3 \partial s} = \frac{\partial^2 v_3}{\partial y_2 \partial s} - \frac{\partial}{\partial s}\left(\frac{2(y_2 - y_1)v_3}{y_1^2 + y_2^2 + y_3^2}\right) \tag{8.109}$$

where N is defined by,

$$N = \frac{\partial v_1}{\partial s} - \frac{\partial v_2}{\partial s} \tag{8.110}$$

8.13 DISCUSSION AND CONCLUSION

Comparing solutions for Eqs. (8.74–8.75) and Eqs. (8.36, 8.55) (for $v_3 = F_4(s)F_5(y_1, y_2, y_3)$) it is evident that the former one has as solution for u_z a time component that is smooth(except at infinity) opposite to that of the smooth force and spatially non-smooth y_1 and y_2 velocities whereas in the latter equations u_z has a finite time blowup(with the first derivative and higher) for f_0, f_1 and y_1 and y_2 smooth velocity function inputs [27]. Equation (8.101) is the full non-separable reduced PDE. Oscillations of arbitrary height can occur at spatial infinity. For (PNS) it is shown in [57] that there exists a vortex in each cell of the lattice associated with $\mathbb{T}^3$ using the decomposition of pure rotation(Liutex), antisymmetric shear and compression and stretching. A cusp bifurcation for vorticity shows the birth and destruction of vorticies. Here it is known that streaklines can be used to give an idea of where the vorticity in a flow resides. The question of no-finite time blowup for the new equations [18] replacing the Navier–Stokes equations is left for future study.

8.14 APPENDIX 1

The second term in Eq. (8.18) involving the tensor product expression for the solution of v_3 given in terms of F_5 is independent of s ($\Phi(s) = H_2(y_1, y_2, y_3)$), and is given as,

$$1/3 \frac{(G_2(y_1,y_2,y_3))^2 \left(\sqrt[3]{s-s_0}\frac{\partial}{\partial y_1}G_2(y_1,y_2,y_3) + \sqrt[3]{s-s_0}\frac{\partial}{\partial y_2}G_2(y_1,y_2,y_3) + \sqrt[3]{s-s_0}\frac{\partial}{\partial y_3}G_2(y_1,y_2,y_3)\right)}{\sqrt[3]{s-s_0}}$$

$$= 1/3\,(G_2(y_1,y_2,y_3))^2 \left(\frac{\partial}{\partial y_1}G_2(y_1,y_2,y_3) + \frac{\partial}{\partial y_2}G_2(y_1,y_2,y_3) + \frac{\partial}{\partial y_3}G_2(y_1,y_2,y_3)\right) \tag{8.111}$$

The surface integral in Eq. (8.18) involving the pressure terms Q is either zero or non-zero depending on if $\Phi(s) = 0$ which occurs when $v_1 = -v_2$ or $\Phi(s) \neq 0$ which occurs when $v_1 = v_2$. In general we have the term $K_1 - \Phi(s) = \lambda_1$, where $\lambda_1 \neq 0$. The term $K_1 - \Phi(s)$ is associated with taking the gradient of the extended expression $\nabla_D^{-1}\left(\mathcal{G}_{\delta 1} + \mathcal{G}_{\delta 2} + \mathcal{G}_{\delta 3} + \mathcal{G}_{\delta 4}\right) = \vec{r}$.

8.15 APPENDIX 2

```
###Maple Program to solve the PDE Equation 8.29 in Section 8.2.2
### Input Equation 8.29(%) and assign it a variable name Y
Y=%
### Solve for expression $\frac{\partial^2 v_1}{\partial s \partial y3}$ as part of $Y$;
###Assign new variable M1
M1=solve(Y,$\frac{\partial^2 v_1}{\partial s \partial y3}$);
### Solve for expression $\frac{\partial^2 v_2}{\partial s \partial y3}$ as part of $Y$;
### Assign new variable M2
M2=solve(Y,$\frac{\partial^2 v_2}{\partial s \partial y3}$);
simplify(%)
### Form the difference between M1 and M2
M1-M2;
### Next use the definition of Vorticity and set M1-M2 as the following:
M1-M2=diff(Kappa(y1,y2,y3,s),s)-diff(diff(v3(y1,y2,y3,s),y2),s)+diff(diff(v3(y1,y2,y3,s),
y1),s);
### Simplify expression
simplify(%);
### Substitute Ansatz for velocity for v1 and v2
subs(v1(y1,y2,y3,s)=(A+Uy1(y1,y2,y3))*f1(s),v2(y1,y2,y3,s)=(A+Uy2(y1,y2,y3))*f1(s),%)
###Substitute the vorticity in terms of angular velocity
PL:=subs(Kappa(y1,y2,y3,s)=(2*(y2*v3(y1,y2,y3,s)-y3*(Uy2(y1,y2,y3)+A)*f1(s) -y1*v3(y1,y2,
y3,s)+y_3*(Uy1(x,y,z)+A)*f1(s))*(y1^2+y2^2+y3^2)^(-1);
### Substituting forcing terms into PL in terms of WeierstrassP functions
subs(FT1(y1,y2,y3,s)=((f0(s)*F(y1,y2,y3)),FT2(y1,y2,y3,s)=((f0(s)*G(y1,y2,y3)),PL);
subs(F(y1,y2,y3)=A+WeierstrassP(y1,3m^2,m^3)^(-1)*WeierstrassP(y2,3m^2,m^3)^(-1)
*WeierstrassP(y3,3m^2,m^3)^(-1),%)
subs(G(y1,y2,y3)=A+WeierstrassP(y1,3m^2,m^3)^(-1)*WeierstrassP(y2,3m^2,m^3)^(-1)
*WeierstrassP(y3,3m^2,m^3)^(-1),%)
### Substitute Uy1 and Uy2 in terms of reciprocal WeierstrassP functions matching
### the smooth form of forcing functions.
subs(Uy1(y1,y2,y3)=A+WeierstrassP(y1,3m^2,m^3)^(-1)*WeierstrassP(y2,3m^2,m^3)^(-1)
*WeierstrassP(y3,3m^2,m^3)^(-1),%)
subs(Uy2(y1,y2,y3)=A+WeierstrassP(y1,3m^2,m^3)^(-1)*WeierstrassP(y2,3m^2,m^3)^(-1)
*WeierstrassP(y3,3m^2,m^3)^(-1),%)
### Note in the previous few lines consecutive substitutions were used in previous step
###Substitute delta =1 or delta = 1-alpha where alpha is <<1
simplify(%)
###Based on the definition of Newton's second law relating the acceleration to the
###force we equate the following in the ###previous equation.
subs(\frac{d}{ds}=-f0(s),%);
###Substitute Lambda(y1,y2,y3,s) in Eq. (8.14) and it's partial derivative with respect
to y3
###For the first solution set Lambda(y1,y2,y3,s) to approach a constant(or zero)
### Next use the Maple pdsolve command to obtain the separable solution
```

```
pdsolve(%,v3(y1,y2,y3,s))
simplify(%)
###End of Program
```

EXERCISES

8.1 Prove that for each $s > 3/2$ there exists a constant C_s such that whenever $u \in H^s(\mathbb{T}^3)$ then $u \in C^0(\mathbb{T}^3)$ with,

$$\|u\|_{L^\infty} \leq C_s \|u\|_{H^s}.$$

8.2 Helmholtz-Weyl decomposition on the torus. In the case of a 3-torus we decompose only the homogeneous space $\dot{L}^2$ (rather than all of L^2) since we will always include the zero-average condition when considering the Navier--Stokes equations in this setting. To prove $\dot{L}^2 = H \oplus G$ we need to define the appropriate spaces H and G in the present case. Starting with the definition $H(\mathbb{T}^3)$. If u is given by $u(x) = \hat{u}_k e^{ik\cdot x}$ then computing we have,

$$\text{div } \mathbf{u}(x) = i\,(k \cdot \hat{u}_k)\,e^{ik\cdot x}.$$

Definition, $\mathbb{L}^2 := [L^2]^3$ and $\mathbb{H}^s := [H^s]^3$.

So this function is divergence free iff $\hat{u}_k$ is orthogonal to k. We thus define the space $H = H(\mathbb{T}^3)$ as,

$$\{u \in \dot{\mathbb{L}}^2 : u = \sum_{k \in \dot{\mathbb{Z}}^3} \hat{u}_k e^{ik\cdot x}, \hat{u}_{-k} = \overline{\hat{u}} \text{ and } k \cdot \hat{u}_k = 0 \text{ for all } k \in \dot{\mathbb{Z}}^3\}$$

where H is equipped with the $\mathbb{L}^2$ - norm. A function $u \in L^1(\Omega)$ is weakly divergence free if $\langle u, \nabla\phi \rangle = 0$ for every $\phi \in C_c^\infty(\Omega)$. See Lemma in [60] that each $u \in H(\mathbb{T}^3)$ is weakly divergence free and moreover $\langle u, \nabla\phi \rangle = 0$ for all $\phi \in H^1(\mathbb{T}^3)$.

The space G on the torus $\mathbb{T}^3$ is defined as,

$$G(\mathbb{T}^3) = \{u \in \dot{\mathbb{L}}^2 : u = \nabla g, \text{ where } g \in \dot{H}^1(\mathbb{T}^3)\}.$$

Here it is assumed that g belongs to $\dot{H}^1$ instead of H^1 in order to obtain the uniqueness of g in the Helmholtz-Weyl decomposition. It is straightforward to show that the spaces H and G are orthogonal, that is $\langle h, \nabla g \rangle = 0$ for every $h \in H$ and $\nabla g \in G$.

Prove that the space $\dot{L}^2$ can be written as,

$$\dot{L}^2 = H \oplus G,$$

that is every function $u \in \dot{L}^2$ can be written in a unique way as,

$$u = h + \nabla g$$

where the vector-valued function h belongs to H and scalar function g belongs to $\dot{H}^1$. We also have that the functions h and ∇g are orthogonal in $\mathbb{L}^2$, $\langle h, \nabla g \rangle = 0$, and if in addition u belongs to $\dot{\mathbb{H}}^s, s > 0$, then $h \in \dot{\mathbb{H}}^s$ and $g \in \dot{H}^{s+1}$.

8.3 Prove the Prékopa Leindler inequality.

8.4 Prove the Gagliardo-Nirenberg inequality.

8.5 Prove that the Cantor set C is totally disconnected and perfect . A Perfect set is a closed set in which every point is an accumulation point in topology. A space X is disconnected then there are some points $x, y \in X$ and disjoint open neighborhoods $U \ni x$ and $V \ni y$ so that $X = U \cup V$. Now X is totally disconnected if for every x and y there are U and V as above.

8.6 a) Define $f_0(x) = x$ for all $x \in [0, 1]$. Now, let,

$$f_1(x) = \begin{cases} (3/2)x & \text{for } 0 \le x \le 1/3 \\ 1/2 & \text{for } 1/3 < x < 2/3 \\ (3/2)x - 1/2 & \text{for } 2/3 \le x \le 1 \end{cases}$$

Sketch f_0 and f_1 on $[0, 1]$ and notice that f_1 is continuous, increasing and constant on the middle third set $(1/3, 2/3) = [0, 1] \setminus C_1$.

b) Construct $f_2(x)$ by flattening out the middle third of each nonconstant segment of $f_1(x)$. That is, set

$$f_2(x) = \begin{cases} (1/2)f_1(3x) & \text{for } 0 \le x \le 1/3 \\ f_1(x) & \text{for } 1/3 < x < 2/3 \\ (1/2)f_1(3x - 2) + 1/2 & \text{for } 2/3 \le x \le 1 \end{cases}$$

Continuing the process proves that the resulting sequence (f_n) converges uniformly on $[0, 1]$.

8.7 Solve the equation,

$$e^{cx}\frac{x - t}{x - s} = a$$

for x where the remaining terms are constants.

## CHAPTER 9

Introduction to Fractional Calculus

Terry E. Moschandreou

TVDSB

9.1 INTRODUCTION

In this part we first consider Euler's gamma function $\Gamma(z)$, which is a generalization of the factorial $n!$. Here n can be such that $n \in \mathbb{C}$. The following is true for the gamma function.

9.1.1 The Gamma Function

The gamma function $\Gamma(z)$ is defined as,

$$\Gamma(z) = \int_0^\infty e^{-t} t^{z-1} dt \tag{9.1}$$

which converges for $Re(z) > 0$.
The following is true,

$$
\begin{aligned}
\Gamma(x + iy) &= \int_0^\infty e^{-t} t^{x-1+iy} dt \\
&= \int_0^\infty e^{-t} t^{x-1} t^{iy} dt \\
&= \int_0^\infty e^{-t} t^{x-1} e^{\log t^{iy}} dt \qquad \because x = e^{\log x}
\end{aligned}
$$

DOI: 10.1201/9781003452256-9

$$= \int_0^\infty e^{-t} t^{x-1} e^{iy \log(t)} dt \qquad\qquad \because \log m^n = n \log m$$

$$\Gamma(x + iy) = \int_0^\infty e^{-t} t^{x-1} \{\cos[y \log(t)] + i \sin[y \log(t)]\} dt \qquad (9.2)$$

The expression in the square brackets in (Eq. 9.2) is bounded for all t. There is convergence at $t = 0$ and at infinity. Also $x = \mathrm{Re}(z) > 1$.

9.1.2   Properties of Gamma Function

The following equation must be satisfied by the gamma function,

$$\Gamma(z + 1) = z\Gamma(z) \qquad (9.3)$$

integrating by parts we have,

$$\Gamma(z + 1) = \int_0^\infty e^{-t} t^z dt$$

$$= \left[-e^{-t} t^z\right]_{t=0}^{t=\infty} + z \int_0^\infty e^{-t} t^{z-1} dt$$

$$= z\Gamma(z)$$

Now, $\Gamma(1) = 1$, and using Eq. (9.3) we have for $z = 1, 2, 3, \ldots$:

$$\Gamma(2) = 1 \cdot \Gamma(1) = 1 = 1!,$$
$$\Gamma(3) = 2 \cdot \Gamma(2) = 2 \cdot 1! = 2!,$$
$$\Gamma(4) = 3 \cdot \Gamma(3) = 3 \cdot 2! = 3!,$$
$$\cdots\cdots \quad \cdots \quad \cdots$$
$$\Gamma(n + 1) = n \cdot \Gamma(n) = n \cdot (n - 1)! = n!$$

9.1.3   An Important Representation of the Gamma Function

In the limit we have,

$$\Gamma(z) = \lim_{n \to \infty} \frac{n! n^z}{z(z + 1) \ldots (z + n)} \qquad (9.4)$$

$\mathrm{Re}(z) > 0$.

9.2 BETA FUNCTION

The beta function $B(z, w)$ is defined by

$$B(z, w) = \int_0^1 T^{z-1}(1 - T)^{w-1}dT, \quad (\text{Re}(z) > 0, \quad \text{Re}(w) > 0) \quad (9.5)$$

The connection between the gamma function defined by Eq. (9.1) and the beta function Eq. (9.5) is observed using the Laplace transform. We have the following,

$$h_{z,w}(t) = \int_0^t T^{z-1}(1 - T)^{w-1}dT \quad (9.6)$$

Now, $h_{z,w}(1)$ is a convolution of the functions t^{z-1} and t^{w-1} and $h_{z,w}(1) = B(z, w)$.

Recall that the Laplace transform of a convolution of two functions is equal to the product of their Laplace transforms, we obtain:

$$H_{z,w}(s) = \frac{\Gamma(z)}{s^z} \cdot \frac{\Gamma(w)}{s^w} = \frac{\Gamma(z)\Gamma(w)}{s^{z+w}} \quad (9.7)$$

where $H_{z,w}(s)$ is the Laplace transform of the function $h_{z,w}(t)$.

Since $\Gamma(z)\Gamma(w)$ is a constant, we get the function $h_{z,w}(t)$ by taking the inverse Laplace transform of the right-hand side of (9.7). We obtain,

$$h_{z,w}(t) = \frac{\Gamma(z)\Gamma(w)}{\Gamma(z + w)}t^{z+w-1} \quad (9.8)$$

and letting $t = 1$ we obtain the following for the beta function:

$$B(z, w) = \frac{\Gamma(z)\Gamma(w)}{\Gamma(z + w)} \quad (9.9)$$

and

$$B(z, w) = B(w, z) \quad (9.10)$$

The beta function Eq. (9.5) is valid only for $\text{Re}(z) > 0$, $\text{Re}(w) > 0$. Equation (9.9) gives the analytical continuation in the complex plane of the beta function.

Using the beta function we can establish the following two properties of gamma functions.

$$\Gamma(z)\Gamma(1 - z) = \frac{\pi}{\sin(\pi z)} \quad (9.11)$$

The Legendre duplication formula,

$$\Gamma(z)\Gamma(z + \frac{1}{2}) = \sqrt{\pi}2^{1-2z}\Gamma(2z), \quad (2z \neq 0, -1, -2, ...) \tag{9.12}$$

9.3 THE MITTAG-LEFFLER FUNCTION

In terms of power series the two representations of the Mittag-Leffler function are,

$$E_A(x) = \sum_{k=0}^{\infty} \frac{x^k}{\Gamma(Ak + 1)}, \quad A > 0 \tag{9.13}$$

$$E_{A,B}(x) = \sum_{k=0}^{\infty} \frac{x^k}{\Gamma(Ak + B)}, \quad A > 0, B > 0 \tag{9.14}$$

The exponential series defined by Eq. (9.14) has a generalization in Eq. (9.13). The following relations follow,

$$E_{A,B}(x) = \frac{1}{\Gamma(B)} + xE_{A,A+B}(x) \tag{9.15}$$

and

$$E_{A,B}(x) = BE_{A,B+1}(x) + Ax\frac{d}{dx}E_{A,B+1}(x) \tag{9.16}$$

Now (9.16) shows that,

$$\frac{d}{dx}E_{A,B+1}(x) = \frac{1}{Ax}\left[E_{A,B}(x) - BE_{A,B+1}(x)\right].$$

So

$$\frac{d}{dx}E_{A,B}(x) = \frac{1}{Ax}\left[E_{A,B-1}(x) - (B - 1)E_{A,B}(x)\right] \tag{9.17}$$

Now to prove (9.15). By definition given by Eq. (9.14),

$$
\begin{aligned}
E_{A,B}(x) &= \sum_{k=0}^{\infty} \frac{x^k}{\Gamma(Ak+B)} \\
&= \sum_{k=-1}^{\infty} \frac{x^{k+1}}{\Gamma(A(k+1)+B)} \\
&= \sum_{k=-1}^{\infty} \frac{xx^k}{\Gamma(Ak+(A+B))} \\
&= \frac{1}{\Gamma(B)} + x \sum_{k=0}^{\infty} \frac{x^k}{\Gamma(Ak+(A+B))} \\
&= \frac{1}{\Gamma(B)} + x E_{A,A+B}(x).
\end{aligned}
$$

$E_{A,B}(0) = \frac{1}{\Gamma(B)}$. For certain values of A and B we have,

$$
E_{1,1}(x) = \sum_{k=0}^{\infty} \frac{x^k}{\Gamma(k+1)} = \sum_{k=0}^{\infty} \frac{x^k}{k!} = e^x \tag{9.18}
$$

$$
E_{1/2,1}(x) = \sum_{k=0}^{\infty} \frac{x^k}{\Gamma(k/2+1)} = e^{x^2} \, \text{Erfc}(-x) \tag{9.19}
$$

$$
E_{1,2}(x) = \sum_{k=0}^{\infty} \frac{x^k}{\Gamma(k+2)} = \frac{1}{x} \sum_{k=0}^{\infty} \frac{x^{k+1}}{(k+1)!} = \frac{e^x - 1}{x} \tag{9.20}
$$

9.4   RIEMANN-LIOUVILLE FRACTIONAL DERIVATIVES

Consider the integro-differential expression,

$$
{}_a D_t^p f(t) = \left(\frac{d}{dt} \right)^{m+1} \int_a^t (t-T)^{m-p} f(T) dT, \quad (m \leq p < m+1) \tag{9.21}
$$

The expression in (9.21) is the Riemann-Liouville definition of fractional derivative. By repeated integration by parts we have,

$$
{}_a D_t^p f(t) = \left(\frac{d}{dt} \right)^{m+1} \int_a^t (t-T)^{m-p} f(T) dT
$$

$$= \sum_{k=0}^{m} \frac{f^{(k)}(a)(t-a)^{-p+k}}{\Gamma(-p+k+1)}$$

$$+ \frac{1}{\Gamma(-p+m+1)} \int_{a}^{t} (t-T)^{m-p} f^{(m+1)}(T)dT$$

$$_a\mathbf{D}_t^p f(t) = {}_a\mathbf{D}_t^p f(t), \quad (m \le p < m+1) \tag{9.22}$$

It is sufficient that the integral of $f(t)$ exists. Then the integral in the previous expression exists for $t > a$ and can be differentiated $m+1$ times. Conditions on the function $f(t)$ are necessary to obtain the solution of the Abel integral equation. For the combination of integer order integration and differentiation,

9.5 INTEGER ORDER INTEGRATION-DIFFERENTIATION

Suppose that the function $f(T)$ is continuous and integrable in every finite interval (a, t); the function $f(t)$ can have an integrable singularity of order $r < 1$ at the point $T = a$:

$$\lim_{T \to a} (T-a)^r f(t) = \text{ const } (\ne 0).$$

Then,

$$f^{(-1)}(t) = \int_{a}^{t} f(T)dT \tag{9.23}$$

exists and has a finite value, namely equal to 0, as $t \to a$. For $T = a + y(t-a)$ and setting $\epsilon = t - a$,

$$\lim_{t \to a} f^{(-1)}(t) = \lim_{t \to a} \int_{a}^{t} f(T)dT$$

$$= \lim_{t \to a} (t-a) \int_{0}^{1} f(a + y(t-a))dy$$

$$= \lim_{\epsilon \to 0} \epsilon^{1-r} \int_{0}^{1} (\epsilon y)^r f(a + y\epsilon)y^{-r}dy = 0 \tag{9.24}$$

since $r < 1$. Considering the following,

$$f^{(-2)}(t) = \int_a^t dT_1 \int_a^{T_1} f(T)dT = \int_a^t f(T)dT \int_T^t dT_1$$

$$= \int_a^t (t - T)f(T)dT \tag{9.25}$$

Integration now gives the following integral of $f(T)$:

$$f^{(-3)}(t) = \int_a^t dT_1 \int_a^{T_1} dT_2 \int_a^{T_2} f(T_3)\, dT_3$$

$$= \int_a^t dT_1 \int_a^{T_1} (T_1 - T)\, f(T)dT$$

$$= \frac{1}{2} \int_a^t (t - T)^2 f(T)dT \tag{9.26}$$

By mathematical induction, it is now evident that,

$$f^{(-n)}(t) = \frac{1}{\Gamma(n)} \int_a^t (t - T)^{n-1} f(T)dT \tag{9.27}$$

For $n \geq 1$ is fixed and take integer $k \geq 0$. We obtain,

$$f^{(-k-n)}(t) = \frac{1}{\Gamma(n)} D^{-k} \int_a^t (t - T)^{n-1} f(T)dT \tag{9.28}$$

where the symbol $D^{-k}(k \geq 0)$ represents k iterated integrations. Now for a fixed $n \geq 1$ and integer $k \geq n$ the $(k - n)$-th derivative of the function $f(t)$ can be expressed as,

$$f^{(k-n)}(t) = \frac{1}{\Gamma(n)} D^k \int_a^t (t - T)^{n-1} f(T)dT \tag{9.29}$$

where the symbol $D^k(k \geq 0)$ denotes k iterated differentiations.

Both (9.28) and (9.29) can be considered as a case for (9.29), in which $n(n \geq 1)$ is fixed and D^k refers to k integrations if $k \leq 0$ and k differentiations if $k > 0$. If $k = n - 1, n - 2, \ldots$, then , (9.29) gives iterated integrals of $f(t)$; for $k = n$ it gives the function $f(t)$; for $k = n+1, n+2, n+3, \ldots$ it also gives derivatives of order $k - n = 1, 2, 3, \ldots$ of the function $f(t)$.

9.6   INTEGRALS OF ARBITRARY ORDER

Extending the idea of n-fold integration to non-integer values of n, we start with the Cauchy formula (9.27) and replace the integer n by a real value $p > 0$:

$$_a\mathbf{D}_t^{-p} f(t) = \frac{1}{\Gamma(p)} \int_a^t (t-T)^{p-1} f(T) dT \qquad (9.30)$$

In (9.27) the integer n must satisfy the condition $n \geq 1$; For the existence of the integral in (9.30) we must have $p > 0$. We have that,

$$\lim_{p \to 0} {}_a\mathbf{D}_t^{-p} f(t) = f(t) \qquad (9.31)$$

so,

$$_a\mathbf{D}_t^0 f(t) = f(t) \qquad (9.32)$$

If $f(t)$ has continuous derivatives for $t \geq 0$ then integration by parts and (9.3) gives

$$_a\mathbf{D}_t^{-p} f(t) = \frac{(t-a)^p f(a)}{\Gamma(p+1)} + \frac{1}{\Gamma(p+1)} \int_a^t (t-T)^p f'(T) dT$$

and

$$\lim_{p \to 0} {}_a\mathbf{D}_t^{-p} f(t) = f(a) + \int_a^t f'(T) dT = f(a) + (f(t) - f(a)) = f(t).$$

If $f(t)$ is only continuous for $t \geq a$, we write $_a\mathbf{D}_t^{-p} f(t)$ as,

$$_a\mathbf{D}_t^{-p} f(t) = \frac{1}{\Gamma(p)} \int_a^t (t-T)^{p-1} (f(T) - f(t)) dT + \frac{f(t)}{\Gamma(p)} \int_a^t (t-T)^{p-1} dT$$

$$= \frac{1}{\Gamma(p)} \int_a^{t-\delta} (t-T)^{p-1} (f(T) - f(t)) dT \qquad (9.33)$$

$$+ \frac{1}{\Gamma(p)} \int_{t-\delta}^t (t-T)^{p-1} (f(T) - f(t)) dT \qquad (9.34)$$

$$+ \frac{f(t)(t-a)^p}{\Gamma(p+1)} \qquad (9.35)$$

Now consider the integral (9.34). By analysis definition, since $f(t)$ is continuous, for every $\delta > 0$ there exists $\epsilon > 0$ such that

$$|f(T) - f(t)| < \epsilon.$$

The following occurs for (9.34):

$$|I_2| < \frac{\epsilon}{\Gamma(p)} \int_{t-\delta}^{t} (t-T)^{p-1} dT < \frac{\epsilon \delta^p}{\Gamma(p+1)} \qquad (9.36)$$

and since $\epsilon \to 0$ as $\delta \to 0$ we have that for all $p \geq 0$,

$$\lim_{\delta \to 0} |I_2| = 0 \qquad (9.37)$$

Given $\epsilon > 0$ we choose δ such that

$$|I_2| < \epsilon \qquad (9.38)$$

for all $p \geq 0$. For this fixed δ we obtain the following estimate for (9.33):

$$|I_1| \leq \frac{M}{\Gamma(p)} \int_{a}^{t-\delta} (t-T)^{p-1} dT \leq \frac{M}{\Gamma(p+1)} \left(\delta^p - (t-a)^p\right) \qquad (9.39)$$

and for fixed $\delta > 0$

$$\lim_{p \to 0} |I_1| = 0 \qquad (9.40)$$

Consider,

$$\left|{}_aD_t^{-p} f(t) - f(t)\right| \leq |I_1| + |I_2| + |f(t)| \cdot \left|\frac{(t-a)^p}{\Gamma(p+1)} - 1\right|$$

and considering the limits (9.37), (9.38) and (9.40) it follows that,

$$\lim_{p \to 0} \sup \left|{}_aD_t^{-p} f(t) - f(t)\right| \leq \epsilon,$$

where ϵ is arbitrarily small. Hence,

$$\lim_{p \to 0} \sup \left|{}_aD_t^{-p} f(t) - f(t)\right| = 0,$$

and (9.31) is valid if $f(t)$ is continuous for $t \geq a$. If $f(t)$ is continuous for $t \geq a$, then integration of arbitrary real order defined by (9.30) has the following property:

$${}_aD_t^{-p} \left({}_aD_t^{-q} f(t)\right) = {}_aD_t^{-p-q} f(t) \qquad (9.41)$$

thus,

$$
\begin{aligned}
{}_a\mathbf{D}_t^{-p}\left({}_a\mathbf{D}_t^{-q}f(t)\right) &= \frac{1}{\Gamma(q)} \int_a^t (t-T)^{q-1}\, {}_a\mathbf{D}_T^{-p}f(T)dT \\
&= \frac{1}{\Gamma(p)\Gamma(q)} \int_a^t (t-T)^{q-1}dT \int_a^T (T-\xi)^{p-1}f(\xi)d\xi \\
&= \frac{1}{\Gamma(p)\Gamma(q)} \int_a^t f(\xi)d\xi \int_\xi^t (t-T)^{q-1}(T-\xi)^{p-1}dT \\
&= \frac{1}{\Gamma(p+q)} \int_a^t (t-\xi)^{p+q-1}f(\xi)d\xi \\
&= {}_a\mathbf{D}_t^{-p-q}f(t)
\end{aligned}
$$

(To evaluate the integral from ξ to t use the substitution $T = \xi + \zeta(t-\xi)$ so that we can express it in terms of the beta function.) Interchanging p and q, we have

$$
{}_a\mathbf{D}_t^{-p}\left({}_a\mathbf{D}_t^{-q}f(t)\right) = {}_a\mathbf{D}_t^{-q}\left({}_a\mathbf{D}_l^{-p}f(t)\right) = {}_a\mathbf{D}_t^{-p-q}f(t) \qquad (9.42)
$$

We note that,

$$
\frac{d^m}{dt^m}\left(\frac{d^n f(t)}{dt^n}\right) = \frac{d^n}{dt^n}\left(\frac{d^m f(t)}{dt^m}\right) = \frac{d^{m+n} f(t)}{dt^{m+n}} \qquad (9.43)
$$

9.7 DERIVATIVES OF ARBITRARY ORDER

The representation (9.29) for the derivative of an integer order $k - n$ allows us to extend the idea of differentiation to non-integer order. Keeping k and replacing integer n with $A \in \mathbb{R}$ so that $k - A > 0$ gives,

$$
{}_a\mathbf{D}_t^{k-A}f(t) = \frac{1}{\Gamma(A)} \frac{d^k}{dt^k} \int_a^t (t-T)^{A-1}f(T)dT, \quad (0 < A \le 1) \quad (9.44)
$$

with restriction for A is $A > 0$, which is necessary for the convergence in (9.44). WLOG we consider $0 < A \le 1$. Denoting $p = k - A$ we can write (9.44) as,

$$
{}_a\mathbf{D}_t^{p}f(t) = \frac{1}{\Gamma(k-p)} \frac{d^k}{dt^k} \int_a^t (t-T)^{k-p-1}f(T)dT, \quad (k-1 \le p < k)
$$

$$
(9.45)
$$

or

$$_a\mathbf{D}_t^p f(t) = \frac{d^k}{dt^k}\left(_a\mathbf{D}_t^{-(k-p)}f(t)\right), \quad (k-1 \le p < k) \qquad (9.46)$$

If $p = k - 1$, then we obtain an integer-order derivative of order $k - 1$:

$$_a\mathbf{D}_t^{k-1}f(t) = \frac{d^k}{dt^k}\left(_a\mathbf{D}_t^{-(k-(k-1))}f(t)\right)$$
$$= \frac{d^k}{dt^k}\left(_a\mathbf{D}_l^{-1}f(t)\right)$$
$$= f^{(k-1)}(t).$$

Moreover, using (9.32) we see that for $p = k \ge 1$ and $t > a$

$$_a\mathbf{D}_t^p f(t) = \frac{d^k}{dt^k}\left(_a\mathbf{D}_t^0 f(t)\right) = \frac{d^k f(t)}{dt^k} = f^{(k)}(t) \qquad (9.47)$$

so that for $t > a$ the Riemann Liouville fractional derivative (9.45) of order $p = k > 1$ is identical to the conventional derivative of order k.

Some properties of the Riemann Liouville fractional derivative are: First for $p > 0$ and $t > a$,

$$_a\mathbf{D}_t^p\left(_a\mathbf{D}_t^{-p}f(t)\right) = f(t) \qquad (9.48)$$

which says that the Riemann Liouville fractional differentiation operator is a left inverse to the Riemann Liouville fractional integration operator of the same order p. To prove the property in (9.48), consider the case of integer $p = n \ge 1$:

$$_a\mathbf{D}_t^n\left(_a\mathbf{D}_t^{-n}f(t)\right) = \frac{d^n}{dt^n}\int_a^t (t-T)^{n-1}f(T)dT$$
$$= \frac{d}{dt}\int_a^t f(T)dT = f(t).$$

Letting $k - 1 \le p < k$ and incorporating the composition rule (9.42) for the Riemann-Liouville fractional integral, we can write

$$_a\mathbf{D}_l^{-k}f(t) = _a\mathbf{D}_t^{-(k-p)}\left(_a\mathbf{D}_t^{-p}f(t)\right) \qquad (9.49)$$

and, therefore,

$$_a\mathbf{D}_t^p\left(_a\mathbf{D}_t^{-p}f(t)\right) = \frac{d^k}{dt^k}\left\{_a\mathbf{D}_t^{-(k-p)}\left(_a\mathbf{D}_t^{-p}f(t)\right)\right\}$$
$$= \frac{d^k}{dt^k}\left\{_a\mathbf{D}_t^{-p}f(t)\right\} = f(t),$$

and this completes the property of (9.48). Now fractional differentiation and integration do not commute. If the fractional derivative $_a\mathbf{D}_t^p f(t), (k - 1 \leq p < k)$, is integrable, then,

$$_a\mathbf{D}_t^{-p} \left(_a\mathbf{D}_t^p f(t)\right) = f(t) - \sum_{j=1}^{k} \left[_a\mathbf{D}_t^{p-j} f(t)\right]_{t=a} \frac{(t-a)^{p-j}}{\Gamma(p-j+1)} \quad (9.50)$$

We have that,

$$_a\mathbf{D}_t^{-p} \left(_a\mathbf{D}_t^p f(t)\right) = \frac{1}{\Gamma(p)} \int_a^t (t-T)^{p-1} {}_a\mathbf{D}_T^p f(T) dT$$

$$= \frac{d}{dt} \left\{ \frac{1}{\Gamma(p+1)} \int_a^t (t-T)^p {}_a\mathbf{D}_T^p f(T) dT \right\} \quad (9.51)$$

Repeated integration by parts and using (9.42) we obtain

$$\frac{1}{\Gamma(p+1)} \int_a^t (t-T)^p {}_a\mathbf{D}_T^p f(T) dT$$

$$= \frac{1}{\Gamma(p+1)} \int_a^t (t-T)^p \frac{d^k}{dT^k} \left\{ {}_a\mathbf{D}_T^{-(k-p)} f(T) \right\} dT$$

$$= \frac{1}{\Gamma(p-k+1)} \int_a^t (t-T)^{p-k} \left\{ {}_a\mathbf{D}_T^{-(k-p)} f(T) \right\} dT$$

$$- \sum_{j=1}^{k} \left[\frac{d^{k-j}}{dt^{k-j}} \left({}_a\mathbf{D}_t^{-(k-p)} f(t) \right) \right]_{t=a} \frac{(t-a)^{p-j+1}}{\Gamma(2+p-j)}$$

$$= \frac{1}{\Gamma(p-k+1)} \int_a^t (t-T)^{p-k} \left\{ {}_a\mathbf{D}_T^{-(k-p)} f(T) \right\} dT$$

$$- \sum_{j=1}^{k} \left[_a\mathbf{D}_t^{p-j} f(t)\right]_{t=a} \frac{(t-a)^{p-j+1}}{\Gamma(2+p-j)} \quad (9.52)$$

$$= {}_a\mathbf{D}_t^{-(p-k+1)} \left({}_a\mathbf{D}_t^{-(k-p)} f(t) \right) - \sum_{j=1}^{k} \left[_a\mathbf{D}_t^{p-j} f(t)\right]_{t=a} \frac{(t-a)^{p-j+1}}{\Gamma(2+p-j)}$$

$$(9.53)$$

$$= {}_a\mathbf{D}_t^{-1} f(t) - \sum_{j=1}^{k} \left[_a\mathbf{D}_t^{p-j} f(t)\right]_{t=a} \frac{(t-a)^{p-j+1}}{\Gamma(2+p-j)} \quad (9.54)$$

The existence of all terms in Eq. (9.52) follows from the integrability of $_a\mathbf{D}_t^p f(t)$, since the fractional derivatives $_a\mathbf{D}_t^{p-j} f(t), (j = 1, 2, \ldots, k)$, are all bounded at $t = a$. Putting together Eqs. (9.51) and (9.54) completes the proof.

Now If $0 < p < 1$, then

$$_a\mathbf{D}_t^{-p} \left(_a\mathbf{D}_t^p f(t)\right) = f(t) - \left[_a\mathbf{D}_t^{p-1} f(t)\right]_{t=a} \frac{(t-a)^{p-1}}{\Gamma(p)} \tag{9.55}$$

The property (9.48) is a particular case of the general property,

$$_a\mathbf{D}_t^p \left(_a\mathbf{D}_t^{-q} f(t)\right) = {}_a\mathbf{D}_t^{p-q} f(t) \tag{9.56}$$

where we assume that $f(t)$ is continuous and if $p \geq q \geq 0$, that the derivative $_a\mathbf{D}_t^{p-q} f(t)$ exists. Two cases are considered: $q \geq p \geq 0$ and $p > q \geq 0$. If $q \geq p \geq 0$, then using the properties (9.42) and (9.48) we obtain,

$$_a\mathbf{D}_t^p \left(_a\mathbf{D}_t^{-q} f(t)\right) = {}_a\mathbf{D}_t^p \left(_a\mathbf{D}_t^{-p}{}_a\mathbf{D}_t^{-(q-p)}\right)$$

$$= {}_a\mathbf{D}_t^{-(q-p)}$$

$$= {}_a\mathbf{D}_t^{p-q} f(t)$$

Now let us consider the case $p > q \geq 0$. Denoting by m and n integers such that $0 \leq m - 1 \leq p < m$ and $0 \leq n \leq p - q < n$. Then, $n \leq m$. Finally, using the definition (9.45) and the property (9.42) we have that,

$$_a\mathbf{D}_t^p \left(_a\mathbf{D}_t^{-q} f(t)\right) = \frac{d^m}{dt^m} \left\{_a\mathbf{D}_t^{-(m-p)} \left(_a\mathbf{D}_t^{-q} f(t)\right)\right\}$$

$$= \frac{d^m}{dt^m} \left\{_a\mathbf{D}_t^{p-q-m} f(t)\right\}$$

$$= \frac{d^n}{dt^n} \left\{_a\mathbf{D}_t^{p-q-n} f(t)\right\} = {}_a\mathbf{D}_t^{p-q} f(t)$$

The property (9.50) is a particular case of the more general property

$$_a\mathbf{D}_t^{-p} \left(_a\mathbf{D}_t^q f(t)\right) = {}_a\mathbf{D}_t^{q-p} f(t) - \sum_{j=1}^{k} \left[_a\mathbf{D}_t^{q-j} f(t)\right]_{t=a} \frac{(t-a)^{p-j}}{\Gamma(1+p-j)} \tag{9.57}$$

$$(0 \leq k - 1 \leq q < k).$$

To prove the formula (9.57) we first use property (9.42) (if $q \le p$) or property 9.56 (if $q \ge p$) and then property (9.50). We arrive at the following,

$$
{}_a\mathbf{D}_t^{-p}\left({}_a\mathbf{D}_t^q f(t)\right) = {}_a\mathbf{D}_t^{q-p}\left\{{}_a\mathbf{D}_t^{-q}\left({}_a\mathbf{D}_t^q f(t)\right)\right\}
$$

$$
= {}_a\mathbf{D}_t^{q-p}\left\{f(t) - \sum_{j=1}^{k}\left[{}_a\mathbf{D}_t^{q-j}f(t)\right]_{t=a}\frac{(t-a)^{q-j}}{\Gamma(p-j+1)}\right\}
$$

$$
= {}_a\mathbf{D}_t^{q-p}f(t) - \sum_{j=1}^{k}\left[{}_a\mathbf{D}_t^{q-j}f(t)\right]_{t=a}\frac{(t-a)^{p-j}}{\Gamma(1+p-j)}
$$

where we incorporated the derivative of the power function Eq.(9.59):

$$
{}_a\mathbf{D}_t^{q-p}\left\{\frac{(t-a)^{q-j}}{\Gamma(1+q-j)}\right\} = \frac{(t-a)^{p-j}}{\Gamma(1+p-j)}.
$$

9.8 FRACTIONAL DERIVATIVE EXAMPLE

Now as an example we consider the Riemann-Liouville fractional derivative ${}_a\mathbf{D}_t^p f(t)$ of the power function

$$
f(t) = (t-a)^\mu,
$$

where μ is a real number. Let us assume that $n - 1 \le p < n$ and since by the definition of the Riemann-Liouville derivative,

$$
{}_a\mathbf{D}_t^p f(t) = \frac{d^n}{dt^n}\left({}_a\mathbf{D}_t^{-(n-p)}f(t)\right), \quad (n-1 \le p < n) \tag{9.58}
$$

Substitution into the formula (9.58) the fractional integral of order $A = n - p$ of this function, is,

$$
{}_a\mathbf{D}_t^{-A}\left((t-a)^\mu\right) = \frac{\Gamma(1+\mu)}{\Gamma(1+\mu+A)}(t-a)^{\mu+A},
$$

we have,

$$
{}_a\mathbf{D}_t^p\left((t-a)^\mu\right) = \frac{\Gamma(1+\mu)}{\Gamma(1+\mu-p)}(t-a)^{\mu-p} \tag{9.59}
$$

and the restriction for $f(t) = (t-a)^\mu$ is $v > -1$.

9.9   COMPOSITION WITH INTEGER-ORDER DERIVATIVES

Considering the n-th derivative of the Riemann Liouville fractional derivative of real order p and using the definition (9.44) of the Riemann-Liouville derivative we see that,

$$\frac{d^n}{dt^n}\left({}_a\mathbf{D}_t^{k-A}f(t)\right) = \frac{1}{\Gamma(A)}\frac{d^{n+k}}{dt^{n+k}}\int_a^t (t-T)^{A-1}f(T)dT$$
$$= {}_a\mathbf{D}_t^{n+k-A}f(t) \quad (0 < A \le 1) \tag{9.60}$$

denoting $p = k - A$ we have that

$$\frac{d^n}{dt^n}\left({}_a\mathbf{D}_t^p f(t)\right) = {}_a\mathbf{D}_t^{n+p}f(t) \tag{9.61}$$

To consider a reversed order of operations, we note that,

$$_a\mathbf{D}_t^{-n}f^{(n)}(t) = \frac{1}{(n-1)!}\int_a^t (t-T)^{n-1}f^{(n)}(T)dT$$
$$= f(t) - \sum_{j=0}^{n-1}\frac{f^{(j)}(a)(t-a)^j}{\Gamma(j+1)} \tag{9.62}$$

and

$$_a\mathbf{D}_t^p g(t) = {}_a\mathbf{D}_t^{p+n}\left({}_a\mathbf{D}_t^{-n}g(t)\right) \tag{9.63}$$

Using (9.62), (9.63) and (9.59) we obtain:

$$_a\mathbf{D}_t^p\left(\frac{d^n f(t)}{dt^n}\right) = {}_a\mathbf{D}_t^{p+n}\left({}_a\mathbf{D}_t^{-n}f^{(n)}(t)\right)$$
$$= {}_a\mathbf{D}_t^{p+n}\left(f(t) - \sum_{j=0}^{n-1}\frac{f^{(j)}(a)(t-a)^j}{\Gamma(j+1)}\right)$$
$$= {}_a\mathbf{D}_t^{p+n}f(t) - \sum_{j=0}^{n-1}\frac{f^{(j)}(a)(t-a)^{j-p-n}}{\Gamma(1+j-p-n)} \tag{9.64}$$

It is evident that the Riemann-Liouville fractional derivative operator ${}_a\mathbf{D}_t^p$ commutes with $\frac{d^n}{dt^n}$, that is,

$$\frac{d^n}{dt^n}\left({}_a\mathbf{D}_t^p f(t)\right) = {}_a\mathbf{D}_t^p\left(\frac{d^n f(t)}{dt^n}\right) = {}_a\mathbf{D}_t^{p+n}f(t) \tag{9.65}$$

only if at $t = a$ of the fractional differentiation the function $f(t)$ satisfies the following,

$$f^{(k)}(a) = 0, \quad (k = 0, 1, 2, \ldots, n - 1) \tag{9.66}$$

9.10 COMPOSITION WITH FRACTIONAL DERIVATIVES

Considering the composition of two fractional Riemann-Liouville derivative operators: $_a\mathbf{D}_t^p, (m - 1 \le p < m)$ and $_a\mathbf{D}_t^q, (n - 1 \le q < n)$. and incorporating the definition of the Riemann Liouville fractional derivative (9.46), equation (9.50) and the composition with integer order derivatives (9.61) respectively we see that,

$$_a\mathbf{D}_t^p \left(_a\mathbf{D}_t^q f(t) \right) = \frac{d^m}{dt^m} \left\{ _a\mathbf{D}_t^{-(m-p)} \left(_a\mathbf{D}_t^q f(t) \right) \right\}$$

$$= \frac{d^m}{dt^m} \left\{ _a\mathbf{D}_t^{p+q-m} f(t) - \sum_{j=1}^{n} \left[_a\mathbf{D}_t^{q-j} f(t) \right]_{t=a} \frac{(t-a)^{m-p-j}}{\Gamma(1+m-p-j)} \right\}$$

$$= _a\mathbf{D}_t^{p+q} f(t) - \sum_{j=1}^{n} \left[_a\mathbf{D}_t^{q-j} f(t) \right]_{t=a} \frac{(t-a)^{-p-j}}{\Gamma(1-p-j)} \tag{9.67}$$

Next Interchanging p and q (and therefore m and n), we have that,

$$_a\mathbf{D}_t^q \left(_a\mathbf{D}_t^p f(t) \right) = _a\mathbf{D}_t^{p+q} f(t) - \sum_{j=1}^{m} \left[_a\mathbf{D}_t^{p-j} f(t) \right]_{t=a} \frac{(t-a)^{-q-j}}{\Gamma(1-q-j)} \tag{9.68}$$

Comparing (9.67) and (9.68) shows that in general the Riemann-Liouville fractional derivative operators $_a\mathbf{D}_t^p$ and $_a\mathbf{D}_t^p$ do not commute.

$$_a\mathbf{D}_t^p \left(_a\mathbf{D}_t^q f(t) \right) = _a\mathbf{D}_t^q \left(_a\mathbf{D}_t^p f(t) \right) = _a\mathbf{D}_t^{p+q} f(t) \tag{9.69}$$

The following conditions must hold,

$$\left[_a\mathbf{D}_t^{p-j} f(t) \right]_{t=a} = 0, \quad (j = 1, 2, \ldots, m) \tag{9.70}$$

and,

$$\left[_a\mathbf{D}_t^{q-J} f(t) \right]_{t=a} = 0, \quad (j = 1, 2, \ldots, n) \tag{9.71}$$

If $f(t)$ has a sufficient number of continuous derivatives, then (9.70) is equivalent to,

$$f^{(j)}(a) = 0, \quad (j = 0, 1, 2, \ldots, m - 1) \tag{9.72}$$

and the conditions (9.71) are equivalent to

$$f^{(j)}(a) = 0, \quad (j = 0, 1, 2, \ldots, n - 1) \tag{9.73}$$

and the relationship (9.69) holds (i.e. the p-th and q-th derivatives commute) if

$$f^{(j)}(a) = 0, \quad (j = 0, 1, 2, \ldots, r - 1) \tag{9.74}$$

where $r = \max(n, m)$.

9.11 CAPUTO'S FRACTIONAL DERIVATIVE

The Caputo derivative is the most appropriate and useful fractional operator that can be used in modeling real world problems including those encompassing fluid mechanics. The Caputo derivative is of use in modeling phenomena which takes account of interactions within the past and also problems with non-local properties.

9.12 CAPUTO FRACTIONAL DIFFERENTIAL OPERATOR

In this section an alternative operator to the Riemann-Liouville operator (9.21) is considered (see Caputo [64]).

$$D_t^A f(t) = \begin{cases} \frac{1}{\Gamma(n-A)} \int_a^t \frac{f^{(n)}(T)}{(t-T)^{A+1-n}} dT, & 0 \leq n - 1 < A < n \in \mathbb{N}, \ t \geq a \\ \frac{d^n}{dt^n} f(t), & A = n \in \mathbb{N} \end{cases}$$

$$\tag{9.75}$$

is called the Caputo fractional derivative, or Caputo fractional differential operator of order A. Let $a = 0, A = 1/2, (n = 1), f(t) = t$. Then, applying formula (9.75) we get

$$D_t^{1/2} t = \frac{1}{\Gamma(1/2)} \int_0^t \frac{1}{(t - T)^{1/2}} dT.$$

We use the gamma function again and the substitution $u = (t - T)^{1/2}$ and the Caputo fractional derivative of the function $f(t) = t$ is,

$$D_t^{1/2} t = -\frac{1}{\Gamma(1 - \frac{1}{2})} \int_0^t \frac{1}{(t - T)^{1/2}} d(T)$$

$$= -\frac{1}{\sqrt{\pi}} \int_t^0 \frac{1}{\sqrt{u}} du$$

$$= \frac{1}{\sqrt{\pi}} \int_0^t \frac{1}{\sqrt{u}} du$$

$$= \frac{2}{\sqrt{\pi}} (\sqrt{t} - 0).$$

Therefore,

$$D_t^{1/2} t = \frac{2\sqrt{t}}{\sqrt{\pi}}. \tag{9.76}$$

The benefit of using the Caputo derivative is shown here, Podlubny [68]

$$D^A y(t) - \lambda y(t) = 0, \quad t > 0, \quad n - 1 < A < n$$
$$\left[D^{A-k-1} y(t) \right] t = 0 = bk, \quad k = 0, \dots, n - 1 \tag{9.77}$$

and

$$D_t^A y(t) - \lambda y(t) = 0, \quad t > 0, \quad n - 1 < A < n,$$
$$y^{(k)}(0) = b_k, \quad k = 0, \dots, n - 1 \tag{9.78}$$

The Caputo fractional derivative requires the existence of the n-th derivative of the function. Unlike the Riemann-Liouville case, for which

$$\lim_{A \to n-1} D^A f(t) = f^{(n-1)}(t),$$

for the Caputo case

$$\lim_{A \to n-1} D_t^A f(t) = f^{(n-1)}(t) - f^{(n-1)}(0)$$

9.13   MAIN PROPERTIES

Suppose $f(t)$ is continuous and integrable in every finite interval $(0, x), x \in \mathbb{R}$. Suppose, in addition, (see Podlubny [68], p. 63) that these functions may have an integrable singularity of order $r < 1$ at the point $t = 0$, i.e.,

$$\lim_{t \to 0} t^r f(t) = \text{const} \neq 0.$$

Here we have well-defined functions belonging to this class. The operator $D^n, n \in \mathbb{N}$ used is the standard integer-order differentiation operator, that is $D^n = \frac{d^n}{dt^n}$.

9.13.1 Representation of Caputo Fractional Derivative

Lemma 9.1 *Let $n - 1 < A < n, n \in \mathbb{N}, A \in \mathbb{R}$ and $f(t)$ be such that $D_t^A f(t)$ exists. Then*

$$D_t^A f(t) = J^{n-A} D^n f(t) \tag{9.79}$$

Here the Caputo fractional operator is equivalent to $(n - A)$-fold integration after n-th order differentiation. Equation (9.79) follows from (9.75) (See also Gorenflo and Mainardi [67]). The Riemann-Liouville fractional derivative is equivalent to the composition of the same operators $((n - A)$-fold integration and n-th order differentiation) but in reverse order, that is,

$$D^A f(t) = D^n J^{n-A} f(t) \tag{9.80}$$

From (9.79) and (9.80) since $J^{n-A} D^n \neq D^n J^{n-A}$ the result follows.

Proposition 9.1 *In general, the two operators, Riemann-Liouville and Caputo, are different, that is,*

$$D^A f(t) \neq D_t^A f(t)$$

There exists a class of functions for which the two operators are identical

9.13.2 Interpolation Theory

Lemma 9.2 *Let $n - 1 < A < n, n \in \mathbb{N}, A \in \mathbb{R}$ and $f(t)$ be such that $D_t^A f(t)$ exists. Then we have,*

$$\lim_{A \to n} D_t^A f(t) = f^{(n)}(t),$$
$$\lim_{A \to n-1} D_t^A f(t) = f^{(n-1)}(t) - f^{(n-1)}(0) \tag{9.81}$$

Proof 9.1 *The proof uses integration by parts (Podlubny [68], p. 79).*

$$D_t^A f(t) = \frac{1}{\Gamma(n-A)} \int_0^t \frac{f^{(n)}(T)}{(t-T)^{A+1-n}} dT$$

$$= \frac{1}{\Gamma(n-A)} \left(-f^{(n)}(T) \frac{(t-T)^{n-A}}{n-A} \Big|_{T=0}^t \right.$$

$$\left. - \int_0^t -f^{(n+1)}(T) \frac{(t-T)^{n-A}}{n-A} dT \right)$$

$$= \frac{1}{\Gamma(n-A+1)} \left(f^{(n)}(0) t^{n-A} + \int_0^t f^{(n+1)}(T)(t-T)^{n-A} dT \right)$$

Taking the limit for $A \to n$ and $A \to n-1$, respectively, we have that,

$$\lim_{A \to n} D_t^A f(t) = \left(f^{(n)}(0) + f^{(n)}(T) \right) \Big|_{T=0}^t = f^{(n)}(t)$$

and

$$\lim_{A \to n-1} D_t^A f(t) = \left(f^{(n)}(0)t + f^{(n)}(T)(t-T) \right) \Big|_{T=0}^t - \int_0^t -f^{(n)}(T) dT$$

$$= f^{(n-1)}(T) \Big|_{T=0}^t$$

$$= f^{(n-1)}(t) - f^{(n-1)}(0)$$

For the Riemann-Liouville fractional differential operator a corresponding interpolation property holds,

$$\lim_{A \to n} D^A f(t) = f^{(n)}(t),$$

$$\lim_{A \to n-1} D^A f(t) = f^{(n-1)}(t)$$

9.13.3 Linearity of Operator

Lemma 9.3 *Let $n-1 < A < n, n \in \mathbb{N}, A, \lambda \in \mathbb{C}$ and $f(t)$ and $g(t)$ are such that both $D_t^A f(t)$ and $D_t^A g(t)$ exist. The Caputo fractional derivative is a linear operator, that is,*

$$D_t^A(\lambda f(t) + g(t)) = \lambda D_t^A f(t) + D_t^A g(t) \tag{9.82}$$

Proof 9.2 *The proof follows from the definition of fractional integration and the fact that the integral and the classical integer-order derivative are linear operators.*

In a similar fashion, the Riemann-Liouville operator satisfies

$$D^A(\lambda f(t) + g(t)) = \lambda D^A f(t) + D^A g(t)$$

9.13.4 Non-Commutation Property

Lemma 9.4 *Suppose that $n - 1 < A < n, m, n \in \mathbb{N}, A \in \mathbb{R}$ and the functions $f(t)$ is such that $D_t^A f(t)$ exists. Then we have that*

$$D_t^A D^m f(t) = D_t^{A+m} f(t) \neq D^m D_t^A f(t) \qquad (9.83)$$

Corollary 9.1 *Suppose that $n - 1 < A < n, B = A - (n - 1), (0 < B < 1), n \in \mathbb{N}, A, B \in \mathbb{R}$ and $f(t)$ is such that $D_t^A f(t)$ exists. Then we have that,*

$$D_t^A f(t) = D_t^B D^{n-1} f(t)$$

Proof 9.3 *Substitution of B for A and $n - 1$ for m in (9.83) leads to,*

$$D_t^B D^{n-1} f(t) = D_t^{B+n-1} f(t) = D_t^{A-(n-1)+n-1} f(t) = D_t^A f(t).$$

To find the Caputo fractional derivative of arbitrary order A, $(n - 1 < A < n)$ of $f(t)$, it is sufficient to find the Caputo derivative of order $B = A - (n - 1)$ of the $(n - 1)$-th derivative of the function. We note that $A - (n - 1)$ is a real number between 0 and 1. Thus the behavior of the Caputo derivative of order $B \in (0, 1)$ is sufficient for finding the Caputo derivatives of any arbitrary order. In general, the Riemann-Liouville operator is non-commutative and abides by,

$$D^m D^A f(t) = D^{A+m} f(t) \neq D^A D^m f(t), \quad n-1 < A < n, m, n \in \mathbb{N}, A \in \mathbb{R}. \qquad (9.84)$$

The inequalities in Eqs. (9.83) and (9.84) become equalities under the following additional conditions (Podlubny [68], p. 81)

$$f^{(s)}(0) = 0, s = n, n + 1, \ldots, m, \quad \text{for } D_t^A \text{ and}$$
$$f^{(s)}(0) = 0, s = 0, 1, 2, \ldots, m, \quad \text{for } D^A.$$

The Caputo derivative has no restrictions on the values $f^{(s)}(0), s = 0, 1, 2 \ldots, n - 1$. For $m = 3, n = 2$ the following,

$$f(t) = a_0 + a_1 t + a_4 t^4 + a_5 t^5 + \cdots$$

satisfies the condition for Caputo but not for Riemann-Liouville derivatives.

9.14 THE LAPLACE TRANSFORM

If the function

$$F(s) = L\{f(t); s\} = \int_0^\infty e^{-st} f(t) dt, \quad s \in \mathbb{C}, \quad (9.85)$$

exists, it is called the Laplace transform of $f(t)$. The sufficient conditions for the existence of the Laplace transform are that we let $f(t)$ (the function we are taking the Laplace transform of) be (see Greenberg [66], p. 103)

  i. piecewise smooth over every finite interval in $[0, \infty)$ and

 ii. of exponential order A, i.e., there exist constants $M > 0$ and $T > 0$ such that $|f(t)| \le M e^{At}$ for all $t > T$. then the Laplace transform $L\{f(t); s\}$ of $f(t)$ exists.

The Laplace transform applies to initial value problems on semi-infinite regions.

(The inverse Laplace transform)

The function $f(t)$ can be obtained from its Laplace transform $F(s)$ using the inverse Laplace transform

$$f(t) = L^{-1}\{F(s); t\} = \frac{1}{2\pi i} \int_{c-i\infty}^{c+i\infty} e^{st} F(s) ds, \quad c = \mathrm{Re}(s) > c_0, \quad (9.86)$$

where c_0 lies in the right half plane of the absolute convergence of the Laplace integral (9.85). The integral in (9.86) is named the Bromwich integral.

Lemma 9.5 *Basic properties of the Laplace transform:*

Suppose that $f(t)$ and $g(t)$ are two functions, which are zero for $t < 0$ and for which the Laplace transforms $F(s)$ and $G(s)$ exist. The following statements hold (see Greenberg [66], pp. 105–115):

a. *The Laplace transform and its inverse are linear operators, i.e., suppose that $\lambda \in \mathbb{R}$, then*

$$L\{\lambda f(t) + g(t); s\} = \lambda L\{f(t); s\} + L\{g(t); s\} = \lambda F(s) + G(s)$$
$$L^{-1}\{\lambda F(s) + G(s); t\} = \lambda L^{-1}\{F(s); t\} + L^{-1}\{G(s); t\} = \lambda f(t) + g(t)$$
$$(9.87)$$

b. *For the Laplace transform of the convolution of $f(t)$ and $g(t)$ it follows*

$$L\{f(t) * g(t); s\} = F(s)G(s), \qquad (9.88)$$

where the convolution is,

$$f(t)tg(t) = \int_0^t f(t-T)g(T)dT = \int_0^t f(T)g(t-T)dT$$

c. *The limit of the function $sF(s)$ for $s \to \infty$ is given by*

$$\lim_{s \to \infty} sF(s) = f(0). \qquad (9.89)$$

d. *The Laplace transform of the $n-th$ derivative $(n \in \mathbb{N})$ of $f(t)$ is,*

$$L\left\{f^{(n)}(t); s\right\} = s^n F(s) - \sum_{k=0}^{n-1} s^{n-k-1} f^{(k)}(0) = s^n F(s) - \sum_{k=0}^{n-1} s^k f^{(n-k-1)}(0).$$
$$(9.90)$$

Lemma 9.6 *Laplace transforms of fractional operators:*
Suppose that $p > 0$ and $F(s)$ is the Laplace transform of $f(t)$. Then the following statements hold (Podlubny [68], pp. 104–105, p. 21):

a. *The Laplace transform of the fractional integral of order A is,*

$$L\left\{J^A f(t); s\right\} = s^{-A} F(s). \qquad (9.91)$$

b. *The Laplace transform of the Riemann-Liouville fractional differential operator of order A is,*

$$L\left\{D^A f(t); s\right\} = s^A F(s) - \sum_{k=0}^{n-1} s^k \left[D^{A-k-1} f(t)\right]_{t=0}$$

$$= s^A F(s) - \sum_{k=0}^{n-1} s^{n-k-1} \left[D^k J^{n-A} f(t)\right]_{t=0}, \quad n-1 < A < n.$$
$$(9.92)$$

    c. Let $A, B, \lambda \in \mathbb{R}, A, B > 0, p \in \mathbb{N}$. Then the Laplace transform of the two-parameter function of Mittag-Leffler type is given by

$$L\left\{t^{Ap+B-1}E_{A,B}^{(p)}\left(\pm\lambda t^A\right);s\right\} = \frac{p!s^{A-B}}{(s^A \mp \lambda)^{p+1}}, \quad Re(s) > |\lambda|^{1/A}.$$

$$(9.93)$$

Now we take the Laplace transform of the Caputo fractional derivative operator of $f(t)$. We prove the following,

Theorem 9.1 *Suppose that $p > 0$ and $F(s)$ is the Laplace transform of $f(t)$. Then the Laplace transform of the Caputo fractional differential operator of order A is,*

$$L\left\{D_t^A f(t); s\right\} = s^A F(s) - \sum_{k=0}^{n-1} s^{A-k-1} f^{(k)}(0), \quad n-1 < A < n \quad (9.94)$$

Proof 9.4 *To prove (9.94) we consider equation (9.76)*

$$D_t^A f(t) = J^{n-A} D^n f(t).$$

Let $g(t) = D^n f(t)$. Then (9.76) becomes

$$D_t^A f(t) = J^{n-A} g(t).$$

$$(9.95)$$

Using the Laplace transform of the fractional integral (9.91) of order $n - A$ of $g(t)$ and equation (9.95) (Podlubny [68], p. 106), the Laplace transform of the Caputo fractional operator is written as,

$$L\left\{D_t^A f(t); s\right\} = L\left\{J^{n-A} g(t); s\right\} = s^{-(n-A)} G(s) \quad (9.96)$$

where $G(s) = L\{g(t); s\}$ can be expressed using (9.90) as follows,

$$G(s) = s^n F(s) - \sum_{k=0}^{n-1} s^{n-k-1} f^{(k)}(0) \quad (9.97)$$

Finally, substituting (9.97) into (9.92),

$$L\left\{D_t^A f(t); s\right\} = s^{-(n-A)}\left(s^n F(s) - \sum_{k=0}^{n-1} s^{n-k-1} f^{(k)}(0)\right)$$

$$= s^A F(s) - \sum_{k=0}^{n-1} s^{A-k-1} f^{(k)}(0)$$

is proven

Suppose that both Laplace transforms of the Riemann-Liouville and of the Caputo fractional derivatives exist for the function $f(t)$. For the Riemann-Liouville equation (9.96) initial values of the fractional integral J^{n-A} and of its integer derivatives of order $k = 1, 2, \ldots, n-1$ are required (see also Gorenflo and Mainardi [67]). For Caputo case in equation (9.98), the initial values of the function and its integer derivatives of order $k = 1, 2, \ldots, n-1$ are only required. The Laplace transform of the Caputo fractional derivative (9.98) is a generalization of the Laplace transform of the integer-order derivative (9.94) where n is replaced by A. The same is not true for the Riemann-Liouville case.

9.15 CAPUTO VERSUS RIEMANN-LIOUVILLE OPERATOR

Recall that the Riemann-Liouville and the Caputo fractional differential operators do not coincide.

Let $f(t)$ be a function for which both $D^A f(t)$ and $D_t^A f(t)$ exist and $n - 1 < A < n \in \mathbb{N}$. Then,

$$D^A f(t) \neq D_t^A f(t).$$

They can be represented as a composition of the same operators but in a reverse order. Consider $n - 1 < A < n, n \in \mathbb{N}$, in the interpolation property there are also some differences for the values of the parameter $A \to n - 1$, although for $A \to n$ the same is true.

Proposition 9.2 *Let $n - 1 < A < n$. Then it holds*

$$\lim_{A \to \pi} D^A f(t) = \lim_{A \to n} D_t^A f(t) = f^{(n)}(t)$$

Concerning the commutation for functions $f(t)$ such that $f^{(s)}(0) = 0, s = 0, 1, 2 \ldots, m$ each of the two fractional derivatives commutes with the mth order derivative ($m \in \mathbb{N}$), that is,

$$D^m D^A f(t) = D^{A+m} f(t) = D^A D^m f(t)$$

and

$$D_t^A D^m f(t) = D_t^{A+m} f(t) = D^m D_t^A f(t).$$

Next,

Proposition 9.3 *Let $f(t)$ be such that $f^{(s)}(0) = 0, s = 0, 1, 2 \ldots, n - 1$. Then the Riemann-Liouville and the Caputo fractional derivatives coincide, i.e.,*

$$D_t^A f(t) = D^A f(t)$$

For both cases derivatives of order B in the interval $(0, 1)$ can only be considered, since (recall formulas (9.83) and (9.84) for $n - 1 < A < n$

$$D_t^A f(t) = D_t^{A-(n-1)} D^{n-1} f(t)$$
$$D^A f(t) = D^{n-1} D^{A-(n-1)} f(t)$$

where $B = A - (n - 1) \in (0, 1)$ and D^{n-1} is the classical integer-order derivative.

9.16 THE CONSTANT FUNCTION

For Caputo case, we have that,

$$D_t^A c = 0, \quad c = \text{const}$$

whereas for Riemann-Liouville

$$D^A c = \frac{c}{\Gamma(1 - A)} t^{-A} \neq 0, \quad c = \text{const}$$

9.17 CONNECTION WITH THE RIEMANN-LIOUVILLE OPERATOR

We consider the following theorem,

Theorem 9.2 *Let $t > 0, A \in \mathbb{R}, n - 1 < A < n \in \mathbb{N}$. Then the following relation between the Riemann-Liouville and the Caputo Eq. operators holds,*

$$D_t^A f(t) = D^A f(t) - \sum_{k=0}^{n-1} \frac{t^{k-A}}{\Gamma(k + 1 - A)} f^{(k)}(0) \qquad (9.98)$$

Proof 9.5 *Taking the Taylor series about the point 0 gives,*

$$f(t) = f(0) + t f'(0) + \frac{t^2}{2!} f''(0) + \frac{t^3}{3!} f'''(0) + \cdots + \frac{t^{n-1}}{(n-1)!} f^{(n-1)}(0)$$

$$+ R_{n-1} = \sum_{k=0}^{n-1} \frac{t^k}{\Gamma(k+1)} f^{(k)}(0) + R_{n-1}$$

where

$$R_{n-1} = \int_0^t \frac{f^{(n)}(T)(t-T)^{n-1}}{(n-1)!} dT = \frac{1}{\Gamma(n)}$$

$$\int_0^t f^{(n)}(T)(t-T)^{n-1} dT = J^n f^{(n)}(t)$$

Now, using the linearity property of the Riemann-Liouville fractional derivative, the Riemann-Liouville fractional derivative of the power function,

$$D^A f(t) = D^A \left(\sum_{k=0}^{n-1} \frac{t^k}{\Gamma(k+1)} f^{(k)}(0) + R_{n-1} \right)$$

$$= \sum_{k=0}^{n-1} \frac{D^A t^k}{\Gamma(k+1)} f^{(k)}(0) + D^A R_{n-1}$$

$$= \sum_{k=0}^{n-1} \frac{\Gamma(k+1)}{\Gamma(k-A+1)} \frac{t^{k-A}}{\Gamma(k+1)} f^{(k)}(0) + D^A J^n f^{(n)}(t)$$

$$= \sum_{k=0}^{n-1} \frac{t^{k-A}}{\Gamma(k-A+1)} f^{(k)}(0) + J^{n-A} f^{(n)}(t)$$

$$= \sum_{k=0}^{n-1} \frac{t^{k-A}}{\Gamma(k-A+1)} f^{(k)}(0) + D_t^A f(t).$$

Therefore,

$$D_t^A f(t) = D^A f(t) - \sum_{k=0}^{n-1} \frac{t^{k-A}}{\Gamma(k+1-A)} f^{(k)}(0).$$

This conclusively shows that the Caputo and the Riemann-Liouville fractional operator coincide iff $f(t)$ together with its first $n-1$ derivatives vanish at $t = 0$. The following relation between the two fractional derivatives holds,

$$D_t^A f(t) = D^A \left(f(t) - \sum_{k=0}^{n-1} \frac{t^k}{k!} f^{(k)}(0) \right).$$

Proof 9.6 *This formula is proved using relation (9.98), the Riemann-Liouville fractional derivative of the power function and the linearity* *property of the Riemann-Liouville operator respectively, that is,*

$$D_t^A f(t) = D^A f(t) - \sum_{k=0}^{n-1} \frac{t^{k-A}}{\Gamma(k+1-A)} f^{(k)}(0)$$

$$= D^A f(t) - \sum_{k=0}^{n-1} \frac{D^A t^k}{\Gamma(k+1)} f^{(k)}(0)$$

$$= D^A \left(f(t) - \sum_{k=0}^{n-1} \frac{t^k}{k!} f^{(k)}(0) \right).$$

(Leibniz Rule) Let $t > 0, A \in \mathbb{R}, n - 1 < A < n \in \mathbb{N}$. If $f(T)$ and $g(T)$ and all its derivatives are continuous in $[0, t]$ then the following holds

$$D_t^A(f(t)g(t)) = \sum_{k=0}^{\infty} \binom{A}{k} \left(D^{A-k} f(t) \right) g^{(k)}(t)$$

$$- \sum_{k=0}^{n-1} \frac{t^{k-A}}{\Gamma(k+1-A)} \left((f(t)g(t))^{(k)}(0) \right). \tag{9.99}$$

Proof 9.7 *Applying consecutively (9.98 and the Leibniz Rule for Riemann-Liouville derivatives (Podlubny [68], p. 96)*

$$D^A(f(t)g(t)) = \sum_{k=0}^{\infty} \binom{A}{k} \left(D^{A-k} f(t) \right) g^{(k)}(t),$$

then the Leibniz rule for the Caputo derivative is,

$$D_t^A(f(t)g(t)) = D^A(f(t)g(t)) - \sum_{k=0}^{n-1} \frac{t^{k-A}}{\Gamma(k+1-A)} \left((f(t)g(t))^{(k)}(0) \right)$$

$$= \sum_{k=0}^{\infty} \binom{A}{k} \left(D^{A-k} f(t) \right) g^{(k)}(t)$$

$$- \sum_{k=0}^{n-1} \frac{t^{k-A}}{\Gamma(k+1-A)} \left((f(t)g(t))^{(k)}(0) \right).$$

9.18 EXAMPLES OF FRACTIONAL DERIVATIVES

9.19 THE CONSTANT FUNCTION

For the Riemann-Liouville operator, it is true that

$$D^A c = \frac{c}{\Gamma(1-A)} t^{-A} \neq 0, \quad c = \text{const.}$$

To see the importance of using Caputo operator over Riemann-Liouville operator is due to the following,

Lemma 9.7 *For the Caputo fractional derivative,*

$$D_t^A c = 0, \quad c = const.$$

Proof 9.8 *We have that $0 < n-1 < A < n, n \in \mathbb{N}$, and $n \geq 1$. Applying the definition of the Caputo derivative and since the n-th derivative $c^{(n)}(n \in N, n \geq 1)$ is of a constant equals 0, it follows*

$$D_t^A c = \frac{1}{\Gamma(n-A)} \int_a^t \frac{c^{(n)}}{(t-T)^{A+1-n}} dT = 0$$

9.20 THE POWER FUNCTION

Consider the following Taylor series,

$$f(t) = f(0) + f'(0)t + \frac{f''(0)}{2!} t^2 + \frac{f'''(0)}{3!} t^3 + \cdots.$$

Now the Caputo fractional derivative is linear. So then if $D_t^A t^p$ is known, then the Caputo fractional derivative for an arbitrary function can be represented as,

$$D_t^A f(t) = D_t^A \sum_{k=0}^{\infty} \frac{f^{(k)}(0)}{k!} t^k = \sum_{k=0}^{\infty} \frac{f^{(k)}(0)}{k!} D_t^A t^k \qquad (9.100)$$

Theorem 9.3 *The Riemann-Liouville fractional derivative of the power function satisfies*

$$D^A t^p = \frac{\Gamma(p+1)}{\Gamma(p-A+1)} t^{p-A}, \quad n-1 < A < n, p > -1, \quad p \in \mathbb{R}.$$

$$(9.101)$$

Theorem 9.4 *The Caputo fractional derivative of the power function satisfies*

$$D_t^A t^p = \begin{cases} \frac{\Gamma(p+1)}{\Gamma(p-A+1)} t^{p-A} = D^A t^p, & n-1 < A < n, p > n-1, p \in \mathbb{R}, \\ \\ 0, & n-1 < A < n, p \le n-1, p \in \mathbb{N}. \end{cases}$$

$$(9.102)$$

For $p > n - 1$ the Caputo fractional derivative of the power function (9.102) is a generalization of the integer-order derivative of the power function. Recall,

$$(t^p)^{(n)} = \left(pt^{p-1}\right)^{(n-1)} = \left(p(p-1)t^{p-2}\right)^{(n-2)} = p(p-1)\ldots(p-n+1)t^{p-n}$$

$$= \frac{\Gamma(p+1)}{\Gamma(p-n+1)} t^{p-n}, n \in \mathbb{N}, p \in \mathbb{R}$$

For $n = 1$, i.e., $0 < A < 1$, $D_t^A t^p = D^A t^p, p > 0, p \in \mathbb{R}$. The Caputo fractional derivative for any function $f(t)$ can be computed by the formula

$$D_t^A f(t) = \sum_{k=n}^{\infty} \frac{f^{(k)}(0)}{\Gamma(k-A+1)} t^{k-A}$$

Proof 9.9 *Considering (9.100) and (9.102) the following equalities hold*

$$D_t^A f(t) = \sum_{k=0}^{\infty} \frac{f^{(k)}(0)}{k!} D_t^A t^k$$

$$= \sum_{k=n}^{\infty} \frac{f^{(k)}(0)}{k!} \frac{\Gamma(k+1)}{\Gamma(k-A+1)} t^{k-A}$$

$$= \sum_{k=n}^{\infty} \frac{f^{(k)}(0)}{\Gamma(k-A+1)} t^{k-A}.$$

Suppose that $A \in \mathbb{R}$ but $A \notin \mathbb{N}$ (A is the order of differentiation). We consider fractional derivatives of the functions t^2 and t, i.e., $p = 2$ and $p = 1$ respectively. More examples are given in Appendix A. $p = 2$ The function t^2 is considered here. Suppose that $0 < A < 1$. The fractional derivative is given as follows,

$$D_t^A t^2 = \frac{\Gamma(2+1)}{\Gamma(2-A+1)} t^{2-A} = \frac{2}{\Gamma(3-A)} t^{2-A}, n-1 < A < n < 3.$$

9.21 THE EXPONENTIAL FUNCTION

We now consider the exponential function $e^{\lambda t}$. For the Caputo operator,

Theorem 9.5 *Let $A \in \mathbb{R}, n - 1 < A < n, n \in \mathbb{N}, \lambda \in \mathbb{C}$. The the following form is considered,*

$$D_t^A e^{\lambda t} = \sum_{k=0}^{\infty} \frac{\lambda^{k+n} t^{k+n-A}}{\Gamma(k+1+n-A)} = \lambda^n t^{n-A} E_{1,n-A+1}(\lambda t), \quad (9.103)$$

where $E_{A,B}(z)$ is the two-parameter function of Mittag-Leffler type.

Proof 9.10 *To prove the theorem of the relation between Caputo and Riemann-Liouville fractional derivatives (9.98) together with the Riemann-Liouville fractional derivative of the exponential function, we have that,*

$$D^A e^{\lambda t} = t^{-A} E_{1,1-A}(\lambda t)$$

Then for the Caputo derivative,

$$D_t^A e^{\lambda t} = D^A e^{\lambda t} - \sum_{k=0}^{n-1} \frac{t^{k-A}}{\Gamma(k+1-A)} \left(e^{\lambda t} \right)^{(k)}(0)$$

$$= t^{-A} E_{1,1-A}(\lambda t) - \sum_{k=0}^{n-1} \frac{t^{k-A}}{\Gamma(k+1-A)} \cdot \lambda^k$$

$$= \sum_{k=0}^{\infty} \frac{(\lambda t)^k t^{-A}}{\Gamma(k+1-A)} - \sum_{k=0}^{n-1} \frac{\lambda^k t^{k-A}}{\Gamma(k+1-A)}$$

$$= \sum_{k=n}^{\infty} \frac{\lambda^k t^{k-A}}{\Gamma(k+1-A)}$$

$$= \sum_{k=0}^{\infty} \frac{\lambda^{k+n} t^{k+n-A}}{\Gamma(k+n+1-A)}$$

$$= \lambda^n t^{n-A} E_{1,n-A+1}(\lambda t)$$

9.22 SOME OTHER FUNCTIONS

The Caputo derivative of sin and cos trigonometric functions are considered. Similar representations are given by Diethelm, Ford, Freed, and Luchko [65], which are mentioned here,

Theorem 9.6 *Let $\lambda \in \mathbb{C}, A \in \mathbb{R}, n \in \mathbb{N}, n - 1 < A < n$. Then*

$$D_t^A \sin \lambda t = -\frac{1}{2} i (i\lambda)^n t^{n-A} \left(E_{1,n-A+1}(i\lambda t) - (-1)^n E_{1,n-A+1}(-i\lambda t) \right). \tag{9.104}$$

The following representation of the sine function is used

$$\sin z = \frac{e^{iz} - e^{-iz}}{2i}, \quad z \in \mathbb{C}.$$

Now, using the linearity property of the Caputo fractional derivative and formula (9.104) for the exponential function it follows that,

$$
\begin{aligned}
D_t^A \sin \lambda t &= D_t^A \frac{e^{i\lambda t} - e^{-i\lambda t}}{2i} \\
&= \frac{1}{2i} \left(D_t^A e^{i\lambda t} - D_t^A e^{-i\lambda t} \right) \\
&= \frac{1}{2i} \left((i\lambda)^n t^{n-A} E_{1,n-A+1}(i\lambda t) - (-i\lambda)^n t^{n-A} E_{1,n-A+1}(-i\lambda t) \right) \\
&= -\frac{1}{2} i (i\lambda)^n t^{n-A} \left(E_{1,n-A+1}(i\lambda t) - (-1)^n E_{1,n-A+1}(-i\lambda t) \right).
\end{aligned}
$$

The corresponding case for the cosine function is,

$$\cos z = \frac{e^{iz} + e^{-iz}}{2}, \quad z \in \mathbb{C}.$$

We have the following theorem,

Theorem 9.7 *Let $\lambda \in \mathbb{C}, A \in \mathbb{R}, n \in \mathbb{N}, n - 1 < A < n$. Then*

$$D_t^A \cos \lambda t = \frac{1}{2} (i\lambda)^n t^{n-A} \left(E_{1,n-A+1}(i\lambda t) + (-1)^n E_{1,n-A+1}(-i\lambda t) \right). \tag{9.105}$$

Let $\lambda, A \in \mathbb{R}, n \in \mathbb{N}, n - 1 < A < n$. Then $\vec{(x)}$

$$D_t^A \sin \lambda t = \begin{cases} -\frac{i\lambda^n (-1)^{n/2} t^{n-A}}{2\Gamma(n-A+1)} \left[{}_1F_1(1, n - A + 1; i\lambda t) - {}_1F_1(1, n - A + 1; -i\lambda t) \right], \\ \qquad\qquad\qquad\qquad\qquad\qquad\qquad\qquad \text{if } n \text{ is even,} \\ \frac{\lambda^n (-1)^{(n-1)/2} t^{n-A}}{2\Gamma(n-A+1)} \left[{}_1F_1(1, n - A + 1; i\lambda t) + {}_1F_1(1, n - A + 1; -i\lambda t) \right], \\ \qquad\qquad\qquad\qquad\qquad\qquad\qquad\qquad\qquad \text{if } n \text{ is odd.} \end{cases} \tag{9.106}$$

and

$$
D_t^A \cos \lambda t =
\begin{cases}
\frac{\lambda^n (-1)^{n/2} t^{n-A}}{2\Gamma(n-A+1)} \left[{}_1F_1(1, n-A+1; i\lambda t) + {}_1F_1(1, n-A+1; -i\lambda t) \right], \\
\qquad\qquad\qquad\qquad\qquad\qquad \text{if } n \text{ is even}, \\
\frac{i\lambda^n (-1)^{(n-1)/2} t^{n-A}}{2\Gamma(n-A+1)} \left[{}_1F_1(1, n-A+1; i\lambda t) - {}_1F_1(1, n-A+1; -i\lambda t) \right], \\
\qquad\qquad\qquad\qquad\qquad\qquad \text{if } n \text{ is odd}.
\end{cases}
$$

$$(9.107)$$

9.23 FRACTIONAL TIME AND MULTI-FRACTIONAL SPACE INCOMPRESSIBLE AND COMPRESSIBLE NAVIER–STOKES EQUATIONS

Fractional flow and transport models show superiority in describing anomalous diffusion, intermittent turbulence, chaos-induced turbulence diffusion and multifractal behavior of velocity fields of turbulent fluids at low viscosity. When the fractional powers in time and in multi-fractional space are unit integer values, the fractional equations of continuity and momentum for incompressible and compressible fluid flow reduce to the classical Navier–Stokes equations. Recall the Caputo fractional derivative in Eq. (9.75). For $0 < A \leq 1$, a first-order approximation of Caputo's fractional time derivative over a given time interval $[0, T]$, which is divided into N equal subintervals of increment $dt = T/N$ by using nodes $t_n = ndt, n = 0, 1, 2, \ldots, N$, can be written as,

$$
D_t^A f^n = \frac{1}{\Gamma(2-A)} \frac{1}{dt^A} \sum_{j=1}^{n} w^{(A)} \left(f^{n-j+1} - f^{n-j} \right)
\tag{9.108}
$$

where $f^n = f(t_n)$ and the weight $w_j^A = j^{1-A} - (j-1)^{1-A}$.

9.24 CONTINUITY EQUATION OF UNSTEADY FLUID FLOW IN FRACTIONAL TIME AND MULTI-FRACTIONAL SPACE

This section's derivations of the Continuity equation are due to [61] and are mostly repeated here for completeness. The full version of the Navier–Stokes equations including continuity equation in the fractional calculus setting is discussed in [62] on a compactness argument and existence of solutions to fractional time NS equations. For β-order the Caputo fractional derivative $aD_x^\beta f(x)$ of a function $f(x)$ may be defined as (references)

$$aD_x^\beta f(x) = \frac{1}{\Gamma(1-\beta)} \int_a^x \frac{f'(\xi)}{(x-\xi)^\beta} d\xi, \quad 0 < \beta < 1, \quad x \geq a \quad (9.109)$$

where ξ is a dummy variable. A β_i-order($i = 1,2,3$) approximation to a function $f(x_i)$ around $x_i - \triangle x_i$ exists in the following form,

$$f(x_i) = f(x_i - \triangle x_i) + \frac{(\triangle x_i)^{\beta_1}}{\Gamma(\beta_i + 1)} x_i - \triangle x_i D_{x_i}^{\beta_i} f(x_i), \quad i = 1,2,3. \quad (9.110)$$

In the previous equation, for $f(x_i) = x_i$ an analytical relationship between $\triangle x_i$ and $(\triangle x_i)^\beta$ $(i = 1,2,3.)$ that will be applicable throughout the modeling domain of function. This is possible when the lower limit in the Caputo derivative is taken as zero, that is, $\triangle x_i = x_i$. Now the net mass outflow rate (NMOR) from the control volume is expressed as,

$$\text{NMOR} = \left[\rho u_1(x_1, x_2, x_3; t) - \rho u_1(x_1 - \triangle x_1, x_2, x_3; t) \right] \triangle x_2 \triangle x_3 + \\ \left[\rho u_2(x_1, x_2, x_3; t) - \rho u_2(x_1, x_2 - \triangle x_2, x_3; t) \right] \triangle x_1 \triangle x_3 + \\ \left[\rho u_3(x_1, x_2, x_3; t) - \rho u_3(x_1, x_2, x_3 - \triangle x_3; t) \right] \triangle x_1 \triangle x_2. \quad (9.111)$$

Substituting Eq. (9.110) into Eq. (9.111) with $\triangle x_i = x_i$, $i = 1,2,3$. and expressing the resulting Caputo derivative $0D_x^\beta f(x)$, where we take $\triangle x = x$ and make the lower limit in the Caputo derivative 0, by the expression $\frac{\partial^\beta f(x)}{\partial x}^\beta$, the net mass flux (NMF) from the control volume can be expressed to β_i-order, $i = 1,2,3$ as,

$$NMF = \frac{(\triangle x_1)^{\beta_1}}{\Gamma(\beta_1 + 1)} \left(\frac{\partial}{\partial x_1} \right)^{\beta_1} (\rho u_1(\vec{x}; t)) \triangle x_2 \triangle x_3 + \\ \frac{(\triangle x_2)^{\beta_2}}{\Gamma(\beta_2 + 1)} \left(\frac{\partial}{\partial x_2} \right)^{\beta_2} (\rho u_2(\vec{x}; t)) \triangle x_1 \triangle x_3 + \quad (9.112) \\ \frac{(\triangle x_3)^{\beta_3}}{\Gamma(\beta_3 + 1)} \left(\frac{\partial}{\partial x_3} \right)^{\beta_3} (\rho u_3(\vec{x}; t)) \triangle x_1 \triangle x_2$$

where $\vec{x} = (x_1, x_2, x_3)$, ρ is the density of fluid and $u_i(\vec{x}; t)$ is the component of the flow velocity vector in the x_i direction for $i = 1,2,3$. Now it is also true that from Eq. (9.110) substituting $f(x_i) = x_i$ that,

$$(\triangle x_i)_i^\beta = \frac{\Gamma(\beta_i + 1)\Gamma(2 - \beta_i)}{x_i^{1-\beta_i}} (\triangle x_i), \quad i = 1,2,3 \quad (9.113)$$

with respect to β_i-order fractional space in the i-th direction, $i = 1,2,3$.

Introducing Eq. (9.113) into Eq. (9.112) yields for the NMOR through the control volume,

$$
\begin{aligned}
\text{NMOR} = {} & \frac{\Gamma(2-\beta_1)}{x_1^{1-\beta_1}} \left(\frac{\partial}{\partial x_1}\right)^{\beta_1} (\rho u_1(\vec{x};t)) \, \triangle x_1 \triangle x_2 \triangle x_3 + \\
& \frac{\Gamma(2-\beta_2)}{x_2^{1-\beta_2}} \left(\frac{\partial}{\partial x_2}\right)^{\beta_2} (\rho u_2(\vec{x};t)) \, \triangle x_1 \triangle x_2 \triangle x_3 + \\
& \frac{\Gamma(2-\beta_3)}{x_3^{1-\beta_3}} \left(\frac{\partial}{\partial x_3}\right)^{\beta_3} (\rho u_3(\vec{x};t)) \, \triangle x_1 \triangle x_2 \triangle x_3, \quad \vec{x} = (x_1, x_2, x_3)
\end{aligned}
\tag{9.114}
$$

to β_i- order, $i = 1, 2, 3$.

Now we look at the time rates of change. The time rate of change of mass within the control volume may be expressed as,

$$
\frac{\rho(\vec{x}, t) - \rho(\vec{x}, t - \triangle t)}{\triangle t} \triangle x_1 \triangle x_2 \triangle x_3
\tag{9.115}
$$

Substituting Eq. (9.110) into Eq. (9.115) and replacing the fractional power β_i with α, and x by t, and expressing the resulting Caputo derivative operator with its lower limit as 0, by $\frac{\partial^\alpha}{(\partial t)^\alpha}$, yields the following. The time rate of change of mass within the control volume with respect to α-fractional time increments is,

$$
\frac{(\triangle t)^\alpha}{\triangle t \Gamma(\alpha + 1)} \left(\frac{\partial}{\partial t}\right)^\alpha \rho(\vec{x}, t) \triangle x_1 \triangle x_2 \triangle x_3
\tag{9.116}
$$

to $\alpha-$ order. Now we calculate with the Caputo derivative, $0 D_t^\alpha t = \frac{\partial^\alpha t}{(\partial t)^\alpha}$ which is,

$$
\frac{\partial^\alpha t}{(\partial t)^\alpha} = \frac{t^{1-\alpha}}{\Gamma(2 - \alpha)}
\tag{9.117}
$$

Combining with Eq. (9.110) with x replaced by t and β_i by α gives the approximation,

$$
(\triangle t)^\alpha = \frac{\Gamma(\alpha + 1)\Gamma(2 - \alpha)}{t^{1-\alpha}} (\triangle t)
\tag{9.118}
$$

to α- order. Substituting Eq. (9.118) into Eq. (9.110) yields for the time rate of change of mass expression, TRCM, with respect to α-order fractional time increments:

$$
TRCM = \frac{\Gamma(2 - \alpha)}{t^{1-\alpha}} \frac{\partial^\alpha \rho(\vec{x}, t)}{(\partial t)^\alpha} \triangle x_1 \triangle x_2 \triangle x_3
\tag{9.119}
$$

Equating $| TRCM |$ and $| NMOR |$ and furthermore the following inverse relationship is true,

$$\frac{\Gamma(2-\alpha)}{t^{1-\alpha}} \frac{\partial^\alpha \rho(\vec{x},t)}{(\partial t)^\alpha} =$$
$$-\sum_{i=1}^3 \frac{\Gamma(2-\beta_i)}{x_i^{1-\beta_i}} \left(\frac{\partial}{\partial x_i}\right)^{\beta_i} (\rho(\vec{x},t)u_i(\vec{x},t)), \quad \vec{x} = (x_1, x_2, x_3), \tag{9.120}$$

This is the general continuity equation of unsteady, multidimensional fluid flow in fractional time and multi-fractional space which holds for compressible and incompressible flows.

If ρ is a constant we have the fractional version of the continuity equation for incompressible flow.

9.25 MOMENTUM EQUATIONS OF UNSTEADY FLOW IN FRACTIONAL TIME AND MULTI-FRACTIONAL SPACE

This section's derivations are due to [61]. The net momentum flux, (NMF) through the control volume can be expressed as,

$$NMF = \left[\rho u_i u_1 \big|_{(x_1,x_2,x_3;t)} -\rho u_i u_1 \big|_{(x_1-\triangle x_1,x_2,x_3;t)} \right]\triangle x_2 \triangle x_3$$
$$+\left[\rho u_i u_2 \big|_{(x_1,x_2,x_3;t)} -\rho u_i u_2 \big|_{(x_1,x_2-\triangle x_2,x_3;t)} \right]\triangle x_1 \triangle x_3$$
$$+\left[\rho u_i u_3 \big|_{(x_1,x_2,x_3;t)} -\rho u_i u_3 \big|_{(x_1,x_2,x_3-\triangle x_3;t)} \right]\triangle x_1 \triangle x_2$$
$$\tag{9.121}$$

along directions $x_i, i = 1, 2, 3$. Next the net pressure forces on the control volume surface are,

$$\left(P \big|_{(x_1,x_2,x_3;t)} -P \big|_{(x_1-\triangle x_1,x_2,x_3;t)} \right)\triangle x_2 \triangle x_3 \tag{9.122}$$

$$\left(P \big|_{(x_1,x_2,x_3;t)} -P \big|_{(x_1,x_2-\triangle x_2,x_3;t)} \right)\triangle x_1 \triangle x_3 \tag{9.123}$$

$$\left(P \big|_{(x_1,x_2,x_3;t)} -P \big|_{(x_1,x_2,x_3-\triangle x_3;t)} \right)\triangle x_1 \triangle x_2 \tag{9.124}$$

Now referring to Figure 9.1, above, the net stress for viscous forces (NSF)on the control volume surface may be written as,

$$NSF = \left(\tau_{i1} \big|_{(x_1,x_2,x_3;t)} -\tau_{i1} \big|_{(x_1-\triangle x_1,x_2,x_3;t)} \right)\triangle x_2 \triangle x_3 +$$
$$\left(\tau_{i2} \big|_{(x_1,x_2,x_3;t)} -\tau_{i2} \big|_{(x_1,x_2-\triangle x_2,x_3;t)} \right)\triangle x_1 \triangle x_3 + \tag{9.125}$$
$$\left(\tau_{i3} \big|_{(x_1,x_2,x_3;t)} -\tau_{i3} \big|_{(x_1,x_2,x_3-\triangle x_3;t)} \right)\triangle x_1 \triangle x_2$$

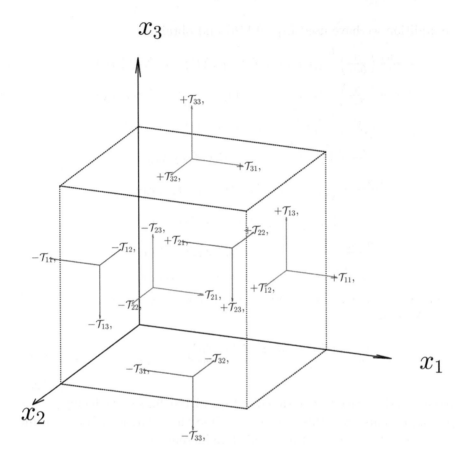

Figure 9.1 Net stress for viscous forces (NSF) on a control volume surface.

along directions $x_i, i = 1, 2, 3$. Finally, the body force on the control volume may be expressed as,

$$-\rho F_i \triangle x_1 \triangle x_2 \triangle x_3, \quad i = 1, 2, 3. \tag{9.126}$$

Substitute Eq. (9.110) into expression for NMF, equation of net pressure forces and expression for NSF with $\triangle x_i = x_i, \quad i = 1, 2, 3$. and follow the same idea as for the continuity equation derivation. Express the resulting Caputo derivative $0D_x^\beta f(x)$ by $\frac{\partial^\beta f(x)}{(\partial x)^\beta}$, the net momentum flux in the x_i direction can be expressed to β_i- order, $i = 1, 2, 3$, where in addition we have used Eq. (9.116) and obtain,

$$
\begin{aligned}
&\frac{\Gamma(2-\beta_1)}{x_1^{1-\beta_1}}\left(\frac{\partial}{\partial x_1}\right)^{\beta_1}\left(\rho(\vec{x};t)u_i(\vec{x};t)u_1(\vec{x};t)\right)\triangle x_1\triangle x_2\triangle x_3+ \\
&\frac{\Gamma(2-\beta_2)}{x_2^{1-\beta_2}}\left(\frac{\partial}{\partial x_2}\right)^{\beta_2}\left(\rho(\vec{x};t)u_i(\vec{x};t)u_2(\vec{x};t)\right)\triangle x_1\triangle x_2\triangle x_3+ \\
&\frac{\Gamma(2-\beta_3)}{x_3^{1-\beta_3}}\left(\frac{\partial}{\partial x_3}\right)^{\beta_3}\left(\rho(\vec{x};t)u_i(\vec{x};t)u_3(\vec{x};t)\right)\triangle x_1\triangle x_2\triangle x_3+ \\
&\frac{\Gamma(2-\beta_i)}{x_i^{1-\beta_i}}\left(\frac{\partial}{\partial x_i}\right)^{\beta_i}\left(P(\vec{x};t)\right)\triangle x_1\triangle x_2\triangle x_3+ \\
&\frac{\Gamma(2-\beta_1)}{x_1^{1-\beta_1}}\left(\frac{\partial}{\partial x_1}\right)^{\beta_1}\left(\tau_{i1}(\vec{x};t)\right)\triangle x_1\triangle x_2\triangle x_3+ \\
&\frac{\Gamma(2-\beta_2)}{x_2^{1-\beta_2}}\left(\frac{\partial}{\partial x_2}\right)^{\beta_2}\left(\tau_{i2}(\vec{x};t)\right)\triangle x_1\triangle x_2\triangle x_3+ \\
&\frac{\Gamma(2-\beta_3)}{x_3^{1-\beta_3}}\left(\frac{\partial}{\partial x_3}\right)^{\beta_3}\left(\tau_{i3}(\vec{x};t)\right)\triangle x_1\triangle x_2\triangle x_3- \\
&\rho(\vec{x};t)F_i(\vec{x};t)\triangle x_1\triangle x_2\triangle x_3,\quad i=1,2,3.
\end{aligned}
\tag{9.127}
$$

The rate of change of momentum within control volume on the interval $(t-\triangle t,t)$ is,

$$
\frac{\left[\rho(\vec{x},t)u_i(\vec{x};t)-\rho(\vec{x},t-\triangle t)u_i(\vec{x};t-\triangle t)\right]}{\triangle t}\triangle x_1\triangle x_2\triangle x_3,\quad i=1,2,3.
\tag{9.128}
$$

Recalling the procedure outlined in Eqs. (9.116–9.118), replacing ρ with ρu_i in previous equations leads to the time rate of change of momentum with respect to α-order fractional time increments,

$$
\frac{\Gamma(2-\alpha)}{t^{1-\alpha}}\frac{\partial^\alpha \rho(\vec{x},t)u_i(\vec{x};t)}{(\partial t)^\alpha}\triangle x_1\triangle x_2\triangle x_3,\quad i=1,2,3.
\tag{9.129}
$$

Now the time rate of momentum change within the control volume is inversely related to the net momentum flux, the previous equations lead to a general momentum equation of unsteady fluid flow in fractional time and multi-fractional space. When the body force is the gravitational force the equation is,

$$
\begin{aligned}
\frac{\partial^\alpha \rho(\vec{x},t)u_i(\vec{x};t)}{(\partial t)^\alpha}=-\ & \\
\sum_{j=1}^{3}\frac{\Gamma(2-\beta_j)}{x_j^{1-\beta_j}}\frac{t^{1-\alpha}}{\Gamma(2-\alpha)}\left(\frac{\partial}{\partial x_j}\right)^{\beta_j}&\left(\rho(\vec{x},t)u_i(\vec{x};t)u_j(\vec{x};t)\right) \\
-\frac{\Gamma(2-\beta_i)}{x_i^{1-\beta_i}}\frac{t^{1-\alpha}}{\Gamma(2-\alpha)}\left(\frac{\partial}{\partial x_i}\right)^{\beta_i}&P(\vec{x},t)+ \\
\sum_{j=1}^{3}\frac{\Gamma(2-\beta_j)}{x_j^{1-\beta_j}}\frac{t^{1-\alpha}}{\Gamma(2-\alpha)}\left(\frac{\partial}{\partial x_j}\right)^{\beta_j}&\left(\tau_{ij}(\vec{x};t)\right)+\frac{t^{1-\alpha}}{\Gamma(2-\alpha)}\rho g_i(\vec{x},t)
\end{aligned}
\tag{9.130}
$$

For incompressible flow, the shear stresses can be expressed using Stoke's viscosity law as in reference [63] where these need to be changed to viscosity relations in multi-fractional space. For ρ constant we can obtain the fractional incompressible equations. (Note that as $\alpha, \beta_i \to n$, the Caputo fractional derivative of a function $f(y)$ to order α or β_i ($i = 1, 2, 3$) is the conventional derivative of the function $f(y)$. Specializing to $n = 1$ we reduce the fractional differential equations to the conventional Navier–Stokes equations.)

Note that for compressible flow the shear stresses remain the same, however the Stokes relations are modified for the effect of volume change in the fluid due to compression [61]. Note that L. Li et al.[62] have proven that the Aubin-Lions lemma and its variants play crucial roles for the existence of weak solutions of nonlinear evolutionary PDEs. They have compactness criteria that are analogies of the Aubin-Lions lemma for the existence of weak solutions to time fractional PDEs. There they show that the time fractional compressible Navier–Stokes equations with constant density have weak solutions in two- and three-dimensional space.

EXERCISES

9.1 With $\alpha > 0$ show that for the Mittag-Leffler function $E_{\alpha,\beta}(t)$

$$t^{\alpha-1} E_{\alpha,\alpha}(-t^\alpha) = -\frac{d}{dt} E_\alpha(-t^\alpha).$$

9.2 Let $x \neq 0$. Show that,

$$E_{-2,1}\left(-\frac{1}{x^2}\right) = 1 - \cos(x)$$

9.3 Show that,

$$\mathcal{L}\left[\frac{F(s)}{s^2}\right] = \int_0^t \int_0^{t_1} f(t_2) dt_2 dt_1$$

where $F(s)$ is the Laplace transform of $f(t)$ admitted to exist.

9.4 Let $f(t)$ be an analytical function and $\mu \in \mathbb{C}$. Show that the Riemann-Liouville fractional derivative of order μ can be written as,

$$(\mathcal{D}^\mu f)(t) = D^\mu f(t) := \sum_{k=0}^{\infty} \binom{\mu}{k} \frac{t^{k-\mu} f^{(k)}(t)}{\Gamma(k - \mu + 1)},$$

with $f^{(k)}(t)$ denoting the derivative of $f(t)$ of order k.

9.5 Set $t > 0$. Consider a function $f(t)$ and admit that the fractional derivatives of Riemann-Liouville and Caputo, of the same order, exist. What must be the condition in order for these two derivatives to coincide?

9.6 Let $\nu \in \mathbb{R}$ such that $n - 1 < \nu \le n$ with $n = 1, 2, 3, \ldots$. Calculate the Caputo fractional derivative of the Mittag-Leffler function. Discuss the case where the order of the derivative is equal to the parameter of the Mittag-Leffler function.

9.7 Let $\nu > 0$, $\lambda \in \mathbb{R}$ and $t \ge 0$. Show that $x(t) = E_\nu(\lambda t^\nu)$ is a solution of the fractional differential equation $_C D^\nu x(t) = \lambda x(t)$ where the derivative is taken in the Caputo sense (C).

9.8 Let $m \in \mathbb{R}$. Calculate the Caputo fractional derivative of order ν, with $n - 1 < \nu \le n$ with $n \in N$ for the following function,

$$f(t) = (t + 1)^m$$

9.9 Let C be a constant. Show that the derivative of $f(t) = C$ is zero, using the Caputo fractional derivative and nonzero, using the Riemann Liouville derivative.

9.10 Calculate the fractional derivative of order $1/2$, in the Riemann-Liouville sense, initialized at zero, of the function $f(t) = \ln(t)$.

9.11 Let $\mu > -1$ and $\nu > 0$. Use the definition of the Caputo fractional derivative to get,

$$_C D_t^\nu t^\mu = \frac{\Gamma(\mu + 1)}{\Gamma(\mu - \nu + 1)} t^{\mu - \nu}$$

## CHAPTER 10

Introduction to Simplicial Complexes, and Discrete Exterior Calculus (DEC)

Keith C. Afas

Western University

10.1 INTRODUCTION

Fluid Mechanics as a mathematical problem has given rise to difficulties in numerical analysis alongside theoretical analysis. Often, for simple configurations of space, the governing equations (Navier Stokes Equations, NSEs) give difficulties for extracting closed-form solutions [63]. These difficulties arise often due to nonlinearities in the equation accounting for inertial effects, prohibiting what would be otherwise solving a vector Poisson's equation [13]. As a consequence, numerical techniques have been developed attempting to solve the NSEs in arbitrary spaces.

10.1.1 Previous Numerical Methods

To find solutions to the NSEs approximately as computers have acquired technological advances, algorithms to numerically solve these equations have naturally risen. Perhaps the oldest of these

DOI: 10.1201/9781003452256-10

techniques, finite-difference (FD), is based on a technique outlined by Brook Taylor in the early 18th century long before the digital revolution. The principle of this method is to impose a grid upon which linear differences in quantities immediately adjacent replace conventional definitions of differentiation. With smaller grids, this method approaches standard differentiation [7,12].

In the 20th century as engineering benefitted from the digital revolution, new numerical methods were developed to solve heat transfer, electrodynamics, and continuum mechanics equations on arbitrary shapes. The general process of these is aimed at decomposing the space upon which the equation is formulated on, into what is referred to as a simplicial mesh [2]. Depending on the treatment of this mesh, large systems of linear or non-linear equations per each mesh element are formulated. These typically produce large sparse equation systems which can be inverted to yield solutions per mesh element of the governing partial differential equation [3].

Techniques which decompose differential equations, into sparse linear systems, and variations, have led to traditional finite-volume (FV) or finite-element (FE) methods [3]. Each of these approaches produces error in accuracy or convergence owing to numerical instabilities and requires increases in either the spatial resolution or time resolution to get stable and accurate. This then leads to laborious computation time, leading to a "no-free-lunch" theorem of a trade-off between spatial/temporal accuracy, and time to produce simulations [63]. This has led to inequalities of restrictions on the time-step of these equations as is outlined in the Courant-Friedrich-Lewy Conditions, and similar conditions on partial differential equations.

10.1.2 Discrete Exterior Calculus as Alternative to FD/FE-Methods

A promising technique that seeks to mitigate the no-free-lunch theorem of FD/FE/FV methods, and prescribes a generalizable method of implementing numerical algorithms to solve partial differential equations on high dimensional manifolds, is the novel field of Discrete Exterior Calculus (DEC). DEC is a framework for computationally implementing equations that are formulated in the language of *exterior calculus*. The central objects of DEC are simplicial meshes, and discrete differential forms, which are discretized analogues of continuous manifolds, and exterior forms seen in exterior calculus [4,5].

This technique has several advantages over other numerical techniques. In FE-type methods, an equation must be formulated in a weak formulation. This may exist for many partial differential equations formed over a manifold, but does not exist for all. In addition, FE methods rely on finding a suitable basis for representing test functions and solutions, which can incur undesired computational time spent on basis regression, spline fitting, least-squared fitting, and floating-point calculations, that scale with resolution.

Structured grid methods like FD are good for avoiding formulating an equation in a weak formulation, or finding a representation basis for decomposing test functions. Using FD-methods, the equation can be discretized on a grid, and an implicit or explicit methods can be readily formulated. This method is generalizable to virtually any form of differential equation, but suffers non-generalizability to non-flat manifolds [7]. Considering the simple example of a 2-sphere in 3-Euclidean space $\mathbb{E}^3$ with radius r_0, which is locally isometric to 2-Euclidean space $\mathbb{E}^2$, its standard UV-parametrization with respect to an azimuthal and polar angle (θ, ϕ) is given by:

$$\mathbf{R} : \Sigma \to \mathbb{E}^3, \mathbf{R} = \begin{bmatrix} r_0 \sin\theta \cos\phi \\ r_0 \sin\theta \sin\phi \\ r_0 \cos\theta \end{bmatrix} \tag{10.1}$$

This parametrization contains points which are degenerate at its poles, since the coordinate $(\theta = 0)$ and $(\theta = \pi)$ correspond to the same point in 3-euclidean space, despite having infinite degrees of freedom due to the unspecified ϕ variable. No matter the parametrization of a sphere, if parametrized as per an orthogonal grid, as per the Hairy-ball theorem, for any genus-0 surface, there will exist points of degeneracy. Similar restrictions exist in non-genus-0 surfaces. These points of degeneracy will induce extra complication in finite-difference algorithms, and methods including Lebedev quadrature to try to circumvent these methods are less obvious to generalize to non-smooth manifolds [8].

Exterior calculus, is a language for calculus, which is formed in a *coordinate independent manner*, circumventing the non-generalizability of FD methods. This makes it well suited to give yield to a generalizable framework, provided there exists an effective discretization that DEC is aimed at providing [11]. This discretization provides a natural language to form differential equations on meshes, without the need to phrase equations in a weak formulation or introduce a decomposition basis as in FE methods [9]. This section will be aimed at providing a brief overview of notation in exterior calculus, its discretization into DEC, implementation for scalar partial differential equations, and application to vector partial differential equations, including the potential for Stokes-flow simulation on arbitrary manifolds.

10.2 EXTERIOR CALCULUS PRELIMINARIES

This section will provide an overview of notation in Exterior calculus; while Exterior calculus is a coordinate-invariant language, certain proofs will require generalized coordinate decompositions of expressions using tensor indices. Therefore where applicable, *einstein summation convention will be used.* Where tensor index notation is used, conventions for levi-civita tensors and covariant derivatives will follow conventions as in Grinfeld's book on the Calculus of Moving Surfaces (CMS) [53]. Exterior calculus is a language for calculus on manifolds, which is formed in a coordinate independent manner, without losing the geometric significance of expressions, and therefore the underlying geometric significance of partial differential equations formed over their respective manifolds. The central algebraic objects in exterior calculus are exterior forms. These are broadly defined as elements of the exterior algebra on a differentiable manifold. Exterior algebra can be classified as an ideal of the tensor algebra on a differentiable manifold, making exterior forms an ideal of tensors, and are similar to tensors in the sense that under a change of coordinates, exterior forms transform in a tensorial manner as per the jacobian of a coordinate space [6,11,53].

Broadly speaking, exterior forms are ideals of elements from a tensor product spaces. As such, like tensors which possess rank, exterior forms are also characterized by rank. By nomenclature, an exterior form of rank κ is referred to as a κ-form, which are defined on a manifold Σ. The space of exterior κ-forms on Σ, is denoted as $\Lambda^{\kappa}\Sigma$. Higher order differential forms are constructed from lower-order differential forms via the exterior product, or wedge product $\wedge$ defined on a manifold, which produces higher order differential forms. For a differential form $\omega \in \Lambda^{\kappa}\Sigma$, and another differential form $\xi \in \Lambda^{n}\Sigma$, the wedge product produces higher-order differential forms $\omega \wedge \xi \in \Lambda^{\kappa+n}\Sigma$. The exterior product is in fact a restriction of the tensor product which defines a tensor algebra, except that it is alternating and antisymmetric. Therefore, the exterior algebra is formed from the ideal of tensor algebra that does not include objects formed from the product itself. The wedge product satisfies the following properties:

1. *Distributivity:* $a \wedge (b + c) = a \wedge b + a \wedge c$

2. *Scalar multiplication:* $a \wedge (kb) = k(a \wedge b)$

3. *Associativity:* $(a \wedge b) \wedge c = a \wedge (b \wedge c)$

4. *Anticommutativity:* $a \wedge a = 0$

For any manifold Σ, there exist tangent $(T\Sigma)$ and co-tangent tensor $(T^*\Sigma)$ bundles, which can be composed using the tensor product $\otimes$ to create the space of tensors on a manifold of rank (m, n) $\mathcal{T}^{(m,n)}\Sigma$. Similarly there exist analogues created from the iterative wedge product of elements from the tangent and co-tangent bundles, referred to as the κ-vector space $\Lambda^\kappa\Sigma$, and κ-form space $\Lambda^{\kappa*}\Sigma$). These are given explicitly using the conventions from Ivancevic and Ivancevic [6] as such:

$$\Lambda^\kappa\Sigma = \bigwedge_{i=1}^{\kappa} T\Sigma, \ \Lambda^{\kappa*}\Sigma = \bigwedge_{i=1}^{\kappa} T^*\Sigma \tag{10.2}$$

In exterior calculus, the basis which spans the κ-form space are the primitive one-forms dS^α, representing a small increment of a co-vector in the direction of S^α. In this manner, κ-forms are represented in component form as:

$$A = A_{\alpha_1\alpha_2...\alpha_\kappa} dS^{\alpha_1} \wedge dS^{\alpha_2} ... \wedge dS^{\alpha_\kappa} \tag{10.3}$$

As mentioned, since the exterior product $\wedge$ is a restriction of the tensor product $\otimes$, the space of exterior forms can be seen to be a subset of the space of tensors such that $\Lambda^\kappa\Sigma \subset \mathcal{T}^{(\kappa,0)}\Sigma$ and $\Lambda^{\kappa*}\Sigma \subset \mathcal{T}^{(0,\kappa)}\Sigma$. Consequentially, κ-vectors and κ-forms can be mapped to the other space by 'lowering' every index of the form/k-vector. This mapping between higher-order κ-vector spaces and κ-form spaces is done via the musical isomorphisms $(\cdot)^\sharp$ and $(\cdot)^\flat$ which satisfy the following:

$$\begin{aligned}(\cdot)^\sharp &: \ \Lambda^{\kappa*}\Sigma \ \rightarrow \ \Lambda^\kappa\Sigma \\ (\cdot)^\flat &: \ \Lambda^\kappa\Sigma \ \rightarrow \ \Lambda^{\kappa*}\Sigma \end{aligned} \tag{10.4}$$

As tensors can be defined over a space to become a tensor field, similarly, exterior forms can be defined everywhere in a space. If an exterior form possesses sufficient continuity, then this allows for the possibility of differentiation on forms, which must account for the curvature of the space upon which the exterior form is defined. Such an exterior form which can be differentiated can be referred to as a differential form and will be elaborated in the subsequent section.

10.2.1 Exterior Derivative on κ-Forms, (d_Σ)

Manifolds, are often restricted to the space of Riemannian manifolds, possessing a metric tensor allowing for a notion of length. This notion allows for differentiation to be performed along its surface through partial differentiation, and can be extended to a derivative which produces higher-rank tensors on a manifold, also referred to as co-variant differentiation. While exterior forms are designed to be coordinate-invariant, the co-variant differentiation machinery on manifolds can be exploited to allow for differentiation on exterior forms that preserves their transformation tensorial nature. The central differential operator is the *exterior derivative* (d_Σ). Similarly to how the covariant-derivative ∇_α extends the partial derivative ∂_α to higher order tensors in a manner that preserves their transformation laws, the exterior derivative extends the role of a differential d on scalar multi-variable functions, to higher-order forms. The action of the exterior derivative can be easily introduced by considering a 0-form/a rank (0,0) tensor (ie. scalar function) $\psi : \Sigma \to \mathbb{R}$. Its exterior derivative is defined as the differential $\mathrm{d}_\Sigma \psi = (\nabla_\alpha \psi) dS^\alpha$. It is seen immediately, that $d\psi$ is a 1-form $\mathrm{d}_\Sigma \psi \in \Lambda^{1*}\Sigma$. Being formed from the co-variant derivative ensures that the differential preserves the tensorial nature of calculus on the manifold Σ. In general, the exterior derivative is a differential operator which maps κ-forms to $(\kappa+1)$-forms, and performs a similar function on κ-vectors, or elements of $\Lambda^\kappa \Sigma$:

$$\begin{aligned}
\mathrm{d}_\Sigma : \quad \Lambda^{\kappa*}\Sigma \quad &\to \quad \Lambda^{(\kappa+1)*}\Sigma \\
\mathrm{d}_\Sigma : \quad \Lambda^{\kappa}\Sigma \quad &\to \quad \Lambda^{\kappa+1}\Sigma
\end{aligned} \tag{10.5}$$

This can be generalized to κ-forms, through iterative application of the differential. In general, for a κ-form $A \in \Lambda^{\kappa*}\Sigma$, given as:

$$A = A_{\alpha_1 \ldots \alpha_\kappa} dS^{\alpha_1} \wedge \ldots \wedge dS^{\alpha_\kappa}$$

Its exterior derivative is given explicitly by:

$$d_\Sigma A = (\nabla_{\beta_1} A_{\alpha_1 \dots}) \, dS^{\beta_1} \wedge dS^{\alpha_1} \wedge \dots \tag{10.6}$$

Taking the exterior derivative a second time yields the following result:

$$d_\Sigma d_\Sigma A = (\nabla_{\beta_2} \nabla_{\beta_1} A_{\alpha_1 \dots \alpha_\kappa}) \, dS^{\beta_2} \wedge dS^{\beta_1} \wedge dS^{\alpha_1} \wedge \dots \wedge dS^{\alpha_\kappa}$$

Through expanding the definition of the co-variant derivatives, it can be shown since $dS^{\beta_2} \wedge dS^{\beta_1}$ places an antisymmetric restriction on the covariant derivatives, this vanishes. This establishes one of the defining properties of exterior derivatives for all κ-forms $\forall A \in \Lambda^{\kappa*}\Sigma$, which is that:

$$\boxed{d_\Sigma^2 = 0} \tag{10.7}$$

The exterior derivative is a powerful operator that allows for the analysis of differential forms. Being an anti-symmetric operator, in order to define higher-order differential operators that do not increase the rank of exterior forms (ie. laplacians, divergences, etc.), an operator is required which decreases the rank of exterior forms. This is accomplished by the Hodge star operator.

10.2.2 Hodge Star Operator, ($\star_\Sigma$)

The next fundamental operator on κ-forms, is the Hodge star operator $\star_\Sigma$. This special operator has the ability to map κ-forms on a m-dimensional manifold, to $(m - \kappa)$-forms:

$$\star_\Sigma : \Lambda^{\kappa*}\Sigma \to \Lambda^{(\dim\Sigma - \kappa)*}\Sigma \tag{10.8}$$

In this sense, it 'completes' the manifold space of exterior forms. Due to this completion, integrating the result over the manifold acts as the 'interior product' or the 'dot product' of the space. To state this explicitly:

$$\forall A, B \in \Lambda^{k*}\Sigma, \quad \int_\Sigma A \wedge \star_\Sigma B = \langle A, B \rangle \star_\Sigma 1 \tag{10.9}$$

where $\star_\Sigma 1$ represents the 'volume form' of the manifold that will be discussed soon. The Hodge star operator therefore has the ability to induce the dot product between κ-forms. To express the effect of the

Hodge star on exterior forms in explicit coordinate representation, this requires the use of the unique fully antisymmetric tensors on a manifold which are scaled permutation symbols, the Levi-Civita tensors on the surface $(\varepsilon_{\alpha_1...\alpha_\kappa\alpha_{\kappa+1}...\alpha_m})$. In general, for a κ-form given by $A = A_{\alpha_1...\alpha_\kappa} dS^{\alpha_1} \wedge ... \wedge dS^{\alpha_\kappa}$, the Hodge star operator takes the following explicit form when applied on a m-dimensional manifold:

$$\star_\Sigma A = \left(\frac{A^{\alpha_1...\alpha_\kappa}\varepsilon_{\alpha_1...\alpha_\kappa\alpha_{\kappa+1}...\alpha_m}}{(m-\kappa)!} \right) dS^{\alpha_{\kappa+1}} \wedge ... \wedge dS^{\alpha_m} \qquad (10.10)$$

This shows that the application of the Hodge star to a κ-form indeed produces a $(m - \kappa)$-form. Applying this to the 0-form scalar value, '1' on an m-manifold we see that:

$$\star_\Sigma 1 = \frac{1}{m!}\varepsilon_{\alpha_1...\alpha_m} dS^{\alpha_1} \wedge ... \wedge dS^{\alpha_m} \qquad (10.11)$$

The definition of the Levi-Civita tensor in terms of the permutation symbol, and executing the iterated sum, this can be expanded to see that:

$$\star_\Sigma 1 = \frac{m!}{m!}\sqrt{|g..|}e_{12...m} dS^1 \wedge dS^2 \wedge ... \wedge dS^m$$

Simplifying, the expression, and integrating over the m-manifold, we see that:

$$\int_\Sigma \star_\Sigma 1 = \int_\Sigma \sqrt{|g..|}dS^1 dS^2 ... dS^m$$

This reveals that $\star_\Sigma 1$ is the area element on 2-manifolds, volume element on 3-manifolds etc. Thus, one may make the equivalence from differential geometry and exterior calculus $\star_\Sigma 1 = d\Sigma$, such that any integration of a 0-form over the manifold Σ can be re-writ in the following form:

$$\int_\Sigma (\cdot)\, d\Sigma = \int_\Sigma \star_\Sigma(\cdot) = \int_\Sigma (\cdot)\, \star_\Sigma 1 \qquad (10.12)$$

In general, a κ-form can be integrated over κ-chain, which is defined defined as a κ-dimensional region of space, that may be a subspace of the manifold upon which a differential form is defined over. In this case, integrating a scalar field ψ over a surface is translated into integrating the m-form $\star_\Sigma \psi$ over the m-chain, Σ. Speaking particularly on 2-manifolds, the results of iterated Hodge star operator application can be observed. When acting on a 0-form and 1-form ψ and $u = u_\alpha dS^\alpha$, we have:

$$\star_\Sigma \psi = \frac{1}{2!}\varepsilon_{\alpha\beta}\psi dS^\alpha \wedge dS^\beta$$

$$\star_\Sigma \star_\Sigma \psi = \frac{1}{2!0!}\varepsilon^{\alpha\beta}\varepsilon_{\alpha\beta}\psi$$

$$\star_\Sigma \star_\Sigma \psi = \frac{2!}{2!0!}\psi$$

$$\star_\Sigma \star_\Sigma \psi = \psi$$

$$\star_\Sigma u = \frac{1}{1!}u^\alpha \varepsilon_{\alpha\beta} dS^\beta$$

$$\star_\Sigma \star_\Sigma u = u_\alpha \varepsilon^{\alpha\beta}\varepsilon_{\beta\gamma}dS^\gamma$$

$$\star_\Sigma \star_\Sigma u = -u_\alpha \delta^\alpha_\gamma dS^\gamma$$

$$\star_\Sigma \star_\Sigma u = -u_\gamma dS^\gamma$$

$$\star_\Sigma \star_\Sigma u = -u$$

And thus it can be seen that $(\star_\Sigma)^2 = 1$ on 0-forms and $(\star_\Sigma)^2 = -1$ on 1-forms in a 2-manifold. Similarly, it can be seen that $(\star_\Sigma)^2 = 1$ on 2-forms in a 2-manifold. Summarizing results:

$$(\star_\Sigma)^2 = (-1)^\kappa, \ \dim \Sigma = 2 \tag{10.13}$$

With the exterior derivative defined, and the Hodge star operator defined on exterior forms, in order to convert partial differential equations into exterior calculus, analogues to differential operators seen in calculus must be identified with equivalent expressions in exterior calculus

10.2.3 Co-Differential Operator and Analogues to the Divergence of Vector Fields on Manifolds

Having defined the exterior derivative, and Hodge star operator, certain specialized differential operators resembling differential operators in vector calculus may also be defined on Σ. The co-differential operator $\tilde{\delta}_\Sigma$ is one which uses the Hodge star operators to modify the exterior derivative. The co-differential is defined explicitly on κ-forms in an m-manifold by $\tilde{\delta}_\Sigma = (-1)^{m(\kappa+1)+1} \star_\Sigma d_\Sigma \star_\Sigma$. By performing form-order analysis on this expression, it can be seen that this operator takes a κ-form to a $(\kappa - 1)$-form, leading to the definition of the co-differential in general as:

$$\tilde{\delta}_\Sigma : \Lambda^{\kappa*}\Sigma \to \Lambda^{(\kappa-1)*}\Sigma \tag{10.14}$$

On a 0-form, the co-differential is undefined, and by convention, it is taken that $\tilde{\delta}_\Sigma \psi = 0$ for a 0-form ψ. On a 2-manifold, the co-differential can be simplified to $\tilde{\delta}_\Sigma = - \star_\Sigma \, d_\Sigma \star_\Sigma$. Evaluating this on a 1-form $u = u_\alpha dS^\alpha$ in a 2-manifold, we see:

$$\tilde{\delta}_\Sigma u = - \star_\Sigma \, d_\Sigma \star_\Sigma u = - \star_\Sigma \, d_\Sigma \star_\Sigma (u_\alpha dS^\alpha)$$

$$\tilde{\delta}_\Sigma u = - \star_\Sigma \, d_\Sigma \left(\frac{1}{1!} u^\alpha \varepsilon_{\alpha\beta} dS^\beta \right)$$

$$\tilde{\delta}_\Sigma u = - \star_\Sigma \left(\frac{1}{1!} \nabla_\gamma \left(u^\alpha \varepsilon_{\alpha\beta} \right) dS^\gamma \wedge dS^\beta \right)$$

$$\tilde{\delta}_\Sigma u = - \frac{1}{1!1!} \epsilon^{\gamma\beta} \varepsilon_{\alpha\beta} \nabla_\gamma u^\alpha = - \frac{1}{1!1!} \delta^\gamma_\alpha \nabla_\gamma u^\alpha$$

$$\tilde{\delta}_\Sigma u = - \nabla_\alpha u^\alpha \tag{10.15}$$

This operator of $\nabla_\alpha u^\alpha$ seen in Eq. (10.15) is equivalent to the divergence operator on a 2-manifold. Therefore, we can see for a surface vector field $\vec{u}$ with a surface 1-form field related through $\{u = \vec{u}^\flat, \vec{u} = u^\sharp\}$, then letting $\nabla_\Sigma \cdot (\cdot)$ be the divergence operator on surface vectors, the following connection between exterior calculus and vector calculus exists:

$$\begin{aligned} \nabla_\Sigma \cdot \vec{u} &= -\tilde{\delta}_\Sigma \vec{u}^\flat \\ \nabla_\Sigma \cdot \vec{u} &= \star_\Sigma d_\Sigma \star_\Sigma \vec{u}^\flat \end{aligned} \tag{10.16}$$

10.2.4 Laplace-deRham Operator and Analogues to the Laplace-Beltrami Operator

On a surface, it is known that the Laplace-Beltrami (LB) Operator may be formed from the co-variant derivatives as $\nabla^2_\Sigma = \nabla_\alpha \nabla^\alpha$. This operator is critical for understanding the surface geometry and for implementing several geometric algorithms on surface including feature smoothing, eigenfunction decomposition, and others. In differential geometry, the LB Operator being defined from the co-variant derivative can be defined on a general tensor. In exterior calculus, a similar second-order differential operator known as the Laplace-deRham (LDR) Operator, or Δ^{dR}_Σ can be defined which takes κ-forms to κ-forms. This operator is defined as:

$$\Delta_\Sigma^{dR} : \Lambda^{\kappa*}\Sigma \to \Lambda^{\kappa*}\Sigma \tag{10.17}$$

Explicitly, the LDR operator can be defined in terms of the co-differential and exterior derivative. Since the co-differential decreases the rank of a κ-form, and the exterior derivative increases the rank, they can be alternated, providing the explicit form of the LDR operator acting on a κ-form:

$$\Delta_\Sigma^{dR} = d_\Sigma \tilde{\delta}_\Sigma + \tilde{\delta}_\Sigma d_\Sigma \tag{10.18}$$

Since the relationship exists on a 2-manifold that $\tilde{\delta}_\Sigma = -\star_\Sigma d_\Sigma \star_\Sigma$, then we can see that:

$$\Delta_\Sigma^{dR} = - \left(d_\Sigma \star_\Sigma d_\Sigma \star_\Sigma + \star_\Sigma d_\Sigma \star_\Sigma d_\Sigma \right) \tag{10.19}$$

This form of the LDR operator will be examined on 0-forms in the subsequent section.

Definition of the Laplacian on 0-Forms

In order to implement numerical methods on a simplicial complex (mesh) representation $\mathcal{K}$ of a 2-manifold Σ, the LB Operator $\nabla_\Sigma^2 = \nabla_\alpha \nabla^\alpha$ must be related to the LDR Operator $\Delta^{dR} : \Lambda^{\kappa*}\Sigma \to \Lambda^{\kappa*}\Sigma$ on 0-forms. All 0-forms $\psi(S) \in \Lambda^{0*}\Sigma$ are co-closed (ie. that is to say, $\tilde{\delta}_\Sigma \Phi = 0$) since there does not exist a lower-Co-homology. Therefore, for 0-forms, the LDR operator is simplified to $\Delta_\Sigma^{dR} = -\star_\Sigma d_\Sigma \star_\Sigma d_\Sigma$. The LDR Operator can be simplified in local coordinates:

$$\Delta_\Sigma^{dR}\psi = -\star_\Sigma d_\Sigma \star_\Sigma d_\Sigma \psi$$
$$\Delta_\Sigma^{dR}\psi = -\star_\Sigma d_\Sigma \star_\Sigma \left(\nabla_\alpha \psi dS^\alpha \right)$$
$$\Delta_\Sigma^{dR}\psi = -\star_\Sigma d_\Sigma \left(\nabla^\alpha \psi \varepsilon_{\alpha\beta} dS^\beta \right)$$
$$\Delta_\Sigma^{dR}\psi = -\star_\Sigma \left(\varepsilon_{\alpha\beta} \nabla_\gamma \nabla^\alpha \psi dS^\gamma \wedge dS^\beta \right)$$
$$\Delta_\Sigma^{dR}\psi = -\varepsilon_{\gamma\beta}\varepsilon^{\alpha\beta} \nabla^\gamma \nabla_\alpha \psi = -\delta_{\gamma\beta}^{\alpha\beta} \nabla^\gamma \nabla_\alpha \psi$$
$$\Delta_\Sigma^{dR}\psi = -\delta_\gamma^\alpha \nabla^\gamma \nabla_\alpha \psi = -\nabla^\alpha \nabla_\alpha \psi$$
$$\Delta_\Sigma^{dR}\psi = -\nabla_\Sigma^2 \psi$$

Therefore, on 0-forms, we see that the two second-order operators are identical to a sign difference. That is to say, the relationships between the Laplace-Beltrami and Laplace-deRham Operator is:

$$\boxed{\Delta_\Sigma^{dR} = -\nabla_\Sigma^2} \tag{10.20}$$

10.2.5 Integration, Stokes Theorem, and Ostrogradsky's Theorem

The last topic to discuss from exterior calculus is integration over κ-chains. In general, chains are regions of κ-dimensional space in the manifold over which the exterior calculus is defined. In general, κ-forms are integrated over κ-chains. For example, considering a 0-form $\psi = \psi(S)$ on the 2-manifold Σ. The 2-manifold can be considered as a curvi-linear 2-chain, and therefore, if we wish to integrate ψ over Σ, ψ must be converted into a 2-chain, which is equivalent to applying the Hodge star to convert it from a 0-form to a 2-form: $\star_\Sigma \psi = \psi \star_\Sigma 1$, and then this can be integrated over Σ:

$$\text{Integral of } \psi(S) \text{ over } \Sigma = \int_\Sigma \psi \star_\Sigma 1 \tag{10.21}$$

Let $\mathcal{M}_\Sigma$ be a κ-dimensional chain on the 2-manifold Σ. $\mathcal{M}_\Sigma$ may possess a boundary given by $\partial\mathcal{M}_\Sigma$ over which a $(\kappa - 1)$-form can be integrated. In general, integrals over chains and their boundaries can be related if the integrand has a particular form. This is known as stokes theorem, and letting $\Phi \in \Lambda^{(\kappa-1)*}\Sigma$ be a $(\kappa - 1)$-form, the theorem is stated as:

$$\int_{\mathcal{M}_\Sigma} d\Phi = \oint_{\partial\mathcal{M}_\Sigma} \Phi \tag{10.22}$$

This states that the integral of the exterior derivative of a $(\kappa - 1)$-form over the κ-chain $\mathcal{M}_\Sigma$, is equivalent to the integral of the $(\kappa - 1)$-form over the boundary of $\mathcal{M}_\Sigma$. In this theorem, it can be seen explicitly that letting Σ be a 2-manifold, $\mathcal{M}_\Sigma$ be a κ-chain subset of Σ with boundary $\partial\mathcal{M}_\Sigma$, and letting u be a 1-form defined by $u = u_\alpha dS^\alpha$, the integral of u's exterior derivative is an integral of a 2-form over a 2-chain. Integrating over $\mathcal{M}_\Sigma$, this is given by:

$$\int_{\mathcal{M}_\Sigma} du = \int_{\mathcal{M}_\Sigma} \nabla_\alpha u_\beta dS^\alpha \wedge dS^\beta$$

Expanding this explicitly, the equation becomes:

$$\int_{\mathcal{M}_\Sigma} du = \int_{\mathcal{M}_\Sigma} e^{\alpha\beta} \nabla_\alpha u_\beta dS^1 dS^2$$
$$\int_{\mathcal{M}_\Sigma} du = \int_{\mathcal{M}_\Sigma} \frac{1}{\sqrt{|g..|}} e^{\alpha\beta} \nabla_\alpha u_\beta \star_\Sigma 1$$
$$\int_{\mathcal{M}_\Sigma} du = \int_{\mathcal{M}_\Sigma} \nabla_\alpha \left(\varepsilon^{\alpha\beta} u_\beta \right) \star_\Sigma 1$$

Letting n_α be the normal to the boundary $\partial \mathcal{M}_\Sigma$, using Ostrogradsky's Theorem [1], the integral simplifies to the following:

$$\int_{\mathcal{M}_\Sigma} du = \oint_{\partial \mathcal{M}_\Sigma} n_\alpha \varepsilon^{\alpha\beta} u_\beta \star_{\partial \Sigma} 1$$

It can be shown that $n_\alpha \varepsilon^{\alpha\beta}$ is tangential to the boundary, and so this integral is a line integral of u_β around the perimeter of the boundary, completing Stokes theorem. Conversely, considering integrating the divergence of a vector field $\nabla_\Sigma \cdot \vec{u} = -\tilde{\delta}_\Sigma \vec{u}^b$ within $\mathcal{M}_\Sigma$, the Hodge star must be applied to convert it into a 2-form to be able to integrate around the 2-chain. Therefore, letting $u = \vec{u}^b$ be the 1-form corresponding to the vector field, the 2-form integrand to integrate is $- \star_\Sigma \tilde{\delta}_\Sigma u$. Simplifying this, we see that since the dual application of the Hodge star results in a scalar, the integrand to integrate is $(d_\Sigma \star_\Sigma u)$. Expressing this in local coordinates:

$$\int_{\mathcal{M}_\Sigma} \nabla_\Sigma \cdot \vec{u} \, d\Sigma = \int_{\mathcal{M}_\Sigma} d_\Sigma \star_\Sigma u$$

$$\int_{\mathcal{M}_\Sigma} \nabla_\Sigma \cdot \vec{u} \, d\Sigma = \int_{\mathcal{M}_\Sigma} d_\Sigma \left(u^\alpha \varepsilon_{\alpha\beta} dS^\beta \right)$$

$$\int_{\mathcal{M}_\Sigma} \nabla_\Sigma \cdot \vec{u} \, d\Sigma = \int_{\mathcal{M}_\Sigma} \varepsilon_{\alpha\beta} \nabla_\gamma u^\alpha dS^\gamma \wedge dS^\beta$$

$$\int_{\mathcal{M}_\Sigma} \nabla_\Sigma \cdot \vec{u} \, d\Sigma = \int_{\mathcal{M}_\Sigma} e^{\gamma\beta} \varepsilon_{\alpha\beta} \nabla_\gamma u^\alpha dS^1 dS^2$$

$$\int_{\mathcal{M}_\Sigma} \nabla_\Sigma \cdot \vec{u} \, d\Sigma = \int_{\mathcal{M}_\Sigma} \varepsilon^{\gamma\beta} \varepsilon_{\alpha\beta} \nabla_\gamma u^\alpha \star_\Sigma 1$$

$$\int_{\mathcal{M}_\Sigma} \nabla_\Sigma \cdot \vec{u} \, d\Sigma = \int_{\mathcal{M}_\Sigma} \nabla_\alpha u^\alpha \star_\Sigma 1$$

$$\int_{\mathcal{M}_\Sigma} \nabla_\Sigma \cdot \vec{u} \, d\Sigma = \oint_{\partial \mathcal{M}_\Sigma} n_\alpha u^\alpha \star_{\partial \Sigma} 1$$

This last equality is exactly the 1-form integrated outwards around the boundary, which demonstrates the Divergence theorem on manifolds.

With preliminaries of exterior calculus covered, discretizations on simplicial complexes will be discussed.

10.3 EXTERIOR CALCULUS DISCRETIZATION

In this section, a brief introduction will be provided on simplicial complexes, signed incidence matrices, and basic geometrical metrics on them, identifying these metrics with their continuous analogues on manifolds through exterior calculus.

10.3.1 Simplicial Complexes

Simplicial complexes are informally a collection of primitive geometrical objects, which are connected topologically, in a manner to approximate continuous analogues in calculus. These objects are individually referred to as an n-simplex, where n described their dimensions. They are informally the shape which uniquely contains $(n + 1)$ points. For example, a 0-simplex is a point, a 1-simplex is a line defined by two points, a 2-simplex is a triangle defined with three points, and a 3-simplex is a tetrahedron defined from four points. In general, a simplicial complex is denoted as $\mathcal{K} = \{\Omega^0(\mathcal{K}), \Omega^1(\mathcal{K}), \Omega^2(\mathcal{K}), \ldots\}$, where $\Omega^n(\mathcal{K})$ is the set of n-simplices in the simplicial complex (set of points, edges, faces, cells etc). The cardinality, or size of each set of simplices, is represented by the Betti numbers $\{b_0, b_1, b_2, \ldots\}$, where $b_0 = |\Omega^0(\mathcal{K})|$, $b_1 = |\Omega^1(\mathcal{K})|$. The Betti numbers represent the number of vertices (b_0), edges (b_1), faces (b_2), and so forth. The topology of a simplicial complex is defined via matrices named "incidence matrices" $\hat{c}_{\mathcal{K}}^\kappa$ that denote which higher-order $(\kappa + 1)$-simplices are formed from which lower-order κ-simplices on $\mathcal{K}$, and are therefore defined for each lower-order set of κ-simplices, where $1 \leq \kappa \leq \dim \Sigma - 1$. This can be defined as such:

Definition 10.1 *Let $\mathcal{K}$ be a simplicial complex, and let b_κ be the Betti numbers on $\mathcal{K}$ indicating the number of κ-simplices in $\mathcal{K}$. Let $\Omega^\kappa(\mathcal{K})$ be a canonical ordering of the set of κ-simplices in $\mathcal{K}$, assigning a number $(1..b_\kappa)$ to each κ-simplex, such that $b_k = |\Omega^k(\mathcal{K})|$. The κ'th incidence matrix $\hat{c}_{\mathcal{K}}^\kappa \in \mathbb{R}^{|\Omega^{\kappa+1}| \times |\Omega^\kappa|}$ is a sparse matrix with entries $(\hat{c}_{\mathcal{K}}^\kappa)_{ij} = 1$ if the j'th κ-simplex is contained in the i'th $(\kappa + 1)$-simplex.*

Each simplex is associated with an "orientation" depending on how it is constructed from lower-order simplices. For example, an edge is defined from two points. The direction from the first point to the second point defines the orientation of the 1-simplex. If the points are reversed, the 1-simplex attains negative orientation. Similarly, a face is defined from three edges. Each face's orientation is stated to be positive when edges are composed to form a clockwise loop in the face. For the rest of the section, references to discretized quantities on $\mathcal{K}$ are denoted with a hat $(\hat{\cdot})$.

10.3.2 Discretized Exterior Forms

In DEC, a simplicial complex is postulated to approximate a continuous manifold [4], and therefore, objects like continuous forms are integrated with respect to each simplex, and represented as a quantity per simplex. For example, consider a 0-form $\psi : \Sigma \to \mathbb{R}$, and let $\hat{\psi}$ represent its discrete analogue. Then the de-Rham map corresponding to the discretization of ψ over the simplicial complex $\mathcal{K}$ approximating the smooth manifold Σ can be given as the integral of the 0-form $\psi : \Sigma \to \mathbb{R}$ over the 0-chain $\{v_i \in \Omega^0(\mathcal{K})\}$:

$$\hat{\psi}_i = \left\{ \int_{v_i} \psi(\Sigma) = \psi|_{v_i}, \forall v_i \in \Omega^0(\mathcal{K}) \right\} \tag{10.23}$$

This is nothing more than evaluating the 0-form at the vertex v_i belonging to the set of vertices in $\mathcal{K}$. This yields a $b_0 \times 1$ array. Similarly, consider a 1-form $\xi \in \Lambda^{1*}\Sigma$, $\xi = \xi_\alpha dS^\alpha$. This 1-form $\xi : \Sigma \to \Lambda^{1*}\Sigma$ can be integrated along each 1-simplex of the mesh $\{e_i \in \Omega^1(\mathcal{K})\}$ yielding the following discretized 1-form on $\mathcal{K}$

$$\hat{\xi}_i = \left\{ \int_{e_i} \xi_\alpha dS^\alpha, \forall e_i \in \Omega^1(\mathcal{K}) \right\} \tag{10.24}$$

In the special case that $\xi_\alpha dS^\alpha$ is an exact differential form in that it can be expressed as the exterior derivative of a 0-form $\phi : \Sigma \to \mathbb{R}$, or $\xi = d\phi$ ($\xi_\alpha = \nabla_\alpha \phi$), then we can express its discretization as $\hat{\xi}_i = \int_{e_i} d\phi$. Through application of Ostrogradsky's theorem, we see that $\hat{\xi}_i = \int_{\partial e_i} \phi = \phi|_{v \in e_i}$. Letting the set of vertices that composed edge e_i be $e_i = \{v_{e_i,1}, v_{e_i,2}\}$, then we can complete the discretization:

$$\hat{\xi}_i = \int_{\partial e_i} \phi = \phi|_{v_{e_i,2}} - \phi|_{v_{e_i,1}} \tag{10.25}$$

It can be seen therefore that the discretization of the 1-form $d\phi$ is nothing more than evaluating the difference of the 0-form ϕ between the vertices of each edge. It should also be mentioned, the orientation of each simplex comes into play. If the edge was reversed, the integral of the 1-form over the edge would have its sign flipped. In general, the orientation of κ-simplices is critical when discretizing κ-forms. This philosophy of discretizing continuous κ-forms on simplicial complexes forms the basis of DEC.

10.3.3 Exterior Derivative on Simplicial Complexes

In the preliminaries on exterior calculus, the exterior derivative was one of the central operators. On a simplicial complex, this is one of the most fundamental operators. Recalling the exterior derivative mapped κ-forms to $(\kappa + 1)$-forms on continuous manifolds, the exterior derivative will map discretized κ-forms, which are integrated quantities over κ-simplices to discretized $(\kappa + 1)$-forms, which are integrated quantities over $(\kappa+1)$-simplices. Discretized κ-forms are integrated per κ-simplex, to form b_κ-arrays. This means that they are elements of $\mathbb{R}^{b_\kappa}$, or $\mathbb{R}^{|\Omega^\kappa|}$. The exterior derivative on Σ, d_Σ when discretized over $\mathcal{K}$, will in fact result in several matrices that takes discretized κ-forms to $(\kappa + 1)$-forms, or using $\hat{d}_\mathcal{K}^\kappa$ to denote the discretized exterior derivative on a κ-form, $\hat{d}_\mathcal{K}^\kappa \in \mathbb{R}^{|\Omega^{\kappa+1}| \times |\Omega^\kappa|}$.

It may not be immediately apparent, but the form of the matrix per row will depend on the orientation of the simplex upon which the differentiation is taking place. As it turns out, as the exterior derivative is dimension invariant, this matrix is also dimension invariant. It can be seen to be only dependent on the topology of the manifold, and in fact is none other than the κ'th signed incidence matrix on the simplicial complex. Informally, this sparse matrix $\hat{d}_\mathcal{K}^\kappa \in \mathbb{R}^{|\Omega^{\kappa+1}| \times |\Omega^\kappa|}$, is similar to the incidence matrix discussed earlier $c_\mathcal{K}^\kappa$, where each row corresponding to a $(\kappa+1)$-simplex contains non-zero entries at the index corresponding to the κ-simplex. The difference is the signed incidence matrix factors in the orientation (either $+1/-1$) required of the κ-simplex to form a cycle of the $(\kappa + 1)$-simplex.

$$(\hat{d}_\mathcal{K}^\kappa)_{ij} = \begin{cases} +1 & , \quad j\text{'th } \kappa\text{-simplex} \in i\text{'th } (\kappa + 1)\text{-simplex (normal orientation)} \\ -1 & , \quad j\text{'th } \kappa\text{-simplex} \in i\text{'th } (\kappa + 1)\text{-simplex (flipped orientation)} \\ 0 & , \qquad\qquad\qquad\qquad\qquad\qquad \text{otherwise} \end{cases}$$

(10.26)

Therefore, this is a modification on the incidence matrix $\hat{c}_\mathcal{K}^\kappa$, and yields $(\dim\Sigma\text{-}1)$ exterior derivative matrices $\{\hat{d}_\mathcal{K}^0, \hat{d}_\mathcal{K}^1, ..., \hat{d}_\mathcal{K}^{\dim\Sigma-1}\}$. Similarly to exterior derivatives which satisfy $d_\Sigma d_\Sigma = 0$, it can be seen that this theorem holds in the discrete case. In general, we see that:

$$\hat{d}_{\mathcal{K}}^{\kappa+1}\hat{d}_{\mathcal{K}}^{\kappa} = 0, \{\forall \kappa \in \mathbb{Z} | 0 \le \kappa \le \dim\Sigma - 2\} \tag{10.27}$$

Having discussed the discretization of the exterior derivative, a similar treatment will now be provided for the Hodge star operator.

10.3.4 Hodge Star Operator on Simplicial Complexes

As the exterior derivative was a differential operator that mapped κ-forms to $(\kappa + 1)$-forms, and the discrete exterior derivative was a matrix that mapped discretized values of a κ-forms on a κ-simplex to discretized values of a $(\kappa + 1)$-forms on a $(\kappa + 1)$-simplex, a similar treatment will be sought for the Hodge star operator which mapped κ-forms to $(\dim \Sigma - \kappa)$-forms.

A central theme that re-emerges in DEC is the concept of the dual mesh to a simplicial complex $\mathcal{K} \sim \Sigma$. In computer graphics communities, this is well known as the "Voronoi-dual" of a mesh [4], and is formed by taking barycentric centers of each $\dim\Sigma$-simplex, joining them, and creating a new mesh. This results in a new polyhedral mesh $\mathcal{K}^{\dagger}$, which can also have its own set of simplices. These are distinguished from the main, or "primal" simplicial complex $\mathcal{K}$, by the addition of a $\dagger$.

As there were sets of κ-simplices in the primal simplicial complex $\Omega^{\kappa}(\Sigma)$, there are also those in the dual simplicial complex $\Omega^{\kappa\dagger}(\mathcal{K}^{\dagger})$. These dual simplicial complexes have their own betti numbers $\{b_{\kappa}^{\dagger}\}$, and their own exterior derivative operators $\hat{d}_{\mathcal{K}}^{\kappa\dagger}$, which act on discretized exterior forms on the dual simplicial complex, referred to as "dual" exterior forms, to distinguish them from "primal" exterior forms. The challenge is to map elements on the primal simplicial complex $\mathcal{K}$, to the dual simplicial complex $\mathcal{K}^{\dagger}$, and operating back and forth between these two simplicial complexes forms the bases for DEC. While this meaning is not evident in the continuous case, this is the function served by the Hodge star operators.

The significance of this cannot be overstated. As the Hodge stars allowed traversing down in rank of differential forms over Σ, and juggling operators on differential forms to obtain various differential metrics, the existence of a dual simplicial complex is what allows the same for discretized meshes, and for effective analyses to be performed in DEC. Discretized Hodge star operators $\star_{\mathcal{K}}^{\kappa}$ are also matrices which map discrete primal κ-forms to discrete dual $(\dim \Sigma - \kappa)$-forms. It can be shown that the cardinality of the set of simplices on a dual simplicial complex is complimentary to the primal simplicial complex, in that in general, we see:

$$b_\kappa^\dagger = b_{\dim\Sigma-\kappa}, \{\kappa \in \mathbb{Z} | 0 \le \kappa \le \dim\Sigma\} \tag{10.28}$$

In this case, it can be seen that a discretized Hodge star operator which maps a discrete primal κ-form to a discrete dual $(\dim \Sigma - \kappa)$-form is a $b_\kappa \times b_\kappa$ matrix. While the exterior derivatives encoded the topology of the manifold, the Hodge star operators $\star_\Sigma$ encode the physical manifestation of the manifold, describing areas, lengths, volumes, etc. This is demonstrated by recalling that integration over the manifold was possible simply by application of $\star_\Sigma 1$ There has been much research gone into developing suitable definitions of the discretized Hodge star operator, but the most widely used is the diagonal size-ratio matrix definition which compares the 'sizes' of a primal κ-simplex $\sigma^i_{\Omega^\kappa} \in \Omega^\kappa(\mathcal{K})$ with the size of its canonical corresponding dual $(\dim \Sigma - \kappa)$-simplex $\sigma^i_{\Omega^{(\dim\Sigma-\kappa)\dagger}} \in \Omega^{(\dim\Sigma-\kappa)\dagger}(\mathcal{K}^\dagger)$ given by:

$$(\hat{\star}^\kappa_\mathcal{K})_{ij} = \begin{cases} \dfrac{\left|\sigma^i_{\Omega^{(\dim\Sigma-\kappa)\dagger}}\right|}{\left|\sigma^i_{\Omega^\kappa}\right|} & , \quad i = j \\ 0 & , \quad i \ne j \end{cases} \tag{10.29}$$

Here, the magnitude operator $|\cdot|$ when acting on simplices σ^i assesses their size, which can be length, area, volume, depending on the dimension of σ^i, and setting that for a 0-form, $|\sigma^i| = 1$

10.3.5 Hodge-deRham Co-Homology

One of the Fundamental Results in DEC is the Hodge-deRham Co-homology [4,10]. To current, discretized exterior forms, exterior derivatives, and Hodge star operators have been discussed on primal and dual meshes. Exterior derivatives and Hodge stars have been shown to exist on the primal forms, and the exterior derivative has been established to also exist on the dual simplicial complex. The relation between Hodge star operators and exterior derivatives on the primal and dual mesh is not completely clear.

The Hodge-deRham Co-homology is a major result that establishes the relation between the primal and dual mesh through the exterior derivative (since both share the same topology), and through the Hodge star operator (since the operator is defined using objects on both). The relation is fundamental as it allows the liberal application of these operators and formulation of equation on either the primal or dual mesh and allows a normalized expression of operators through their relations. The easiest manner to observe a connection is through the Hodge star operators. It was defined earlier that entries for the discretized Hodge star matrix are formed from ratios between magnitudes of simplicial complexes on both the primal and dual Hodge star operators. This definition was made from the vantage point of the primal simplicial complex. For the dual simplicial complex, it can be seen to be given by:

$$
(\hat{\star}_{\mathcal{K}}^{\kappa\dagger})_{ij} = \begin{cases} \dfrac{\left|\sigma^i_{\Omega\dim\Sigma-\kappa}\right|}{\left|\sigma^i_{\Omega\kappa\dagger}\right|} & , \quad i = j \\ 0 & , \quad i \neq j \end{cases}
$$

If a particular Hodge star operator on the dual simplicial complex is considered $\star_{\mathcal{K}}^{(\dim\Sigma-\kappa)\dagger}$, then we see:

$$
\left(\hat{\star}_{\mathcal{K}}^{(\dim\Sigma-\kappa)\dagger}\right)_{ij} = \begin{cases} \dfrac{\left|\sigma^i_{\Omega\kappa}\right|}{\left|\sigma^i_{\Omega(\dim\Sigma-\kappa)\dagger}\right|} & , \quad i = j \\ 0 & , \quad i \neq j \end{cases}
$$

It can be noticed that each entry is the reciprocal of the discretized Hodge star operator on the primal simplicial complex. Therefore, the equality can be made between Hodge stars on the primal and dual complexes as:

$$
\boxed{\hat{\star}_{\mathcal{K}}^{(\dim\Sigma-\kappa)\dagger} = \left(\hat{\star}_{\mathcal{K}}^{\kappa}\right)^{-1}}
$$

In addition, it can be shown since the topologies of the manifolds are independent, the exterior derivatives defined from their topologies follow a similar relationship:

$$
\boxed{\left(\hat{d}_{\mathcal{K}}^{\kappa}\right)^T = \hat{d}_{\mathcal{K}}^{(\dim\Sigma-\kappa-1)\dagger}}
$$

Therefore, the Hodge-deRham Co-homology can be summarized at the bottom relating topology and extrinsic geometry of the primal and dual simplicial complexes:

$$
\left\{\hat{\star}_{\mathcal{K}}^{(\dim\Sigma-\kappa)\dagger} = \left(\hat{\star}_{\mathcal{K}}^{\kappa}\right)^{-1}, \left(\hat{d}_{\mathcal{K}}^{\kappa}\right)^T = \hat{d}_{\mathcal{K}}^{(\dim\Sigma-\kappa-1)\dagger}\right\} \tag{10.30}
$$

10.3.6 Specialized Differential Operators

With all relevant operators discussed, a small word will be delivered on the discretization of differential operators through their expression in exterior calculus. As mentioned earlier, the dimension of the underlying manifold determined the signs of co-differentials and Laplace-deRham operators. The same is true in the discrete case, and for the purposes of this investigation, cases on dim $\Sigma = 2$ will be focused on.

10.3.6.1 *Discrete Co-Differential* $(\hat{\tilde{\delta}}^{\kappa}_{\mathcal{K}})$

As outlined earlier, the co-differential $\tilde{\delta}_{\Sigma} : \Lambda^{\kappa*}\Sigma \to \Lambda^{(\kappa-1)*}\Sigma$ for $\Sigma \subset \mathbb{R}^2$ was defined in terms of the Hodge star operators and exterior derivatives as $\tilde{\delta}_{\Sigma} = - \star_{\Sigma} d_{\Sigma}\star_{\Sigma}$. This can be discretized easily to the discrete co-differential $\hat{\tilde{\delta}}^{\kappa}_{\mathcal{K}}$, by considering the discretized exterior derivatives and Hodge star operators, and how they operate on primal and dual forms. Consider a 1-form $u = u_{\alpha}dS^{\alpha}$ which is filtered through the deRham-map to become discretized to the primal discrete 1-form

$$\hat{U} = \begin{bmatrix} U_1 \\ U_2 \\ \cdots \\ U_i \end{bmatrix}, \hat{U}_i = \int_{e_i \in \Omega^1(\mathcal{K})} u_{\alpha}dS^{\alpha} \qquad (10.31)$$

The process of applying the discretized co-differential is as follows:

1. Applying the Hodge star to a primal discrete 1-form requires the discretized Hodge star operator $\hat{\star}^1_{\mathcal{K}}$, which maps the primal discrete 1-form to the dual (2-1)-form, or dual 1-form $\hat{\star}^1_{\mathcal{K}}\hat{U}$.

2. Applying the exterior derivative to a dual discrete 1-form requires the discretized exterior derivative $\hat{d}^{1\dagger}_{\mathcal{K}}$, which maps the dual discrete 1-form to the dual 2-form, $\hat{d}^{1\dagger}_{\mathcal{K}}\hat{\star}^1_{\mathcal{K}}\hat{U}$.

3. Applying the Hodge star to a dual discrete 2-form requires the discretized Hodge star operator $\hat{\star}^{2\dagger}_{\mathcal{K}}$, which maps the dual discrete 2-form to the primal (2-2)-form, or primal 0-form $\hat{\star}^{2\dagger}_{\mathcal{K}}\hat{d}^{1\dagger}_{\mathcal{K}}\hat{\star}^1_{\mathcal{K}}\hat{U}$.

4. Imposing the Hodge-deRham Co-homology to express every operator with respect to the primal simplicial complex, we see that this is equivalent to $(\hat{\star}^0_{\mathcal{K}})^{-1}(\hat{d}^0_{\mathcal{K}})^T\hat{\star}^1_{\mathcal{K}}\hat{U}$

Negating this value since the co-differential has a negative leading coefficient, we see the discretized co-differential acting on 1-primal forms has the DEC form of:

$$\hat{\tilde{\delta}}_{\mathcal{K}}^1 = -\left(\hat{\star}_{\mathcal{K}}^0\right)^{-1}\left(\hat{d}_{\mathcal{K}}^0\right)^T \hat{\star}_{\mathcal{K}}^1, \int_{v_i \in \Omega^0(\mathcal{K})} \tilde{\delta}_{\Sigma} u = \int_{v_i \in \Omega^0(\mathcal{K})} -\nabla^\alpha u_\alpha = \hat{\tilde{\delta}}_{\mathcal{K}}^1 \hat{U}$$

Generalizing this by repeating the process for a discrete primal κ-form on a μ-dimensional Manifold, and applying the Hodge-deRham Cohomology, we see that in general:

$$\boxed{\hat{\tilde{\delta}}_{\mathcal{K}}^\kappa = (-1)^{\mu(\kappa+1)+1}\left(\hat{\star}_{\mathcal{K}}^{\kappa-1}\right)^{-1}\left(\hat{d}_{\mathcal{K}}^{\kappa-1}\right)^T \hat{\star}_{\mathcal{K}}^\kappa} \qquad (10.32)$$

10.3.6.2 *Discrete Laplace-deRham Operator* $(\hat{\Delta}_{\mathcal{K}}^\kappa)$

In a similar fashion to the above co-differential discretization, the Laplace-deRham operator can be defined generally acting on a κ-form on an μ-dimensional manifold as $\Delta_{\Sigma}^{\mathrm{dR}} = \tilde{\delta}_{\Sigma} d_{\Sigma} + d_{\Sigma} \tilde{\delta}_{\Sigma}$. Expanding the definition of the co-differential on a κ-form on a μ-dimensional manifold, this simplifies in terms of the exterior derivative and Hodge star operators as:

$$\Delta_{\Sigma}^{\mathrm{dR}} = \begin{cases} (-1)^{2\mu+1} \star_{\Sigma} d_{\Sigma} \star_{\Sigma} d_{\Sigma} & , \quad \kappa = 0 \\ (-1)^{\mu(\kappa+1)+1}\left(d_{\Sigma} \star_{\Sigma} d_{\Sigma} \star_{\Sigma} +(-1)^\mu \star_{\Sigma} d_{\Sigma} \star_{\Sigma} d_{\Sigma}\right) & , \quad 1 < \kappa < \mu \\ (-1)^{\mu(\mu+1)+1} d_{\Sigma} \star_{\Sigma} d_{\Sigma} \star_{\Sigma} & , \quad \kappa = \mu \end{cases}$$
$$(10.33)$$

Through careful application of the discretized differential operators and Hodge stars, this can be discretized onto the primal-dual simplicial complex $\mathcal{K}$:

$$\hat{\Delta}_\kappa^\kappa = \begin{cases} (-1)^{2\mu+1}\hat{\star}_\kappa^{\mu\dagger}\hat{d}_\kappa^{(\mu-1)\dagger}\hat{\star}_\kappa^1\hat{d}_\kappa^0 & , \quad \kappa = 0 \\[2mm] (-1)^{\mu(\kappa+1)+1}\left(\hat{d}_\kappa^{\kappa-1}\hat{\star}_\kappa^{(\mu-\kappa+1)\dagger}\hat{d}_\kappa^{\kappa}\hat{\star}_\kappa^{\kappa} + (-1)^\mu\hat{\star}_\kappa^{(\mu-\kappa)}\hat{d}_\kappa^{(\mu-\kappa-1)\dagger}\hat{\star}_\kappa^{\kappa+1}\hat{d}_\kappa^\kappa\right) & , \quad 1 < \kappa < \mu \\[2mm] (-1)^{\mu(\mu+1)+1}\hat{d}_\kappa^{\mu-1}\hat{\star}_\kappa^{1\dagger}\hat{\star}_\kappa^{0\dagger}\hat{d}_\kappa^\mu & , \quad \kappa = \mu \end{cases} \tag{10.34}$$

Through careful application of the Hodge-deRham Co-homology, we see the discretized Laplace-deRham operator take the following form:

$$\hat{\Delta}_\kappa^\kappa = \begin{cases} (-1)^{2\mu+1}\left(\hat{\star}_\kappa^0\right)^{-1}\left(\hat{d}_\kappa^0\right)^T\hat{\star}_\kappa^1\hat{d}_\kappa^0 & , \quad \kappa = 0 \\[2mm] (-1)^{\mu(\kappa+1)+1}\left(\hat{d}_\kappa^{\kappa-1}\left(\hat{\star}_\kappa^{\kappa-1}\right)^{-1}\left(\hat{d}_\kappa^{\kappa-1}\right)^T\hat{\star}_\kappa^{\kappa} + (-1)^\mu\left(\hat{\star}_\kappa^{\kappa}\right)^{-1}\left(\hat{d}_\kappa^{\kappa}\right)^T\hat{\star}_\kappa^{\kappa+1}\hat{d}_\kappa^\kappa\right) & , \quad 1 < \kappa < \mu \\[2mm] (-1)^{\mu(\mu+1)+1}\hat{d}_\kappa^{\mu-1}\left(\hat{\star}_\kappa^{\mu-1}\right)^{-1}\left(\hat{d}_\kappa^{\mu-1}\right)^T\hat{\star}_\kappa^\mu & , \quad \kappa = \mu \end{cases}$$
$$\tag{10.35}$$

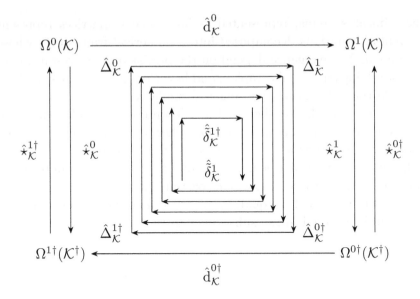

Figure 10.1 Hodge-deRham Co-homology Commutative Diagram on a 1-simplicial complex embedded in $\mathbb{R}^2$

10.3.7 Graphical Hodge-deRham Commutative Diagrams

With the Hodge-dRham Co-homology, and outlining Laplace-deRham operators as cycles in this Co-homology, this can be visually summarized using commutative diagrams. Consider first a 1-dimensional manifold, embedded in two dimensions. This would have a single exterior derivative $\hat{d}^0_\mathcal{K}$, and a single co-differential $\hat{\tilde{d}}^1_\mathcal{K}$, as well as two Hodge star operators $\{\hat{\star}^0_\mathcal{K}, \hat{\star}^1_\mathcal{K}\}$ and two Laplace-deRham operators $\{\hat{\Delta}^0_\mathcal{K}, \hat{\Delta}^1_\mathcal{K}\}$. This would be reflected onto the dual simplicial complex, and arranged as follows:

This can be extended to 2D manifolds where larger cycles involving the Laplace-deRham operators can be observed (Figure 10.1).

10.4 CONCLUSION

In conclusion, Discrete Exterior Calculus is a new and exciting framework which can be used to translate differential operators on manifolds into linear systems on simplicial complex meshes [4].

These linear systems, represented as large sparse matrices represent differential equations decomposed into a numerical framework, whose accuracy and convergence depend on the precision of the resolution of the underlying simplicial complex, and how close it represents the base manifold the differential equations is formulated upon.

This framework can be used to decompose partial differential equations directly into numerical approaches, by expressing it first in a coordinate-independent exterior calculus form, and following through with replacing exterior calculus operators as their DEC counterparts. The applications to solving scalar partial differential equations and vector partial differential equations of use to fluid mechanics and other areas will be covered in the subsequent section.

## CHAPTER 11

Applications of Discrete Exterior Calculus (DEC) to Fluid Mechanics and Fluid-Structure Interactions

Keith C. Afas

Western University

11.1 INTRODUCTION

Fluid Mechanics is a discipline which gives rise to several rigorous analytical and numerical methods to obtain solutions. As discussed, finite-difference (FD) and finite-element (FE) methods contain several drawbacks owing to their lack of generalizability, and limited scope of application.

Discrete Exterior Calculus (DEC) offers an alternative to numerical methods, by supplying a framework that successfully discretizes the machinery of exterior calculus, and allows for differential operators to be naturally expressed in a coordinate-invariant method, that conforms to generalizable meshes, with varying topology, or local geometry. In Chapter 10, the theoretics for notation in exterior calculus, and subsequent discretization into its form acting on discretized exterior

DOI: 10.1201/9781003452256-11

forms defined as arrays on a simplicial mesh were presented. In addition, several differential operators which ext on exterior forms were discretized, into matrices that depended on the topology and extrinsic geometry of the manifold which was approximated as a simplicial complex. In this chapter, the application of these arrays and matrices to numerical techniques which solve partial differential equations (PDEs) is examined and compared against numerical solutions. This will be done for scalar partial differential equations, and vector partial differential equations.

11.1.1 Partial Differential Equation (Scalar): Poisson's and Laplace's Equation

To examine the applicability of DEC to scalar PDEs, Laplace's equation and Poisson's equation will be considered.

11.1.1.1 Poisson's Equation on a Plane

Poisson's equation is a scalar PDE of significance for Harmonic Theory on manifolds, as well as of significance for the Heat and Wave equation. This equation is given by defining ∇^2 as the Laplacian in 2D or 3D space Ω, and $\Phi : \Omega \to \mathbb{R}$ as a scalar field:

$$\nabla^2 \Phi = \Psi \tag{11.1}$$

where $\Psi : \Omega \to \mathbb{R}$ is a function defined throughout space, and is commonly referred to as the "source function", which gives genesis to Φ. Typically, Poisson's equation is in n-dimensional flat space (ie. the square $n = 2$, the cube $n = 3$) and is parametrized by the position vector $\mathbf{R} = (x_1, x_2, \ldots, x_n)$. As such, the Laplacian can be defined as the sum of second partial derivatives in the space.

$$\nabla^2 = \frac{\partial}{\partial x_1^2} + \frac{\partial}{\partial x_2^2} + \ldots + \frac{\partial}{\partial x_n^2} = \sum_{m=1}^{n} \frac{\partial}{\partial x_m^2} \tag{11.2}$$

This expression can be transformed into other general coordinate systems by use of the Jacobian of the transformation. For non-standard shapes, and for unconventional topologies, Poisson's equation can be extended to a manifold Σ, by identifying the Laplacian ∇^2 in n-dimensional space Ω, as the Laplace-Beltrami/Laplace-deRham

Operator Δ_Σ^{dR} on Σ. If it is specified that Σ is a 2-dimensional manifold, then it can be seen that the solution to Poisson's equation is now a scalar function on Σ, or $\Phi : \Sigma \to \mathbb{R}$. The LDR operator acts on exterior forms, where Φ can be thought of as a 0-form. In Chapter 10 recalling that Δ_Σ^{dR} on a 0-form on a 2-dimensional manifold is $\Delta_\Sigma^{dR} = - \star_\Sigma d_\Sigma \star_\Sigma d_\Sigma$ where d_Σ and $\star_\Sigma$ are the exterior derivative and Hodge star operator respectively, then Poisson's equation on a manifold Σ simplifies to:

$$\Delta_\Sigma^{dR}\Phi = - \star_\Sigma d_\Sigma \star_\Sigma d_\Sigma\Phi = \Psi \qquad (11.3)$$

where now Ψ is a function defined on the manifold Σ. With Poisson's equation in this coordinate independent expression, it is well suited for discretization into DEC. In this chapter, Poisson's equation will be considered on manifold Σ, which is the unit plane. A plane and its discretization are displayed in the following Figure 11.1 : Poisson's equation will be considered on this simplicial complex for the source

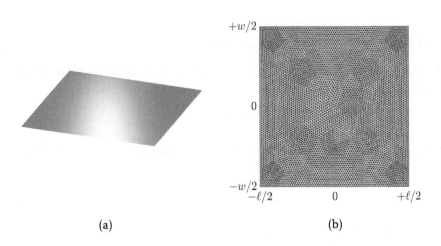

(a)   (b)

Figure 11.1 Geometric representation of a $\ell \times w$ plane (a), and its discretization into a simplicial complex (b) composed of 11,776 elements and 6,033 nodes. Here, the unit plane is considered, so $\{w = 1, \ell = 1\}$. (a) A $w \times \ell$ plane in 3D. (b) Mesh discretization of plane.

function $\Psi = -20e^{-50(x^2+y^2)}$ under Dirichlet Conditions:

$$\Phi|_{x=-\ell/2} = -\frac{1}{5}\sin(\pi y), \Phi|_{x=\ell/2} = \frac{1}{5}\sin(\pi y)$$

$$\Phi|_{y=-w/2} = -\frac{1}{5}\sin(\pi x), \Phi|_{y=w/2} = \frac{1}{5}\sin(\pi x)$$

By identifying the boundary of the plane as $\partial\Sigma$, we can decompose its boundary into the union of the left, right, bottom, and top boundaries, denoted as $\{\partial\Sigma_1, \partial\Sigma_2, \partial\Sigma_3, \partial\Sigma_4\}$. By defining $\Xi = \frac{1}{5}\sin(\pi x)\sin(\pi y)$, then the Dirichlet functions on each of these boundaries can be given as $\{\Xi|_{\partial\Sigma_1} \Xi|_{\partial\Sigma_2}, \Xi|_{\partial\Sigma_3}, \Xi|_{\partial\Sigma_4}\}$, respectively. In this fashion, all the Dirichlet boundary conditions can be summarized by the equation $\Phi|_{\partial\Sigma} = \Xi|_{\partial\Sigma}$. Replacing the boundary conditions with this form, and combining it with Poisson's equation in exterior calculus form, the full coordinate-independent form of the Poisson equation system which will be examined in this section is:

$$\left\{ \overbrace{-\star_\Sigma d_\Sigma \star_\Sigma d_\Sigma \Phi = \Psi}^{\text{Poisson's equation}}, \overbrace{\Phi|_{\partial\Sigma} = \Xi|_{\partial\Sigma}}^{\text{Dirichlet condition}} \right\} \qquad (11.4)$$

11.1.1.2 Laplace's Equation on an Annulus

Laplace's equation is a related PDE to Poisson's equation, if the source equation is identically zero $\Psi = 0$, the Laplace's equation is:

$$\nabla^2\Phi = 0 \qquad (11.5)$$

Similarly to Poisson's equation, this can be expressed on a manifold as the following form, or $\Delta_\Sigma^{\text{dR}}\Phi = 0$:

$$\star_\Sigma d_\Sigma \star_\Sigma d_\Sigma \Phi = 0 \qquad (11.6)$$

In this chapter, Poisson's equation will be considered on manifold Σ, which is an annulus with an inner radius of $r_1 = 2$, and $r_2 = 4$. This annulus and its discretization are displayed in the following Figure 11.2b: Laplace's equation will be considered on this simplicial complex for the source function under Dirichlet Conditions:

$$\Phi|_{r_1} = 0, \Phi|_{r=r_2} = 4\sin(5\theta) \qquad (11.7)$$

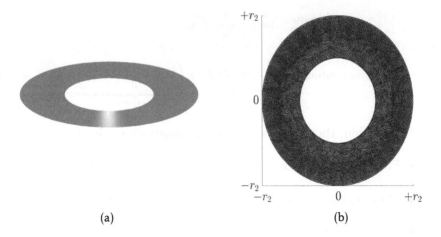

Figure 11.2 Geometric representation of an annulus with inner/outer radii r_1, r_2 (a), and its discretization into a simplicial complex (b) composed of 19,003 elements and 9,773 nodes. Here, the radii $\{r_1 = 2, r_2 = 4\}$ are considered. (a) Annulus with radii r_1, r_2. (b) Mesh discretization of annulus.

Similarly to the Poisson's equation case, letting the inner ring and outer ring boundaries be denoted by $\partial\Sigma_1$ and $\partial\Sigma_2$ respectively, and letting the outer Dirichlet function be denoted by the boundary function $\Xi :$ $\partial\Sigma_2 \to \mathbb{R}$, such that $\Xi(\partial\Sigma_2) = 4\sin(5\theta)$, then the full coordinate-invariant form of the Laplace equation system which will be examined in this section is:

$$\left\{ \overbrace{\star_\Sigma d_\Sigma \star_\Sigma d_\Sigma \Phi = 0}^{\text{Laplace's equation}}, \overbrace{\Phi|_{\partial\Sigma_1} = 0, \Phi|_{\partial\Sigma_2} = \Xi}^{\text{Dirichlet conditions}} \right\} \tag{11.8}$$

11.1.2 Partial Differential Equation (Vector): Poiseuille Flow and Stokes Flow

To examine the applicability of DEC to scalar PDEs, the Navier Stokes Equations (NSEs) will be considered. For a linear-stress fluid, the NSE mass-conservation and momentum-conservation equations take the general form:

Mass Conservation : $\underbrace{\partial_t \rho}_{\substack{\text{density} \\ \text{change}}} + \underbrace{\vec{\nabla} \cdot (\rho \vec{u})}_{\substack{\text{density decrease} \\ \text{due to outward flux}}} = 0$

Momentum Conservation : $\rho \underbrace{\left(\partial_t + \vec{u} \cdot \vec{\nabla} \right) \vec{u}}_{\text{inertial effects}} - \underbrace{\mu \left(\nabla^2 \vec{u} + \frac{1}{3} \vec{\nabla}(\vec{\nabla} \cdot \vec{u}) \right)}_{\text{viscosity effects}} + \underbrace{\vec{\nabla} p - \rho \vec{f}}_{\substack{\text{external} \\ \text{forces}}} = \vec{0}$

In this section, there will be several assumptions made:

- Fluid has negligible inertial effects: $\left(\partial_t + \vec{u} \cdot \vec{\nabla} \right) \vec{u} \approx \vec{0}$

- Flow is at *steady-state*: $\{ \partial_t \vec{u} = 0, \partial_t \rho \}$

- Fluid is *isotropic*: $\{ \vec{\nabla} \rho = 0 \}$

- Fluid is *incompressible*: $\{ \vec{\nabla} \cdot \vec{u} = 0 \}$

- Fluid has *unit density* and *unit viscosity*: $\rho = \mu = 1$

- Fluid has *no body forces*: $\vec{f} = \vec{0}$

Using these assumptions, the NSEs over a space Ω with a vector space $\mathcal{V}$ reduce to the system for the vector fluid velocity field $\vec{u} : \Omega \to \mathcal{V}$, and scalar pressure field $p : \Omega \to \mathbb{R}$, given by:

$$\left\{ \vec{\nabla} \cdot \vec{u} = 0, \vec{\nabla} p - \nabla^2 \vec{u} = \vec{0} \right\} \tag{11.9}$$

In n-dimensions, these equations can be defined explicitly with respect to a cartesian-like flat coordinate system $(x_1, x_2, \ldots, x_n)$. Defining $\vec{u} = [u_1 \quad u_2 \quad \ldots \quad u_n]^T$, and abbreviating $\partial/\partial x_m = \partial_m$, then the set of equations has the following explicit form:

$$\left\{ \sum_{m=1}^{n} \partial_m u_m = 0, \partial_k p - \sum_{m=1}^{n} \partial_m^2 u_k = 0 \right\} \tag{11.10}$$

This form of the NSEs is essentially the equations which define *stokes flow*. These NSEs are defined in a space Ω, but can be extended to a general n-dimensional manifold Σ, by utilizing the Laplace-deRham operator Δ_Σ^{dR}, the co-differential $\tilde{\delta}_\Sigma$, and the exterior derivative d_Σ. In the context of exterior forms, the pressure field $p(\Sigma)$ can be considered to be a 0-form on Σ, where $\vec{u}$ can be identified as the sharp musical isomorphism of a 1-form $u = u_\alpha dS^\alpha$ defined on Σ, such that $\vec{u} = u^\sharp$. In n-dimensional space, recalling that the Laplace-deRham operator on a 1-form takes the following form $\Delta_\Sigma^{dR} u = -(d_\Sigma \star_\Sigma d_\Sigma \star_\Sigma +(-1)^n \star_\Sigma d_\Sigma \star_\Sigma d_\Sigma)$, the co-differential takes the following form $\delta_\Sigma = -\star_\Sigma d_\Sigma \star_\Sigma$, all that remains is for the Laplacian ∇^2 and divergence operator $\vec{\nabla} \cdot (\cdot)$ to be identified with these in explicit form. For the purposes of this chapter, the exterior calculus form of the NSEs will be analyzed for two cases of interest:

- Poiseuille Flow on a Rectangular Manifold

- Stokes Flow on a Rectangular Manifold, around a circular hole.

Both of these cases assume a base manifold Σ which is $(n = 2)$-dimensional. Consequentially, it was established in Chapter 10 that for a 1-form on a 2-dimensional manifold, $\nabla^2 = -\Delta_\Sigma^{dR}$ and $\tilde{\delta}_\Sigma = -\vec{\nabla} \cdot (\cdot)$. Therefore, the NSEs can be expressed in exterior calculus form:

$$\left\{ d_\Sigma p + \Delta_\Sigma^{dR} u = 0, -\tilde{\delta}_\Sigma u = 0 \right\} \tag{11.11}$$

These can be expressed in terms of exterior derivatives and Hodge star operators as the following:

$$\left\{ \overbrace{d_\Sigma p - (d_\Sigma \star_\Sigma d_\Sigma \star_\Sigma + \star_\Sigma d_\Sigma \star_\Sigma d_\Sigma)u = 0}^{\text{Navier-Stokes momentum equations}}, \overbrace{\star_\Sigma d_\Sigma \star_\Sigma u = 0}^{\text{mass conservation}} \right\} \tag{11.12}$$

This equation will have boundary conditions formulated in line with the two cases of interest considered above.

Poiseuille Flow on a Rectangular Manifold

For Poiseuille flow on a rectangular manifold, the manifold represents a cross-section through the midsection of a rectangular pipe. Consequentially, the geometry and a suitable discretization into a simplicial mesh can be considered as shown below:

To obtain a steady-state solution on this discretization of the rectangular manifold, suitable boundary conditions must be imposed. To obtain poiseuille flow, poiseuille-like solutions will be set on the boundary of the mesh. Similarly to the Poisson equation case, the left, right, bottom, and top boundaries will be denoted by $\{\partial\Sigma_1, \partial\Sigma_2, \partial\Sigma_3, \partial\Sigma_4\}$ respectively. The solutions to this system are both the pressure field and velocity field, so both of these must have values prescribed on the boundary.

For the pressure conditions, a pressure differential of $\vec{\nabla}p = -\vec{\mathbf{x}}$ is desired to be obtained. Therefore, at the inlet and outlet, $\{p|_{\partial\Sigma_1} = 4, p|_{\partial\Sigma_2} = -4\}$ will take care of this. For the stability of the solver, a linearly decreasing profile at the top and bottom of the rectangular manifold in line with the inlet and outlet values will also be specified $\{p|_{\partial\Sigma_2} = p|_{\partial\Sigma_3} = 4 - x\}$. Letting $P(\Sigma) = 4 - x$, then the pressure field's Dirichlet boundary conditions can be summarized as $p|_{\partial\Sigma} = P|_{\partial\Sigma}$. This completely fixes the conditions required for the pressure scalar field (Figure 11.3).

Next, for the velocity vector field, a no-slip condition on the top and bottom of the mesh will be considered $\vec{\mathbf{u}}|_{\partial\Sigma_3} = \vec{\mathbf{u}}|_{\partial\Sigma_4} = \vec{\mathbf{0}}$. Finally, the velocity at the inlet and outlet will be set to the steady-state poiseuille solution in advance. $\{\vec{\mathbf{u}}|_{\partial\Sigma_1} = \vec{\mathbf{u}}|_{\partial\Sigma_2} = \frac{1}{2}\left(y^2 - 1\right)\hat{\mathbf{x}}\}$. To specify this in exterior calculus form, special care will need to be taken. The no-slip condition can be easily implemented on the 1-form as $u|_{\partial\Sigma_3\cup\partial\Sigma_4} = 0$,

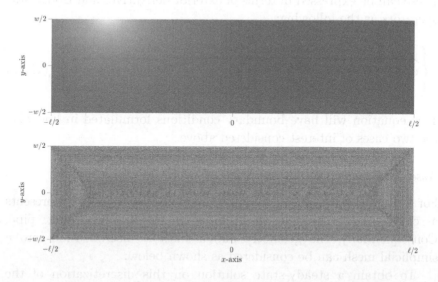

Figure 11.3 Geometry (top) and mesh simplicial complex (bottom) of a $\ell \times w$ plane, with $\{w = 2, \ell = 8\}$. Discretization into a simplicial complex produced a mesh with 38,622 nodes and 76,332 elements.

where since $\hat{\mathbf{x}}^\flat = dx$ in 1-form expression, and letting the 1-form $\xi = \frac{1}{2}(y^2 - 1)dx$, then we can state that $\{u_{\partial \Sigma_1 \cup \partial \Sigma_2} = \xi\}$. Therefore, the final coordinate-independent form of the NSEs on a rectangular manifold Σ is:

$$
\left\{
\begin{array}{ll}
\overbrace{\mathrm{d}_\Sigma p - (\mathrm{d}_\Sigma \star_\Sigma \mathrm{d}_\Sigma \star_\Sigma + \star_\Sigma \mathrm{d}_\Sigma \star_\Sigma \mathrm{d}_\Sigma)u = 0}^{\text{Navier-Stokes momentum equations}} , & \overbrace{\star_\Sigma \mathrm{d}_\Sigma \star_\Sigma u = 0}^{\text{mass conservation}} \\[4pt]
\overbrace{u|_{\partial \Sigma_1 \cup \partial \Sigma_2} = \xi \,,\ u|_{\partial \Sigma_3 \cup \partial \Sigma_4} = 0}^{\text{velocity Dirichlet conditions}} , & \overbrace{p|_{\partial \Sigma} = P}^{\text{pressure Dirichlet conditions}}
\end{array}
\right\}
$$

$$(11.13)$$

Stokes Flow on a Rectangular Manifold with a Hole

For Stokes flow around a hole on a rectangular manifold, the rectangular manifold is taken as in the poiseuille flow case representing a cross-section through the midsection of a rectangular pipe, and a hole is immersed in the center of the manifold with radius r_1 (Figure 11.4). Consequentially, the geometry and a suitable discretization into a simplicial mesh can be considered as shown below:

As for boundary conditions for this equation, they are identical to the poiseuille flow case, with the addition of flow around the hole, which adds a boundary $\partial \Sigma_5$ to the equation which must be dealt with in terms of pressure and velocity. For pressure, it is assumed that the pressure at this hole will be the values of the function $P = 4 - x$ evaluated on the hole, and the velocity will have what is called a free-slip condition on the hole, where the velocity at the hole only requires its normal component to vanish, meaning the velocity is locally tangential to the hole. This is expressed as $\mathbf{n} \cdot \vec{\mathbf{u}}|_{\partial \Sigma_5} = 0$, or in exterior calculus form, where the dot product can be expressed in terms of the edge product and the Hodge star operator, $\star_\Sigma(n \wedge \star_\Sigma u|_{\partial \Sigma_5}) = 0$. Therefore, the final form of the NSEs on this stokes flow problem around a hole on a rectangular manifold is:

$$
\left\{
\begin{array}{ll}
\overbrace{\mathrm{d}_\Sigma p - (\mathrm{d}_\Sigma \star_\Sigma \mathrm{d}_\Sigma \star_\Sigma + \star_\Sigma \mathrm{d}_\Sigma \star_\Sigma \mathrm{d}_\Sigma)u = 0}^{\text{Navier-Stokes momentum equations}} , & \overbrace{\star_\Sigma \mathrm{d}_\Sigma \star_\Sigma u = 0}^{\text{mass conservation}} \\[4pt]
\underbrace{u|_{\partial \Sigma_1 \cup \partial \Sigma_2} = \xi \,,\ u|_{\partial \Sigma_3 \cup \partial \Sigma_4} = 0 \,,\ \star_\Sigma(n \wedge \star u|_{\partial \Sigma_5}) = 0}_{\text{velocity Dirichlet conditions}} , & \overbrace{p|_{\partial \Sigma} = P}^{\text{pressure Dirichlet conditions}}
\end{array}
\right\}
$$

$$(11.14)$$

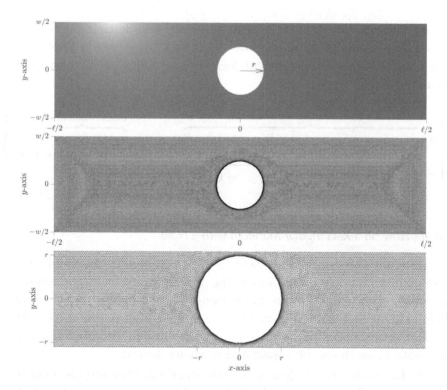

Figure 11.4 Geometry (top row), and mesh simplicial complex (middle) of a $\ell \times w$ plane with an r-radius hole, with $\{w = 2, \ell = 8, r = 1/2\}$. Mesh is divided with further refinement around the hole (bottom). Discretization into a simplicial complex produced a mesh with 41,177 nodes, and 80,051 elements.

11.2   METHODS

As covered in the introduction, all the final forms of the equation will be translated into Discrete Exterior Calculus (DEC) operators, and a linear system will be formulated to solve the systems contained in Eqs. (11.4, 11.8, 11.13, 11.14) on a simplicial complex $\mathcal{K}$ which is approximating their base manifold $\Sigma \sim \mathcal{K}$. In general, the principle behind DEC solvers is to consider translating DEC operators into matrix form directly, formulate the governing equation as a matrix system, and then boundary conditions are treated on a case-by-case basis, usually by either formulating special differential operators to include these conditions as force terms in the system (as in the case for the NSEs), or by separating the problem into boundary and non-boundary nodes, and solving each linear system independently. In general, this can be simplified as the following code:

Data: Σ, BCs, PDE
Result: Solution
%% Generate Mesh simplex structure $\mathcal{K}$ from geometry
Σ %%
$\mathcal{K} \leftarrow$ generateDECMesh(Σ);
%% Get simplices $\Omega^\kappa(\mathcal{K})$, and betti Numbers b_κ %%
$\Omega^\kappa(\mathcal{K}) \leftarrow$ getSimplices($\mathcal{K}$);
$b_\kappa \leftarrow$ getBettiNumbers($\mathcal{K}$);
%% Generate exterior derivatives, hodge star
operators %%
$\hat{\mathrm{d}}_\mathcal{K}^\kappa \leftarrow$ getExteriorDerivatives($\mathcal{K}$);
$\hat{\star}_\mathcal{K}^\kappa \leftarrow$ getHodgeStar($\mathcal{K}$);
%% Find nodes on boundary $\partial\Sigma$ %%
$i_\text{nodes} \leftarrow 1..b_0$;
$\partial\mathcal{K} \leftarrow$ getBoundary($\mathcal{K}$);
if $type(BCs)==$"$dirichlet$" **then**
$\quad | \quad i_{\partial\mathcal{K}} \leftarrow$ findNodeIndices($\partial\Sigma$);
$\quad | \quad i_{\sim\partial\Sigma} \leftarrow i_\text{nodes}(i_{\partial\Sigma})$;
end
Solution $\leftarrow$ solverDEC(PDE, BCs, $i_{\partial\Sigma}, \hat{\mathrm{d}}_\mathcal{K}^\kappa, \hat{\star}_\mathcal{K}^\kappa, i_{\partial\mathcal{K}}$)
    Algorithm 1: Generalized DEC Solving Algorithm

The form of the solverDEC function, will be expanded upon in the subsequent sections per PDE considered.

11.2.1 DEC Poisson Equation on Plane

As covered above, the exterior calculus form of Poisson's equation considered on the plane, along with the Dirichlet boundary conditions considered, are given as:

$$\left\{ \overbrace{- \star_\Sigma \mathrm{d}_\Sigma \star_\Sigma \mathrm{d}_\Sigma \Phi = \Psi,}^{\text{Poisson's equation}} \overbrace{\Phi|_{\partial\Sigma} = \Xi|_{\partial\Sigma}}^{\text{Dirichlet condition}} \right\}$$

The scalar field Φ can be identified as a 0-form on Σ, and therefore can be discretized on the vertices of the approximating simplicial complex $\Omega^0(\mathcal{K})$. This converts the scalar field into an array of scalar values, as a discretized 0-form on $\mathcal{K}$ by $\phi_i = \int_{v_i \in \Omega^0(\mathcal{K})} \Phi$; this can be collected into a $\mathbb{R}^{|\Omega^0(\mathcal{K})| \times 1}$ array, $\vec{\phi} = [\phi_1, \phi_2, \ldots, \phi_{b_0}]^T$. Similarly, the source function Ψ can also be discretized in like into a discrete 0-form on $\mathcal{K}$ as $\psi_i = \int_{v_i \in \Omega^0(\mathcal{K})} \Psi$, and collected into the vertex array $\vec{\psi}$. Finally, replacing the exterior derivative operators with their DEC counterparts, and being careful of using the primal or dual DEC operators, Poisson's equation may be expressed as $(-\hat{\star}_{\mathcal{K}}^{2\dagger} \hat{d}_{\mathcal{K}}^{1\dagger} \hat{\star}_{\mathcal{K}}^1 \hat{d}_{\mathcal{K}}^0) \vec{\phi} = \vec{\psi}$. Applying the Hodge-deRham Cohomology, this can be expressed entirely in terms of DEC matrices with respect to the primal simplicial mesh:

$$\left(- \left(\hat{\star}_{\mathcal{K}}^0 \right)^{-1} \left(\hat{d}_{\mathcal{K}}^0 \right)^T \hat{\star}_{\mathcal{K}}^1 \hat{d}_{\mathcal{K}}^0 \right) \vec{\phi} = \vec{\psi} \tag{11.15}$$

At this point, the Dirichlet boundary conditions will be considered. The condition is set in the continuous manifold upon $\partial\Sigma$. On the simplicial complex, the boundary $\partial\mathcal{K}$, represents a set of node indices. On some of these, the solution is known, and on the non-boundary node indices, the linear system from Eq. (11.15) must be utilized. The array $\vec{\phi}$ can be separated into two components therefore, $\vec{\phi} = \vec{\phi}_{\partial\mathcal{K}} + \vec{\phi}_{\mathcal{K}}$, where letting Ξ be discretized per node into $\xi_i = \int_{v_i \in \Omega^0(\mathcal{K})} \Xi$, we can define $\vec{\phi}_{\partial\mathcal{K}}$ per component $((\phi_{\partial\mathcal{K}})_i)$ as:

$$(\phi_{\partial\mathcal{K}})_i = \begin{cases} \xi_i & , \quad i \in \partial\mathcal{K} \\ 0 & , \quad \text{otherwise} \end{cases} \tag{11.16}$$

In this manner, we see that the full DEC system can be simplified to:

$$\left(- \left(\hat{\star}_{\mathcal{K}}^0 \right)^{-1} \left(\hat{d}_{\mathcal{K}}^0 \right)^T \hat{\star}_{\mathcal{K}}^1 \hat{d}_{\mathcal{K}}^0 \right) \vec{\phi}_{\mathcal{K}} = \vec{\psi} - \left(- \left(\hat{\star}_{\mathcal{K}}^0 \right)^{-1} \left(\hat{d}_{\mathcal{K}}^0 \right)^T \hat{\star}_{\mathcal{K}}^1 \hat{d}_{\mathcal{K}}^0 \right) \vec{\phi}_{\partial\mathcal{K}} \tag{11.17}$$

And therefore, we have the solution for $\vec{\phi}$ on **non-boundary nodes** as:

$$-\left(\left(\hat{\star}_{\mathcal{K}}^0 \right)^{-1} \left(\hat{d}_{\mathcal{K}}^0 \right)^T \hat{\star}_{\mathcal{K}}^1 \hat{d}_{\mathcal{K}}^0 \right) \vec{\phi}_{\mathcal{K}} = \vec{\psi} + \left(\left(\hat{\star}_{\mathcal{K}}^0 \right)^{-1} \left(\hat{d}_{\mathcal{K}}^0 \right)^T \hat{\star}_{\mathcal{K}}^1 \hat{d}_{\mathcal{K}}^0 \right) \vec{\phi}_{\partial\mathcal{K}}$$

Redefining the Laplace-deRham operator as $\hat{\Delta}_{\mathcal{K}}^0 = -\left(\hat{\star}_{\mathcal{K}}^0 \right)^{-1} \left(\hat{d}_{\mathcal{K}}^0 \right)^T \hat{\star}_{\mathcal{K}}^1 \hat{d}_{\mathcal{K}}^0$, and letting its components be given by $(\hat{\Delta}_{\mathcal{K}}^0)_{ij}$ then the full solution to the DEC problem is given by:

$$\phi_i = \begin{cases} (\phi_{\partial \mathcal{K}})_i & , \quad i \in \partial \mathcal{K} \\ \sum_j \left((\hat{\Delta}^0_{\mathcal{K}})^{-1} \right)_{ij} \left(\psi_j - \sum_k (\hat{\Delta}^0_{\mathcal{K}})_{jk} (\phi_{\partial \mathcal{K}})_k \right) & , \quad \text{otherwise} \end{cases}$$

(11.18)

The computational efficiency of solving this equation lies in the efficiency of sparse solvers to invert the sparse matrix $\hat{\Delta}^0_{\mathcal{K}}$. This can then be visualized using a graphing software, and will be visualized in subsequent sections.

11.2.2 DEC Laplace Equation on Annulus

As covered above, the exterior calculus form of Laplace's equation considered on the annulus, along with the Dirichlet boundary conditions considered, are given as:

$$\left\{ \overbrace{\star_\Sigma d_\Sigma \star_\Sigma d_\Sigma \Phi = 0}^{\text{Laplace's equation}}, \overbrace{\Phi|_{\partial \Sigma_1} = 0, \Phi|_{\partial \Sigma_2} = \Xi}^{\text{Dirichlet conditions}} \right\}$$

Similarly to the above Poisson equation example, the discretization of Laplace's equation, and inclusion of the Dirichlet condition are as follows. To implement two Dirichlet conditions, the boundary node array can be defined as:

$$(\phi_{\partial \mathcal{K}})_i = \begin{cases} 0 & , \quad i \in \partial \mathcal{K}_1 \\ \xi_i & , \quad i \in \partial \mathcal{K}_2 \\ 0 & , \quad \text{otherwise} \end{cases}$$

(11.19)

Having defined this object on the boundary, similarly defining $\hat{\Delta}^0_{\mathcal{K}} = -\left(\hat{\star}^0_{\mathcal{K}} \right)^{-1} \left(\hat{d}^0_{\mathcal{K}} \right)^T \hat{\star}^1_{\mathcal{K}} \hat{d}^0_{\mathcal{K}}$, the boundary node term can be built in as a force term, and the solution to Laplace's equation on the annulus is:

$$\phi_i = \begin{cases} 0 & , \quad i \in \partial \mathcal{K}_1 \\ (\phi_{\partial \mathcal{K}})_i & , \quad i \in \partial \mathcal{K}_2 \\ -\sum_j \left((\hat{\Delta}^0_{\mathcal{K}})^{-1} \right)_{ij} \left(\sum_k (\hat{\Delta}^0_{\mathcal{K}})_{jk} (\phi_{\partial \mathcal{K}})_k \right) & , \quad \text{otherwise} \end{cases}$$

(11.20)

As can be seen, from an Annulus to a Plane, nothing has changed in the solution process, aside from the consideration of the boundary definition as an extra case. This can then be visualized using a graphing software and will be visualized in subsequent sections.

11.2.3 DEC Poiseuille Flow with Steady-State Boundary and Initial Conditions

The utility of DEC as a numerical technique shines when considering systems of PDEs as in the NSEs example for Poiseuille flow. As covered above, the exterior calculus form of the NSEs considered on the rectangle, along with assumptions made and Dirichlet boundary conditions considered, is given as:

$$
\left\{
\begin{array}{cc}
\overbrace{d_\Sigma p - (d_\Sigma \star_\Sigma d_\Sigma \star_\Sigma + \star_\Sigma d_\Sigma \star_\Sigma d_\Sigma)u = 0}^{\text{Navier-Stokes momentum equations}} & , \quad \overbrace{\star_\Sigma d_\Sigma \star_\Sigma u = 0}^{\text{mass conservation}} \\[2mm]
\underbrace{u|_{\partial\Sigma_1 \cup \partial\Sigma_2} = \xi \, , \; u|_{\partial\Sigma_3 \cup \partial\Sigma_4} = 0}_{\text{velocity Dirichlet conditions}} & , \quad \underbrace{p|_{\partial\Sigma} = P}_{\text{pressure Dirichlet conditions}}
\end{array}
\right\}
$$

Similarly to the scalar cases, the NSE solver is no different, with the exception of noting that the two fields of interest are now not only a 0-form p(Σ) but also a 1-form $u(\Sigma)$. These are discretized respectfully as:

$$
p_i = \int_{v_i \in \Omega^0(\mathcal{K})} p \, , \; u_i = \int_{e_i \in \Omega^1(\mathcal{K})} u_\alpha \, dS^\alpha \tag{11.21}
$$

Collecting these array-wise definitions into node and edge numerical arrays $\vec{\pi}$ and $\vec{u}^\flat$, respectively, then the velocity-pressure equation can be discretized in DEC as:

$$
\hat{d}_\mathcal{K}^0 \vec{\pi} - \left(\hat{d}_\mathcal{K}^0 \hat{\star}_\mathcal{K}^{2\dagger} \hat{d}_\mathcal{K}^{1\dagger} \hat{\star}_\mathcal{K}^1 + \hat{\star}_\mathcal{K}^{1\dagger} \hat{d}_\mathcal{K}^{0\dagger} \hat{\star}_\mathcal{K}^2 \hat{d}_\mathcal{K}^1 \right) \vec{u}^\flat = 0 \tag{11.22}
$$

Similarly, the incompressibility condition can be expressed in terms of DEC operators as:

$$
\hat{\star}_\mathcal{K}^{2\dagger} \hat{d}_\mathcal{K}^{1\dagger} \hat{\star}_\mathcal{K}^1 \vec{u}^\flat = 0 \tag{11.23}
$$

The Dirichlet conditions for pressure and no-slip conditions for the velocity 1-form are relatively straightforward, and handled in a manner similarly discussed earlier by separating $\vec{\pi}$ and $\vec{u}^\flat$ into the boundary value arrays $\vec{u}^\flat_{\partial\mathcal{K}}$ and $\vec{\pi}^\flat_{\partial\mathcal{K}}$, and the non-boundary value arrays $\vec{u}^\flat_\mathcal{K}$ and $\vec{\pi}^\flat_\mathcal{K}$. The inlet-outlet conditions for the velocity Dirichlet condition must be handled with more care. In exterior calculus, the value for u on $\partial\Sigma_1$ and $\partial\Sigma_2$ are handled through the definition of the primitive one forms dx and dy. These do not exist on a simplicial complex, with numerous edges. Prescriptions of u on the edges of the simplicial manifold will be automatically interpreted as tangential, and therefore setting the value for u on the edge will be interpreted as a tangential velocity, which is not the goal of the condition that is to prescribe the velocity field coming in **normal** to the boundary. This can be remedied through the use of the hodge star operators $\hat{\star}_\mathcal{K}^1$.

Another function of the Hodge star operators is to map discretized forms to their dual forms on the dual simplicial complex. Interestingly enough, by definition of the dual simplicial complex, these edges are completely perpendicular to the primal simplicial complex, and therefore by prescribing values on the dual simplicial edges, we can achieve a similar result to prescribing values normal to the primal simplicial complex edges. Therefore, one can define another vertex array, $(\vec{\star u})_0^\flat$ in addition to $\vec{u}_0^\flat$ which prescribes values tangential to the edges, and if the correct differential operators are found that include this normal vertex array on the boundary, then the normal Dirichlet boundary condition can be implemented.

In order to factor in this normal vertex array $(\vec{\star u})_0^\flat$, correct differential operators must be sourced to be able to operate on it. A differential operator $\hat{d}_{\partial\mathcal{K}}^1$ must be defined such that it is complimentary to the standard 1-form discrete exterior derivative, and is defined as such:

$$\hat{d}_{\partial\mathcal{K}}^1(\vec{\star u}) = \hat{d}_\mathcal{K}^{1\dagger}\hat{\star}_\mathcal{K}^1\vec{u}, \text{ on } \partial\mathcal{K} \tag{11.24}$$

For a primal discretized 1-form $\vec{u}$ and a dual discretized 1-form $(\vec{\star u})$. Using this, defining the Laplace-deRham discretized operator on 1-forms as $\hat{\Delta}_\mathcal{K}^1 = -\left(\hat{d}_\mathcal{K}^0\hat{\star}_\mathcal{K}^{2\dagger}\hat{d}_\mathcal{K}^{1\dagger}\hat{\star}_\mathcal{K}^1 + \hat{\star}_\mathcal{K}^{1\dagger}\hat{d}_\mathcal{K}^{0\dagger}\hat{\star}_\mathcal{K}^2\hat{d}_\mathcal{K}^1\right)$, and defining the complimentary Laplace-deRham discretized operator on 1-forms formed with $\hat{d}_{\partial\mathcal{K}}^1$ as $\hat{\Delta}_{\partial\mathcal{K}}^1 = -\left(\hat{d}_\mathcal{K}^0\hat{\star}_\mathcal{K}^{2\dagger}\hat{d}_{\partial\mathcal{K}}^1\right)$, then the NSE equations can be modified to include the Dirichlet (tangential and normal) boundary conditions as such:

$$\begin{cases} \hat{d}_\mathcal{K}^0\vec{\pi}_\mathcal{K} + \hat{\Delta}_\mathcal{K}^1\vec{u}_\mathcal{K}^\flat = -\hat{d}_\mathcal{K}^0\vec{\pi}_{\partial\mathcal{K}} - \hat{\Delta}_\mathcal{K}^1\vec{u}_{\partial\mathcal{K}}^\flat - \hat{\Delta}_{\partial\mathcal{K}}^1(\vec{\star u})_0^\flat \\ \hat{\star}_\mathcal{K}^{2\dagger}\hat{d}_\mathcal{K}^{1\dagger}\hat{\star}_\mathcal{K}^1\vec{u}_\mathcal{K}^\flat = -\hat{\star}_\mathcal{K}^{2\dagger}\hat{d}_\mathcal{K}^{1\dagger}\hat{\star}_\mathcal{K}^1\vec{u}_{\partial\mathcal{K}}^\flat - \hat{\star}_\mathcal{K}^{2\dagger}\hat{d}_{\partial\mathcal{K}}^1(\vec{\star u})_0^\flat \end{cases} \tag{11.25}$$

This can be formulated into a master system block matrix equation by defining the node-edge solution array vector $\vec{\sigma}_\mathcal{K} = [\vec{u}_\mathcal{K}^\flat \quad \vec{\pi}_\mathcal{K}]^T$. In this fashion, the entire system of NSEs can be summarized by the following master equation system:

$$\begin{bmatrix} \hat{\Delta}^1_{\mathcal{K}} & \hat{d}^0_{\mathcal{K}} \\ \star^{2\dagger}_{\mathcal{K}}\hat{d}^{1\dagger}_{\mathcal{K}}\star^1_{\mathcal{K}} & \bar{0} \end{bmatrix} \vec{\sigma}_{\mathcal{K}} + \begin{bmatrix} \hat{d}^0_{\mathcal{K}}\vec{\pi}_{\partial\mathcal{K}} + \hat{\Delta}^1_{\mathcal{K}}\vec{u}^\flat_{\partial\mathcal{K}} + \hat{\Delta}^1_{\partial\mathcal{K}}(\vec{\star u})^\flat_0 \\ \star^{2\dagger}_{\mathcal{K}}\hat{d}^{1\dagger}_{\mathcal{K}}\star^1_{\mathcal{K}}\vec{u}^\flat_{\partial\mathcal{K}} + \star^{2\dagger}_{\mathcal{K}}\hat{d}^1_{\partial\mathcal{K}}(\vec{\star u})^\flat_0 \end{bmatrix} = 0 \quad (11.26)$$

This can be solved for $\vec{\sigma}_{\mathcal{K}}$ by the following inversion:

$$\vec{\sigma}_{\mathcal{K}} = - \begin{bmatrix} \hat{\Delta}^1_{\mathcal{K}} & \hat{d}^0_{\mathcal{K}} \\ \star^{2\dagger}_{\mathcal{K}}\hat{d}^{1\dagger}_{\mathcal{K}}\star^1_{\mathcal{K}} & \bar{0} \end{bmatrix}^{-1} \begin{bmatrix} \hat{d}^0_{\mathcal{K}}\vec{\pi}_{\partial\mathcal{K}} + \hat{\Delta}^1_{\mathcal{K}}\vec{u}^\flat_{\partial\mathcal{K}} + \hat{\Delta}^1_{\partial\mathcal{K}}(\vec{\star u})^\flat_0 \\ \star^{2\dagger}_{\mathcal{K}}\hat{d}^{1\dagger}_{\mathcal{K}}\star^1_{\mathcal{K}}\vec{u}^\flat_{\partial\mathcal{K}} + \star^{2\dagger}_{\mathcal{K}}\hat{d}^1_{\partial\mathcal{K}}(\vec{\star u})^\flat_0 \end{bmatrix} \quad (11.27)$$

From this, similarly to the past examples, $\vec{\pi}_{\mathcal{K}}$ and $\vec{u}^\flat_{\mathcal{K}}$ can be recovered, and then reconstructed with the boundary values into $\vec{\pi}$ and $\vec{u}^\flat$.

11.2.4 DEC Stokes Flow around Hole with Steady-State Boundary and Initial Conditions

As covered above, the exterior calculus form of the NSEs considered on the rectangle with an immersed hole, along with assumptions made and Dirichlet boundary conditions considered, are given as:

$$\left\{ \begin{array}{c} \overbrace{d_\Sigma p - (d_\Sigma \star_\Sigma d_\Sigma \star_\Sigma + \star_\Sigma d_\Sigma \star_\Sigma d_\Sigma)u = 0}^{\text{Navier-Stokes momentum equations}} \\ \underbrace{u|_{\partial\Sigma_1 \cup \partial\Sigma_2} = \xi\,, \ u|_{\partial\Sigma_3 \cup \partial\Sigma_4} = 0\,, \ \star_\Sigma(n \wedge \star u|_{\partial\Sigma_5}) = 0}_{\text{velocity Dirichlet conditions}} \end{array} \right., \quad \left. \begin{array}{c} \overbrace{\star_\Sigma d_\Sigma \star_\Sigma u = 0}^{\text{mass conservation}} \\ \underbrace{p|_{\partial\Sigma} = P}_{\text{pressure Dirichlet conditions}} \end{array} \right\}$$

Noting how no-slip and inlet/outlet conditions are used as in the previous section, one could start from the master equation system obtained in the Poiseuille flow case:

$$\vec{\sigma}_{\mathcal{K}} = - \begin{bmatrix} \hat{\Delta}^1_{\mathcal{K}} & \hat{d}^0_{\mathcal{K}} \\ \star^{2\dagger}_{\mathcal{K}}\hat{d}^{1\dagger}_{\mathcal{K}}\star^1_{\mathcal{K}} & \bar{0} \end{bmatrix}^{-1} \begin{bmatrix} \hat{d}^0_{\mathcal{K}}\vec{\pi}_{\partial\mathcal{K}} + \hat{\Delta}^1_{\mathcal{K}}\vec{u}^\flat_{\partial\mathcal{K}} + \hat{\Delta}^1_{\partial\mathcal{K}}(\vec{\star u})^\flat_0 \\ \star^{2\dagger}_{\mathcal{K}}\hat{d}^{1\dagger}_{\mathcal{K}}\star^1_{\mathcal{K}}\vec{u}^\flat_{\partial\mathcal{K}} + \star^{2\dagger}_{\mathcal{K}}\hat{d}^1_{\partial\mathcal{K}}(\vec{\star u})^\flat_0 \end{bmatrix}$$

The only difference to the previous is the inclusion of a hole and the existence of a no-slip condition. This is remedied by finding a discretization of the wedge product between 1-forms and 0-forms, and implementing this as an additional constraint in the matrix, representing the discretization of the normal and wedge product together by the matrix $\mathcal{N}^1_{\mathcal{K}}$, such that the final form of the master equation appears as such:

$$\begin{bmatrix} \hat{\Delta}^1_{\mathcal{K}} & \hat{d}^0_{\mathcal{K}} \\ \star^{2\dagger}_{\mathcal{K}}\hat{d}^{1\dagger}_{\mathcal{K}}\star^1_{\mathcal{K}} & \bar{0} \\ \star^{2\dagger}_{\mathcal{K}}\mathcal{N}^1_{\mathcal{K}}\star^1_{\mathcal{K}} & \bar{0} \end{bmatrix} \vec{\sigma}_{\mathcal{K}} + \begin{bmatrix} \hat{d}^0_{\mathcal{K}}\vec{\pi}_{\partial\mathcal{K}} + \hat{\Delta}^1_{\mathcal{K}}\vec{u}^\flat_{\partial\mathcal{K}} + \hat{\Delta}^1_{\partial\mathcal{K}}(\vec{\star u})^\flat_0 \\ \star^{2\dagger}_{\mathcal{K}}\hat{d}^{1\dagger}_{\mathcal{K}}\star^1_{\mathcal{K}}\vec{u}^\flat_{\partial\mathcal{K}} + \star^{2\dagger}_{\mathcal{K}}\hat{d}^1_{\partial\mathcal{K}}(\vec{\star u})^\flat_0 \\ \star^{2\dagger}_{\mathcal{K}}\mathcal{N}^1_{\mathcal{K}}\star^1_{\mathcal{K}}\vec{u}^\flat_{\partial\mathcal{K}} + \star^{2\dagger}_{\mathcal{K}}\mathcal{N}^1_{\mathcal{K}}(\vec{\star u})^\flat_0 \end{bmatrix} = 0 \quad (11.28)$$

It can be noticed that this equation is not a square matrix, and as such, a least-squares linear solver can be used to recover $\vec{\pi}_\mathcal{K}$ and $\vec{u}^\flat_\mathcal{K}$.

11.2.5 Ptačkova Reconstruction

Methods for solving partial differential equations were outlined above using the framework of DEC. The outputs to this are discretized κ-forms on the primal simplicial complex. These values would be ready to be interpolated back into their scalar and vector field expressions. Scalar arrays $\vec{\phi}$ and $\vec{\pi}$ can be separated into ϕ_i and p_i respectively, and node values can be interpolated back into the respective scalar fields $\phi(\Sigma)$ and $p(\Sigma)$. Also covered was the solution to vector field PDEs; this case must be handled with more care.

The vector field solution was given as a solution array $\vec{u}^\flat$ with values u_i. The discretized 1-form u_i represents the value of the vector field integrated along the edges, and as such, must have its direction recovered, and its lengths scaled before reconstructing it into the vector field $\vec{u}(\Sigma)$. This process of recovering continuous vector fields from discretized primal 1-forms has been covered by Ptačkova and colleagues and will be referred to as Ptačkova reconstruction. The details will not be explored, but a reference is left to the reader.

11.3 RESULTS

Using the previous section's methods, solutions are obtained for the four PDE cases of interest. Linear solvers were utilized as supplied through MATLAB and executed on a 2.3 GHz 18-Core Intel Xeon W processor with 128 GB memory, and 16 GB of RAM. Each solution will be discussed in the subsequent subsections.

11.3.1 Poisson Equation Solution on Plane

Here, the solution to the equation $\nabla^2 \Phi = \Psi$ is obtained, subject to the considered Dirichlet conditions:

As can be seen in Figure 11.5a, by analyzing the boundary of the manifold indeed, the sinusoidal Dirichlet boundary conditions are obtained, and the source function of an exponential heat function produces the expected increase in value in the center of the array. The generation of the plane simplicial complex, formulation of the DEC

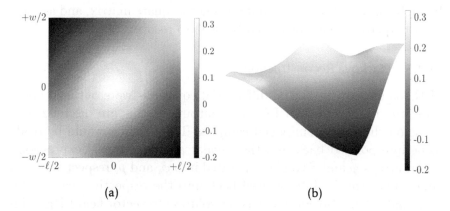

(a)   (b)

Figure 11.5 Solution to Poisson's equation $\Delta_\Sigma^{\mathrm{dR}}\Phi = \psi_\Sigma$, where the manifold Σ is a $w \times \ell$ plane $\{w = 4, \ell = 4\}$ discretized into a simplicial complex $\mathcal{K}$ with 11,776 elements. Here, $\psi_\Sigma : \Sigma \to \mathbb{R}$ is a function on the manifold given by $\psi_\Sigma(x, y) = \mathrm{e}^{-50(x^2+y^2)}$, and the equation is subject to Dirichlet Conditions as covered in Section 11.1.1.1. (a) Flat 2D solution to PDE. (b) Scalar-height solution to PDE.

master equation, and inversion of the system to obtain a solution is obtained approximately in 81 ms, for the simplicial complex containing 11,776 elements. As seen in Figure 11.5b, the solution is sufficient for interpolation into a smooth manifold.

11.3.2   Laplace Equation Solution on Annulus

Here, the solution to the equation $\nabla^2\Phi = 0$ is obtained, subject to the considered Dirichlet conditions:

As can be seen in Figure 11.6a, by analyzing the boundaries of the manifold, the zero and sinusoidal Dirichlet boundary conditions are observed, and in between, an acceptable Laplace equation solution is seen by minimizing the deviation of the inner and outer Dirichlet boundary conditions. The generation of the annulus simplicial complex, formulation of the DEC master equation, and inversion of the system to obtain a solution is obtained approximately in 236 ms, for the simplicial complex containing 19,003 elements. As seen in Figure 11.6b, the solution is sufficient for interpolation into a smooth manifold.

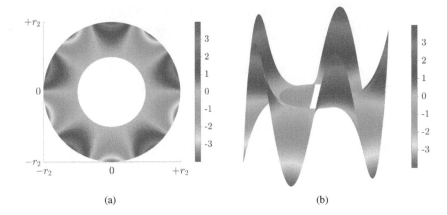

(a)   (b)

Figure 11.6 Solution to Laplace's equation $\Delta_{\Sigma}^{\mathrm{dR}}\Phi = 0$, where the manifold Σ is an annulus with an inner and outer radius of (r_1, r_2) $\{r_1 = 2, r_2 = 4\}$, discretized into a simplicial complex $\mathcal{K}$ with 19,003 elements. Here, the equation is subject to Dirichlet Conditions as covered in Section 11.1.1.2. (a) Flat 2D solution to PDE. (b) Scalar-height solution to PDE.

11.3.3 Poiseuille Flow with Steady-State Boundary and Initial Conditions

Here, the solution to the NSEs for suitable boundary conditions mimicking Poiseuille flow is obtained:

As can be seen in Figure 11.7, the discretized 1-form $u^{\flat}$ is displayed. Values represent the integration of the velocity 1-form along the edges of the simplicial complex. This can undergo Ptačkova Reconstruction, to obtain the final vector field $\vec{u}$ as displayed alongside the obtained pressure field p:

As can be seen in Figure 11.8, both the pressure field and velocity field are displayed. The pressure vield varies linearly from the left of the manifold to the right, while the velocity field increases quadratically from the boundary, to the center of the tube, where the flow is the highest. The boundary conditions are respected at the boundaries and are homogeneous with the middle of the manifold as well as the edges. The generation of the plane simplicial complex, formulation of the DEC NSEs master equation system, and inversion of the system to obtain

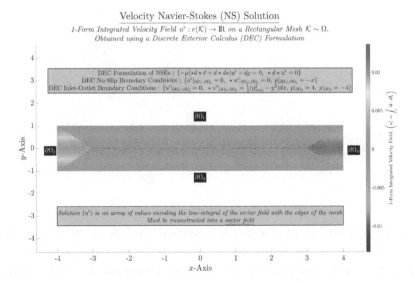

Figure 11.7 Raw solution of primal discrete 1-form $u^\flat$ representing poiseuille solution to NSEs in the Stokes Flow Regime.

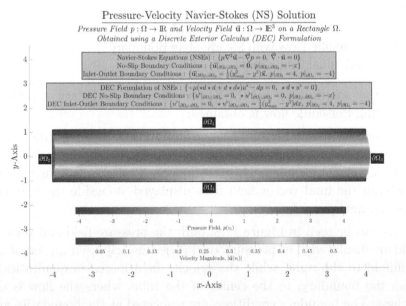

Figure 11.8 Ptačkova Reconstruction of solution of primal vector field $\vec{u}(\Sigma)$ representing poiseuille solution to NSEs in the Stokes Flow Regime.

a solution is obtained approximately in $3,261$ ms, for the simplicial complex containing 76,330 elements.

11.3.4 Stokes Flow around Hole with Steady-State Boundary and Initial Conditions

Here, the solution to the NSEs for suitable boundary conditions mimicking Stokes flow around a hole is obtained:

As can be seen in Figure 11.9, the discretized 1-form $u^{\flat}$ is displayed. Values represent the integration of the velocity 1-form along the edges of the simplicial complex. This can undergo Ptačkova Reconstruction, to obtain the final vector field $\vec{u}$ as displayed alongside the obtained pressure field p:

As can be seen in Figure 11.10, both the pressure field and velocity field are displayed. The pressure field varies linearly from the left of the manifold to the right, while the velocity field increases quadratically from the boundary, to the center of the tube, where the flow is the highest, flows around the hole, attains a velocity increase at the edges

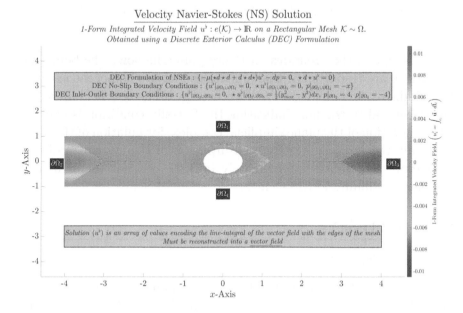

Figure 11.9 Raw solution of primal discrete 1-form $u^{\flat}$ representing stokes flow solution to NSEs around a hole.

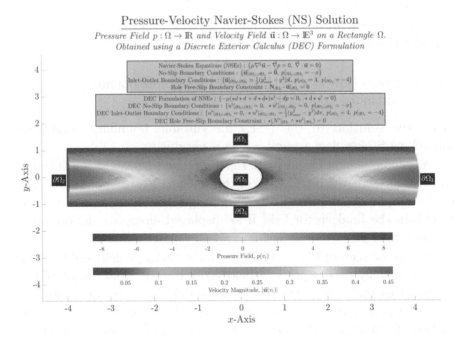

Figure 11.10 Ptačkova Reconstruction of solution of primal vector field $\vec{u}(\Sigma)$ representing stokes flow solution to NSEs around a hole.

of the hole, and then reunites back into poiseuille flow. The boundary conditions are respected at the boundaries and are homogeneous with the middle of the manifold as well as the edges. The flow field lines also are tangential to the hole indicating the free-slip condition is obeyed. The generation of the plane simplicial complex, formulation of the DEC NSEs master equation system, and inversion of the system to obtain a solution is obtained approximately in 3,673 ms, for the simplicial complex containing 78,823 elements.

11.4 DISCUSSION

11.4.1 Solving Scalar PDEs

For the Scalar PDEs considered on the Annulus and on the Plane, both obtained solutions which were reasonable, and obtained in times less than 250ms. The form of the master equations was generalizable and also at the resolution considered, integration into a smooth mesh was possible, indicating that the resolution considered was suitable for interpolation. The application of DEC to solve scalar PDEs is seen to be as effective, if not more robust than the utilization of finite-difference or finite-element methods which would pose difficulties of how to go about solving these equations.

11.4.2 Validation of Poiseuille Flow

It can be seen from Figure 11.8 that the DEC implementation of the NSEs in the Stokes-flow regime reproduces Poiseuille flow, as would be obtained analytically. The magnitude in the center of the tube stays constant at $\max_\Sigma |\vec{u}| = \frac{1}{2}$, as is expected from the analytic Poiseuille flow solution. The discretization of the mesh into $\sim$ 80k faces yields approximately the correct Poiseuille flow solution, and while a full error analysis of lower resolution meshes is not provided, lower resolution and higher resolution meshing can be considered without a large hit to the error of Poiseuille flow to the analytic solution.

11.4.3 Observations on Stokes Flow

It can be seen from Figure 11.10 that the DEC implementation of the NSEs in the Stokes-flow regime with a hole reproduces approximate Stokes flow seen from other numerical studies. Flow is approximately Poiseuille upstream, and diverges at the hole. It can be seen flow is approximately tangential to the hole, increases speed as it hits a bottlenecking of the pathway, and regains Poiseuille flow at the outlet due to the pressure gradient.

Herein lies the advantage of applying DEC to computational fluid dynamics. Such a problem is difficult to consider analytically, though can be easily translated into a coordinate-invariant exterior calculus form, translated almost directly into DEC, and implemented as a powerful numerical technique. The discretization of the mesh into $\sim$ 80k faces yields approximately Stokes flow behavior, and while a full error analysis of lower resolution meshes is not provided, lower resolution and higher resolution meshing can be considered without a large hit to the error of Stokes flow compared to other CFD methods.

11.5 CONCLUSION

In conclusion, Discrete Exterior Calculus is a computational framework, which allows for calculus to be performed on simplicial complex meshes which are approximating continuous manifolds. As seen in Chapter 10, differential operators on manifolds which are discretized into linear operators on simplicial complexes, accurately represent their differential counterparts seen through the solution of Poisson's and Laplace's scalar partial differential equations in efficient time, and in a single time-step.

The application of this framework to the raison d'etre of the book - fluid mechanics - is observed also. A decomposition of the NSEs into exterior calculus form, factoring in Dirichlet-like boundary conditions, and then DEC form subsequently, shows the validity of applying DEC to fluid mechanics problems. Stokes flow around a hole has been presented, and qualitatively yields reasonable results.

The potential application of DEC as a framework is therefore logically sound, directly translated theoretics into computational implementation and is generalizable to higher dimensions. Many of the forms of the equations here remain identical in higher dimension, where only the form of the discretized exterior derivatives and Hodge star operators need to be adjusted. The coordinate-invariant approach of DEC highlights the geometric significance of this equation and yields a generalizable framework where one can consider higher dimensional geometric and physical partial differential equations on manifolds.

The potential of DEC, while young, can be seen to be promising for computational fluid mechanics problems, and can potentially replace finite-difference and finite-element methods in the near future.

Computer Programs

A.1 PRODUCT OF DENSITY AND VELOCITY FOR COMPRESSIBLE FLOW

> > ## *Maple Program to show the relationship between density rho$(r, theta, z, t)$ and velocity $v2(r, theta, z, t)$,*

> > ## *so that Eq3.44 leads to the Hunter $-$ Saxton PDE.*

> > ##

> > ## *Last right hand side term in Eq.3.44 involves the forcing term $F'(T2)$ and is:*

> > $\frac{1}{2} \cdot b2(r, theta, z, t) \cdot diff(b2(r, theta, z, t), theta) \cdot$
$(diff(\text{rho}(r, theta, z, t)^{-2} \cdot b2(r, theta, z, t), theta)):$

> > ## *Next we simplify*

> > *simplify*$(\%):$

> > ## The partial derivative of N wrt to theta in Eq. 3.44 is set equal to $0.5 \cdot$ diff(b2(r,theta,z,t), theta)^2

> > #*Next we simplify*

> > *simplify*$(\%):$

> > ## *Use pdsolve command to solve PDE wrt to rho$(r, theta, z, t)$*

> > *pdsolve*$(\%, \text{rho}(r, theta, z, t)):$

> > ## *The Hunter Saxton PDE is the obtained in the following two lines :*

> > *diff*$(b2(r, theta, z, t), t) + b2(r, theta, z, t) \cdot diff(b2(r, theta, z, t), theta):$

> > $diff(\%, \text{theta}) = \dfrac{1}{2} \cdot diff(b2(r, \text{theta}, z, t), \text{theta})^2:$

> > ##*Next we solve wrt to $b2(r, \text{theta}, z, t)$*

> > $pdsolve(\%, b2(r, \text{theta}, z, t)):$

> > ## *A separable solution arises*

> > ## *The time dependent term is a blowup for $F4(t)$*

> > ## *We have that:*

> > $diff(F4(t), t) = c4 \cdot F4(t)^2:$

> > $dsolve(\%):$

> > ```##The product rho(r,theta,z,t)·b2(r,theta,z,t) is equal to the following two terms multiplied```

> > $P1 := \dfrac{\begin{array}{c}((2 \cdot F1(r,z,t) \cdot (int(\frac{1}{b2 \cdot diff(b2,\text{theta})}, \text{theta})) + F2(r,z,t) \\ + int(\frac{int(diff(b2,\text{theta})^2, \text{theta})}{b2 \cdot diff(b2,\text{theta})}, \text{theta})) \cdot b2)^{\frac{1}{2}}\end{array}}{\begin{array}{c}(2 \cdot F1(r,z,t) \cdot (int(\frac{1}{b2 \cdot diff(b2,\text{theta})}, \text{theta})) + F2(r,z,t) \\ + int(\frac{int(diff(b2,\text{theta})^2, \text{theta})}{b2 \cdot diff(b2,\text{theta})}, \text{theta}))\end{array}}:$

> > $P2 := (\dfrac{1}{B + c4 \cdot t - C1}) \cdot F(r, z) \cdot G(\text{theta}):$

> > ## *Set P2 to b2*

> > ## *And substitute result into P1*

> > $subs(b2(r, \text{theta}, z, t) = \dfrac{1}{B + c4 \cdot t - C1} \cdot F(r, z) \cdot G(\text{theta}), \%):$

> > ## *Next we introduce auxiliary functions F1 and F2 each as functions of (r, z, t)*

> > $subs(F1(r, z, t) = ((B + v4 \cdot t - C1))^{-2}, \%):$

> > $subs(F2(r, z, t) = ((B + c4 \cdot t - C1))^{-3}, \%):$

> > ## *The product is :*

> > $Prod := simplify(\%):$

> > ```##Take the limit at the blowup for F1(r,z,t), F2(r,z,t) at t=(-B+C1)·c4```$^{-1}$

> > ##

> > $limit(Prod, t = (-B + C1) \cdot c4^{-1}):$

> > ##

> > ```## The limit is finite for Prod, indicating that the adiabatic condition rho·b2 is a constant is satisfied```

> > ##*End of Program*

A.2 RAYLEIGH BENARD CONVECTION MAPLE PROGRAM

> > ### *PROGRAM TO SOLVE THE RALEIGH BENARD CONVECTION PROBLEM*

> > $P := diff(diff(uz(x,y,z,t),x),x) + diff(diff(uz(x,y,z,t),y),y) + diff(diff(uz(x,y,z,t),z),z):$

> > $L1 := diff(diff(diff(uz(x,y,z,t),x),x),x)+diff(diff(uz(x,y,z,t),y),y)+ diff(diff(uz(x,y,z,t),z),z),t) - Ra \cdot Pr \cdot((diff(diff(T(x,y,z,t),x),x) + diff(diff(T(x,y,z,t),y),y)+diff(diff(T(x,y,z,t),z),z))-Pr \cdot(diff(diff(P,x),x) + diff(diff(P,y),y) + diff(diff(P,z),z))):$

> > $L2 := diff(T(x,y,z,t),t)-uz(x,y,z,t)-(diff(diff(T(x,y,z,t),x),x)+ diff(diff(T(x,y,z,t),y),y) + diff(diff(T(x,y,z,t),z),z)):$

> >

> > ## *Input Ansatz for velocity in z direction and theta in the same direction.*

> > $A1 := subs(uz(x,y,z,t) = V(z) \cdot X(x,y) \cdot exp(gamma \cdot t), T(x,y,z,t) = \theta(z) \cdot X(x,y) \cdot exp(gamma \cdot t), L1):$

> >

> > $A2 := subs(uz(x,y,z,t) = V(z) \cdot X(x,y) \cdot exp(gamma \cdot t), T(x,y,z,t) = \theta(z) \cdot X(x,y) \cdot exp(gamma \cdot t), L2):$

> > $simplify(\%):$

> > $-(\theta(z)(\frac{\partial^2}{\partial x^2}X(x,y))+\theta(z)(\frac{\partial^2}{\partial y^2}X(x,y))+X(x,y)(-\gamma\theta(z)+\frac{d^2}{dz^2}\theta(z)+ V(z)))e^{\gamma t}:$

> > ## *Cast into a separable form to obtain $X(x,y)$*

> > $\dfrac{\%}{\theta(z) \cdot X(x,y)}:$

> > $expand(\%):$

> > $-\dfrac{\frac{\partial^2}{\partial x^2}X(x,y)}{X(x,y)} - \dfrac{\frac{\partial^2}{\partial y^2}X(x,y)}{X(x,y)} = -\Lambda^2:$

> > $pdsolve(\%):$

> > $\dfrac{d^2}{dx^2} `F1(x) = - `c_1 `F1(x):$

> > $dsolve(\%):$

> > $\dfrac{d^2}{dy^2} \, \dot{}F2(y) = \dot{}F2(y) \, \dot{c_1} + \Lambda^2 \, \dot{}F2(y):$

> > $dsolve(\%):$

> > $\gamma - \dfrac{V(z)}{\theta(z)} - \dfrac{\frac{d^2}{dz^2}\theta(z)}{\theta(z)} = -\Lambda^2:$

> > $A1s := simplify(A1):$

> >

> > $A1sout := simplify(subs(X(x,y) = (\,\dot{}C1\,\sin(\sqrt{\dot{c_1}}x) + \,\dot{}C2\cos(\sqrt{\dot{c_1}}x))\cdot$
$(\,\dot{}C1\sin(\sqrt{-\Lambda^2 - \dot{c_1}}\,y) + \,\dot{}C2\cos(\sqrt{-\Lambda^2 - \dot{c_1}}\,y)), A1s)):$

> > $simplify(\%):$

> > $F1 := (2\Lambda^2 \overset{2}{\Pr} Ra + \gamma)(\dfrac{d^2}{dz^2}V(z)) - \Lambda^2\theta(z)\Pr Ra + \Lambda^2(\Lambda^2 \overset{2}{\Pr} Ra + \gamma)$
$V(z) + \Pr((\dfrac{d^4}{dz^4}V(z))\Pr - \dfrac{d^2}{dz^2}\theta(z))Ra:$

> > $F2 := \gamma - \dfrac{V(z)}{\theta(z)} - \dfrac{\frac{d^2}{dz^2}\theta(z)}{\theta(z)} + \Lambda^2:$

> > $\#\# Use$ the dsolve command to solve for V and theta in the system of
coupled ordinary differential equations

> > $dsolve(\{F1, F2\}):$

> >

> > $\#\# The$ solution for V as a function of z with permuted coefficients

> > $V := C6 \, e^{- z\sqrt{2\gamma \Pr^2 Ra^2 - 2\gamma Ra - 2}\sqrt{\frac{4\Lambda^4 \Pr^4 Ra^2 + 4\Lambda^2 \Pr^4 Ra^2\gamma + \Pr^4 Ra^2\gamma^2 + 4\Lambda^2 \Pr^2 Ra\gamma - 4\Pr^3 Ra^2 + 2\Pr^2 Ra\gamma^2 + \gamma^2 Ra}{2\Pr Ra}}} +$

$C5 \cdot e^{z\sqrt{2\gamma \Pr^2 Ra^2 - 2\gamma Ra - 2}\sqrt{\frac{4\Lambda^4 \Pr^4 Ra^2 + 4\Lambda^2 \Pr^4 Ra^2\gamma + \Pr^4 Ra^2\gamma^2 + 4\Lambda^2 \Pr^2 Ra\gamma - 4\Pr^3 Ra^2 + 2\Pr^2 Ra\gamma^2 + \gamma^2 Ra}{2\Pr Ra}}} +$

$C4 \, e^{- z\sqrt{2\gamma \Pr^2 Ra^2 - 2\gamma Ra + 2}\sqrt{\frac{4\Lambda^4 \Pr^4 Ra^2 + 4\Lambda^2 \Pr^4 Ra^2\gamma + \Pr^4 Ra^2\gamma^2 + 4\Lambda^2 \Pr^2 Ra\gamma - 4\Pr^3 Ra^2 + 2\Pr^2 Ra\gamma^2 + \gamma^2 Ra}{2\Pr Ra}}} +$

$$C3 \cdot e^{\dfrac{z \; \sqrt{2\gamma \Pr^2 Ra^2 - 2\gamma Ra + 2} \; \sqrt{\begin{array}{l} 4\Lambda^4 \Pr^4 Ra^2 + 4\Lambda^2 \Pr^4 Ra^2\gamma \\ + \Pr^4 Ra^2\gamma^2 + 4\Lambda^2 \Pr^2 Ra\gamma \\ -4\Pr^3 Ra^2 + 2\Pr^2 Ra\gamma^2 + \gamma^2 Ra \end{array}}}{2\Pr Ra}}$$
$$+ \; C1 \cdot \sin(\Lambda \cdot z) + C2 \cdot \cos(\Lambda \cdot z):$$

> > $Vout := simplify(\%):$

> > ##$Substitute$ in $Prandtl$ and $Rayleigh$ $numbers$, $together$ $with$ $Lambda$ and use the 6 $boundary$ $conditions$ for V, the $second$ $derivative$ of V and the $fourth$ $derivative$ of V at $z = 0, 1$ $respectively;$

> > $V1 := subs(Ra = 20000, \Pr = 7.2, \Lambda = -0.61, Vout):$

> > $Q3 := subs(z = 0, diff(diff(V1, z), z)) = 0:$

> > $Q2 := subs(z = 0, diff(diff(diff(diff(V1, z), z), z), z)):$

> > $Q1 := subs(z = 0, V1):$

> > $Q6 := subs(z = 1, diff(diff(V1, z), z)):$

> > $Q5 := subs(z = 1, diff(diff(diff(diff(V1, z), z), z), z)):$

> > $Q4 := subs(z = 1, V1):$

> > $fsolve(\{Q1, Q2, Q3, Q4, Q5\}):$

> > $subs(\text{`}C1 = C1, \text{`}C2 = 0, \text{`}C3 = 0.03149953220\ C1,$
$\text{`}C4 = -0.0314995322\ C1, \text{`}C5 = -0.5221495197\ IC1,$
$\text{`}C6 = 0.5221495197\ IC1, Ra = 200000, \Pr = 7.2, \Lambda = -0.61, \theta(z)):$

> > ##$Plot$ $theta$ $versus$ z at $C1 = 1$ and $gamma = 1,$

> > $THETA := subs(\text{gamma} = -20, C1 = 10, \%):$

> > $plot(\Re(THETA), z = 0..1):$

> > $VELOCITYz := subs(C1 = C1, C2 = 0, C3 = 0.03149953220\ C1,$
$C4 = -0.0314995322\ C1, C5 = -0.5221495197\ IC1,$
$C6 = 0.5221495197\ IC1, Ra = 200000, \Pr = 7.2, \Lambda = -0.61, V):$

> > $subs(C1 = 10, \text{gamma} = -20, \%):$

> > $plot(\Re(\%), z = 0..1):$

> > ##$Higher$ $harmonics$ for $velocity$, $four$ $times$ $Lambda$ $Here$ we see $that$ the $behaviour$ of the $velocity$ is $sinusoidal$, $when$ $gamma$ is $sufficiently$ $large$

> > $VELOCITYz := subs(C1 = C1, C2 = 0, C3 = 0.03149953220\ C1,$
$C4 = -0.0314995322\ C1, C5 = -0.5221495197\ IC1,$
$C6 = 0.5221495197\ IC1, Ra = 200000, \Pr = 7.2, \Lambda = 10 \cdot 0.61, V):$

> > $subs(C1 = 10, \text{gamma} = -20, \%):$

> > $plot(\Re(\%), z = 0..1):$

> >

> >

A.3 EQUATION 10.37 SOLUTION SECTION 10.2.2

> > *The governing PDE Equation 8.29 to be solved in Section 8.2.2*

> > $(\frac{\partial}{\partial t} uz(x, y, z, t))^2 \mu(-1 + \delta)(\frac{\partial^3}{\partial x^2 \partial z} uz(x, y, z, t)) + (\frac{\partial}{\partial t} uz(x, y, z, t))^2$
$\mu(-1+\delta)(\frac{\partial^3}{\partial y^2 \partial z} uz(x, y, z, t)) + (\frac{\partial}{\partial t} uz(x, y, z, t))^2 \mu(-1+\delta)(\frac{\partial^3}{\partial z^3} uz(x, y, z, t))$
$+(\frac{\partial}{\partial t} uz(x, y, z, t)) uz(x, y, z, t)^2 (\frac{\partial^3}{\partial t \partial z^2} uz(x, y, z, t)) \delta\rho - uz(x, y, z, t)^2$
$(\frac{\partial^2}{\partial t \partial z} uz(x, y, z, t))^2 \delta\rho - 2\rho((\frac{\delta}{2} - \frac{1}{2})(\frac{\partial}{\partial t} uz(x, y, z, t))^2 - uz(x, y, z, t)$
$(\frac{\partial}{\partial t} uz(x, y, z, t))(\frac{\partial}{\partial z} uz(x, y, z, t))\delta + \delta(uz(x, y, z, t)(FT1(x, y, z, t)+$
$\frac{\partial}{\partial t} ux(x, y, z, t))(\frac{\partial}{\partial x} uz(x, y, z, t)) + uz(x, y, z, t)(FT2(x, y, z, t)+$
$\frac{\partial}{\partial t} uy(x, y, z, t))(\frac{\partial}{\partial y} uz(x, y, z, t)) + \frac{\Lambda(x,y,z,t)}{2} + \frac{\Phi(t)}{2}))(\frac{\partial^2}{\partial t \partial z} uz(x, y, z, t)) +$
$(\frac{\partial}{\partial t} uz(x, y, z, t))(((-1+\delta)(ux(x, y, z, t)\delta - 1)(\frac{\partial}{\partial t} uz(x, y, z, t)) + 2uz(x, y, z, t)$
$\rho\delta(FT1(x, y, z, t) + \frac{\partial}{\partial t} ux(x, y, z, t)))(\frac{\partial^2}{\partial x \partial z} uz(x, y, z, t)) + ((-1+\delta)(uy(x, y, z, t)\delta - 1)(\frac{\partial}{\partial t} uz(x, y, z, t)) + 2uz(x, y, z, t)\rho\delta(FT2(x, y, z, t) + \frac{\partial}{\partial t} uy(x, y, z, t)))$
$t)))(\frac{\partial^2}{\partial y \partial z} uz(x, y, z, t)) + 3(\frac{\partial}{\partial t} uz(x, y, z, t))(-\frac{2}{3} + (\rho + \frac{2}{3})\delta) uz(x, y, z, t)(\frac{\partial^2}{\partial z^2} uz$
$(x, y, z, t)) + 2uz(x, y, z, t)(\frac{\partial}{\partial t} uz(x, y, z, t))(-1 + \delta)(\frac{\partial^2}{\partial x^2} uz(x, y, z, t)) +$
$2uz(x, y, z, t)(\frac{\partial}{\partial t} uz(x, y, z, t))(-1 + \delta)(\frac{\partial^2}{\partial y^2} uz(x, y, z, t)) + 2(\frac{\partial^2}{\partial t \partial z} ux(x, y, z,$
$t))uz(x, y, z, t)(\frac{\partial}{\partial x} uz(x, y, z, t))\rho\delta + 2(\frac{\partial^2}{\partial t \partial z} uy(x, y, z, t))uz(x, y, z, t)(\frac{\partial}{\partial y} uz$
$(x, y, z, t))\rho\delta + ((-1 + (3\rho + 1)\delta)(\frac{\partial}{\partial t} uz(x, y, z, t))^2 + (-1 + \delta)((\frac{\partial}{\partial x} ux(x, y, z,$
$t))^2 + 2(\frac{\partial}{\partial y} ux(x, y, z, t))(\frac{\partial}{\partial x} uy(x, y, z, t)) + (\frac{\partial}{\partial y} uy(x, y, z, t))^2))(\frac{\partial}{\partial t} uz(x, y, z,$
$t)) + 2\rho(((FT1(x, y, z, t) + \frac{\partial}{\partial t} ux(x, y, z, t))(\frac{\partial}{\partial x} uz(x, y, z, t)) + (\frac{\partial}{\partial y} uz(x, y,$
$z, t))(FT2(x, y, z, t) + \frac{\partial}{\partial t} uy(x, y, z, t)))(\frac{\partial}{\partial z} uz(x, y, z, t)) + uz(x, y, z, t)(\frac{\partial}{\partial x} uz$
$(x, y, z, t))(\frac{\partial}{\partial z} FT1(x, y, z, t)) + uz(x, y, z, t)(\frac{\partial}{\partial y} uz(x, y, z, t))(\frac{\partial}{\partial z} FT2(x, y, z,$
$t)) + \frac{(\frac{\partial}{\partial z} \Lambda(x,y,z,t))}{2})\delta)$:

> > *Simplify governing equation and assign to variable Y*

> > $Y := simplify(\%)$:

> > *Solve for expression* $\frac{\partial^2}{\partial t \partial z} ux(x, y, z, t)$ *as part of* Y; *assign new variable M1*

> > $M1 := solve(Y, \frac{\partial^2}{\partial t \partial z} ux(x, y, z, t))$:

> > *Solve for expression* $\frac{\partial^2}{\partial t \partial z} uy(x, y, z, t)$ *as part of* Y; *assign new variable M2*

> > $M2 := solve(Y, \frac{\partial^2}{\partial t \partial z} uy(x, y, z, t))$:

> > $simplify(\%)$:

> > $M1 := \%$:

> > $M2 := \%$:

> > *Form the difference between M1 and M2*

> > $M1 - M2$:

> > *Next use the definition of Vorticity and set M1 − M2 as the following*

> > $\% = diff(K(x, y, z, t), t) - diff(diff(uz(x, y, z, t), y), t) + diff(diff(uz(x, y, z, t), x), t)$:

> > *Simplify expression*

> > *Substitute Ansatz for velocity for ux and uy*

> > $subs(ux(x, y, z, t) = (A + Ux(x, y, z)) \cdot f1(t) \cdot \cos(0 \cdot f1 \cdot t), uy(x, y, z, t) = (A + Uy(x, y, z)) \cdot f1(t) \cdot \cos(0 \cdot f1 \cdot t), \%)$:

> > *Substitute the vorticity in terms of angular velocity*

> > $PL := subs\left(K(x, y, z, t) = \dfrac{\begin{array}{l}2 \cdot (y \cdot uz(x, y, z, t) \\ -z \cdot (Uy(x, y, z) + A) \cdot \\ f1(t) - x \cdot uz(x, y, z, t) \\ +z \cdot (Ux(x, y, z) + A) \cdot f1(t))\end{array}}{x^2 + y^2 + z^2}, \%\right)$:

> >

> > *Substitute forcing terms into PL*

> > $subs(FT1(x, y, z, t) = ((f0(t) \cdot F(x, y, z))), FT2(x, y, z, t) = (f0(t) \cdot G(x, y, z)), PL)$:

> > $subs(F(x, y, z) = A + \mathrm{WeierstrassP}(x, 3 \cdot m^2, m^3)^{-1} \cdot \mathrm{WeierstrassP}(y, 3 \cdot m^2, m^3)^{-1} \cdot \mathrm{WeierstrassP}(z, 3 \cdot m^2, m^3)^{-1}, G(x, y, z) = A + \mathrm{WeierstrassP}(x, 3 \cdot m^2, m^3)^{-1} \cdot \mathrm{WeierstrassP}(y, 3 \cdot m^2, m^3)^{-1} \cdot \mathrm{WeierstrassP}(z, 3 \cdot m^2, m^3)^{-1}, \%)$:

> > *Substitute Ux and Uy in terms of reciprocal WeierstrassP functions (matching the smooth form of the forcing functions)*

> > $subs(Ux(x, y, z) = (\mathrm{WeierstrassP}(x, 3 \cdot m^2, m^3) \cdot \mathrm{WeierstrassP}(y, 3 \cdot m^2, m^3) \cdot \mathrm{WeierstrassP}(z, 3 \cdot m^2, m^3))^{-1}, Uy(x, y, z) = (\mathrm{WeierstrassP}(x, 3 \cdot m^2, m^3) \cdot \mathrm{WeierstrassP}(y, 3 \cdot m^2, m^3) \cdot \mathrm{WeierstrassP}(z, 3 \cdot m^2, m^3))^{-1}, \%)$:

> > *Substitute delta is equal to 1 (see section 8.2.2 for explanation)*

> > $subs(\delta = 1, \%)$:

> > $simplify(\%)$:

> > *Based on the definition of Newton's second law relating the acceleration to the force we equate the following in the previous equation*

> $>subs(\dfrac{d}{dt}f1(t) = -f0(t), \%)$:

> $>$ *Substitute Lambda*(x, y, z, t) *and it's partial derivative with respect to z to zero (see explanation in Section 8.2.2) gives,*

> $>$

> $>$ *Next we use the Maple pdsolve command to obtain the separable solution,*

> $>pdsolve(\%, uz(x, y, z, t))$:

> $>$

A.4 GENERAL FORCING SOLUTION MAPLE SCRIPT

> $>$ #####*Program to compute Solution of General forcing term Lambda*(x, y, z, t) *according to chapter 8 Equation* 8.14

> $>$ ##### *Governing PDE given by Equation* $(8.29) - (8.35)$

> $>$ #####*Input Equation* $(8.29), LP$:

> $>$

> $>$

> $>$ ####*Solve LP for derivative of f1*(t) *wrt to t*

> $>$

> $>solve(\%, (\dfrac{d}{dt}f1(t)))$:

> $>$ #### *Simplify and multiply by constant A and subtract constant A multiplied by derivative of f1 wrt to t*

> $>$

> $>L1 := \% \cdot A - ((\dfrac{d}{dt}f1(t))) \cdot A$:

> $>$ ####

> $>$ #### *Next solve for f0*(t) *in LP*

> $>$ ####

> $>solve(\%, f0(t))$:

> $>simplify(\%)$:

> $>$ #### *Multiply resulting PDE by constant A and subtract A times f0*(t)

> $>$ ####

> $>\% \cdot (1 - A) - f0(t) \cdot (1 - A)$:

> $>$ ####

> > ####

> > $L2 := \%$:

> > $\dfrac{1}{\eta^3} \cdot f0(t) \cdot F(x,y,z) \cdot diff(uz(x,y,z,t)^2, x) + f0(t) \cdot G(x,y,z) \cdot$ $diff(uz(x,y,z,t)^2, y) - uz(x,y,z,t) \cdot diff(uz(x,y,z,t), z) \cdot H(x,y,z,t) +$ $f1(t) \cdot diff(uz(x,y,z,t) \cdot H(x,y,z,t), x) + f1(t) \cdot diff(uz(x,y,z,t) \cdot$ $H(x,y,z,t), y) + f1(t) \cdot diff(uz(x,y,z,t) \cdot H(x,y,z,t), z)$:

> > $subs(\Lambda(x,y,z,t) = \%, L1 + L2)$:

> > $simplify(\%)$:

> > ##### $Substitute$ $separble$ $form$ for $forcing$ $H = H(x,y,z,t)$

> > $subs(H(x,y,z,t) = fa(x) \cdot fb(y) \cdot fc(z) \cdot fd(t), \%)$:

> > #####

> >

> > $subs(F(x,y,z) = G(x,y,z), \%)$:

> >

> >

> > $subs(G(x,y,z) = \text{WeierstrassP}(x, 3 \cdot m^2, m^3)^{-1} \cdot \text{WeierstrassP}(y, 3 \cdot m^2, m^3)^{-1} \cdot \text{WeierstrassP}(z, 3 \cdot m^2, m^3)^{-1}, \%)$:

> >

> >

> > $subs(fd(t) = f0(t), g4(t) = f0(t), \%)$:

> >

> >

> > $subs(fa(x) = \text{WeierstrassP}(x, 3 \cdot m^2, m^3)^{-1}, fb(y) = \text{WeierstrassP}(y, 3 \cdot m^2, m^3)^{-1}, fc(z) = \text{WeierstrassP}(z, 3 \cdot m^2, m^3)^{-1}, g1(x) = \text{WeierstrassP}(x, 3 \cdot m^2, m^3)^{-1}, g2(y) = \text{WeierstrassP}(y, 3 \cdot m^2, m^3)^{-1}, g3(z) = \text{WeierstrassP}(z, 3 \cdot m^2, m^3)^{-1}, \%)$:

> >

> >

> > $simplify(\%)$:

> > ##$SOLVE$ for $x^2 + y^2 + z^2$

> >

> > $solve(\%, x^2 + y^2 + z^2)$:

> > ### simplify

> > $simplify(\%)$:

> > ### Divide by $(\frac{\partial}{\partial t} uz(x, y, z, t))^2 uz(x, y, z, t)(\frac{\partial}{\partial x} uz(x, y, z, t))$ $(\frac{\partial}{\partial y} uz(x, y, z, t))$ on both sides

> > ### Now $x^2 + y^2 + z^2$ is bounded also the product of $\text{WeierstrassP}(y, 3m^2, m^3)^2 \text{WeierstrassP}(x, 3m^2, m^3)^2 \text{WeierstrassP}(z, 3m^2, m^3)^3$ is infinite at a countable sequence.

> > ### So then we have infinity is equal to infinity provided that the denominator of right hand side of equation is equal to zero

> > ###

> > ### Examine Phi from equation 8.14

> > ### If we solve the right hand side of equation in previous two lines then we obtain the following

> > ### First Right hand sdie of Phi equation is :

> > $diff(uz(x, y, z, t), t) \cdot uz(x, y, z, t) \cdot diff(uz(x, y, z, t), x)$: We substitute the proposed solution of RHS of PDE in above analysis

> > $subs(uz(x, y, z, t) = (3\,{}^{\cdot}c_4(\int \Phi(t)dt) + \,{}^{\cdot}C1)^{\frac{1}{3}} \cdot \,{}^{\cdot}F6(z, y + x), \%)$:

> > Now equate to the left side of Φ equation

> > $\,{}^{\cdot}F6(z, y + x)^2 \,{}^{\cdot}c_4\Phi(t)D_2(\,{}^{\cdot}F6)(z, y + x) = uz(x, y, z, t)^2$:

> > $subs(uz(x, y, z, t) = (3\,{}^{\cdot}c_4(\int \Phi(t)dt) + \,{}^{\cdot}C1)^{\frac{1}{3}} \cdot \,{}^{\cdot}F6(z, y + x), \%)$:

> > $simplify(\%)$:

> > $solve(\%, (\int \Phi(t)dt))$:

> > ### Set the following equal :

> > $diff(\dfrac{(\,{}^{\cdot}c_4\Phi(t)D_2(\,{}^{\cdot}F6)(z, y + x))^{\frac{3}{2}} - \,{}^{\cdot}C1}{3\,{}^{\cdot}c_4}, t) = \Phi(t)$:

> > $dsolve(\%)$:

> > $solve(\%, \Phi(t))$:

> > #### Substitute Phi into the following :

$$
> \; > \frac{\begin{array}{l} 8\mathrm{WeierstrassP}(z,3m^2,m^3)^3 \, \dot{}F6(z,y+x)^3 \, \dot{}c_4^2\Phi(t)^2 \\ \mathrm{D}_2(\, \dot{}F6)(z,y+x)^2\mathrm{WeierstrassP}(y,3m^2,m^3)^2 \\ \mathrm{WeierstrassP}(x,3m^2,m^3)^2\eta^3(x-y) \end{array}}{(3\, \dot{}c_4(\int \Phi(t)dt)+\, \dot{}C1)^{\frac{1}{3}}} :
$$

$$
> \; > subs(\Phi(t)=\frac{(t+\, \dot{}C1)^2}{\mathrm{D}_2(\, \dot{}F6)(z,y+x)^3\, \dot{}c_4},\%):
$$

$> \; > simplify(\%):$

$> \; > simplify(\%):$

$> \; > limit(\%,t=-\, \dot{}C1):$

$> \; >$

$> \; > \#\#\#$ *So the limit of this solution is zero and its reciprocal is infinity and we have infinity is equal to infinity.*

$> \; > \#\#\#Back$ *to solving the denominator DE set to zero*

$> \; > \#\#\#$

$> \; > \#\#\#$

$> \; >$

$> \; > \#\#\#$ *Dividing by* $\mathrm{WeierstrassP}(y,3m^2,m^3)\cdot\mathrm{WeierstrassP}(x,3m^2,m^3)$ $\mathrm{WeierstrassP}(z,3m^2,m^3)^3$

$> \; > \#\#\#$ *The expression that survives along coordinate spatial axes is :*

$> \; > \#\#\#$

$$
> \; > \mathrm{WeierstrassP}(y,3m^2,m^3)\Big(\frac{\begin{array}{l} uz(x,y,z,t)^2(\frac{\partial}{\partial t}uz(x,y,z,t)) \\ (\frac{\partial}{\partial x}uz(x,y,z,t)-\frac{\partial}{\partial y}uz(x,y,z,t)) \\ (\frac{\partial^3}{\partial t\partial z^2}uz(x,y,z,t)) \end{array}}{2}-
$$

$$
uz(x,y,z,t)^2(\frac{\partial}{\partial x}uz(x,y,z,t)-\frac{\partial}{\partial y}uz(x,y,z,t))
$$

$$
\frac{(\frac{\partial^2}{\partial t\partial z}uz(x,y,z,t))^2}{2}-(\frac{\partial}{\partial x}uz(x,y,z,t)-
$$

$\frac{\partial}{\partial y}uz(x,y,z,t))(-uz(x,y,z,t)(\frac{\partial}{\partial t}uz(x,y,z,t))(\frac{\partial}{\partial z}uz(x,y,z,t))+Auz(x,y,$
$z,t)(\frac{d}{dt}f1(t)+f0(t))(\frac{\partial}{\partial x}uz(x,y,z,t))+Auz(x,y,z,t)(\frac{d}{dt}f1(t)+f0(t))(\frac{\partial}{\partial y}uz$
$(x,y,z,t)) + \frac{\Phi(t)}{2})(\frac{\partial^2}{\partial t\partial z}uz(x,y,z,t)) + (Auz(x,y,z,t)(\frac{\partial}{\partial y}uz(x,y,z,t) -$
$\frac{\partial}{\partial y}uz(x,y,z,t))(\frac{d}{dt}f1(t)+f0(t))(\frac{\partial^2}{\partial x\partial z}uz(x,y,z,t))+Auz(x,y,z,t)(\frac{\partial}{\partial x}uz(x,$
$y,z,t) - \frac{\partial}{\partial y}uz(x,y,z,t))(\frac{d}{dt}f1(t)+f0(t))(\frac{\partial^2}{\partial y\partial z}uz(x,y,z,t))+$
$\frac{3uz(x,y,z,t)(\frac{\partial}{\partial t}uz(x,y,z,t))(\frac{\partial}{\partial x}uz(x,y,z,t)-\frac{\partial}{\partial y}uz(x,y,z,t))(\frac{\partial^2}{\partial z^2}uz(x,y,z,t))}{2} - uz(x,y,z,$
$t)(\frac{\partial}{\partial x}uz(x,y,z,t))(\frac{\partial}{\partial y}uz(x,y,z,t))(\frac{\partial^2}{\partial t\partial x}uz(x,y,z,t)) + uz(x,y,z,t)(\frac{\partial}{\partial x}uz$
$(x,y,z,t))(\frac{\partial}{\partial y}uz(x,y,z,t))(\frac{\partial^2}{\partial t\partial y}uz(x,y,z,t))+(\frac{\partial}{\partial x}uz(x,y,z,t)-\frac{\partial}{\partial y}uz(x,y,$

$z, t))(\frac{\partial}{\partial z} uz(x, y, z, t))(\frac{3(\frac{\partial}{\partial z} uz(x,y,z,t))(\frac{\partial}{\partial t} uz(x,y,z,t))}{2} + A(\frac{\partial}{\partial x} uz(x, y, z, t) + \frac{\partial}{\partial y}$
$uz(x, y, z, t))(\frac{d}{dt} f1(t) + f0(t))))(\frac{\partial}{\partial t} uz(x, y, z, t)))\text{WeierstrassP}(x, 3m^2, m^3):$

```
>  > simplify(%):
>  > ###Perturbation about A = 0
>  > ###
>  > ###
>  > subs(A = 0, %):
>  > simplify(%):
>  > pdsolve(%, uz(x, y, z, t)):
>  >
```

Bibliography

[1] B. Taylor, *Methodus Incrementorum Directa & Inversa*. Gulielmi Innys, London, p. 108, 1715.

[2] D.N. Arnold, R.S. Falk and R. Winther, Finite element exterior calculus, homological techniques, and applications. *Acta Numerica*, 15:1–155, May 2006.

[3] P.D. Boom, O. Kosmas, L. Margetts and A.P. Jivkov, A geometric formulation of linear elasticity based on discrete exterior calculus. *International Journal of Solids and Structures*, 236–237:111345, February 2022.

[4] K. Crane, F. de Goes, M. Desbrun and P. Schroder, Digital geometry processing with discrete exterior calculus. In *ACM SIGGRAPH 2013 Courses*. ACM, July 2013.

[5] K. Crane, U. Pinkall and P. Schroder, Spin transformations of discrete surfaces. In *ACM SIGGRAPH 2011 papers*. ACM, July 2011.

[6] V.G. Ivancevic and T.T. Ivancevic, *Applied Differential Geometry: A Modern Introduction*. World Scientific, Singapore, May 2007.

[7] P.S. Jensen, Finite difference techniques for variable grids. *Computers & Structures*, 2(1–2):17–29, February 1972.

[8] V.I. Lebedev, Quadratures on a sphere. *USSR Computational Mathematics and Mathematical Physics*, 16(2):10–24, January 1976.

[9] M.S. Mohamed, A.N. Hirani and R. Samtaney, Numerical convergence of discrete exterior calculus on arbitrary surface meshes. *International Journal for Computational Methods in Engineering Science and Mechanics*, 19(3):194–206, April 2018.

[10] L. Ptackova and L. Velho, A simple and complete discrete exterior calculus on general polygonal meshes. *Computer Aided Geometric Design*, 88:102002, June 2021.

[11] M. Spivak. *Calculus on Manifolds: A Modern Approach to Classical Theorems of Advanced Calculus*. CRC Press, Boca Raton, FL, 2018.

[12] J.C. Strikwerda, *Finite Difference Schemes and Partial Differential Equations, Second Edition*. Society for Industrial and Applied Mathematics, Philadelphia, PA, January 2004.

[13] G.W. Bluman and S.C. Anco, *Symmetry and Integration Methods for Differential Equations*. Springer-Verlag, Berlin, 2002.

[14] A.G. Hansen, *Similarity Analyses of Boundary Value Problems in Engineering*, pp. 86–92. Prentice Hall, Englewood Cliffs, NJ, 1964.

[15] H. Jia and V. Šverak, *On Scale-Invariant Solutions of the Navier-Stokes Equations*. Proceedings of the 6th European Congress of Mathematicians, 2012.

[16] VR, Poisson Equation for Pressure, http://www.thevisualroom.com/poisson_for_pressure.html

[17] L. Burns, Maxwell's equations are universal for locally conserved quantities. *Advances in Applied Clifford Algebras*, 29(4):62, 2019. doi:10.1007/s00006-019-0979-7

[18] C. Liu and Z. Liu, New governing equations for fluid dynamics. *AIP Advances*, 11:115025, 2021. doi:10.1063/5.0074615

[19] T.E. Moschandreou, No finite time blowup for 3D incompressible Navier Stokes equations via scaling invariance. *Mathematics and Statistics*, 9(3):386–393, 2021.

[20] R. Benzi and M. Colella, A simple point vortex model for two-dimensional decaying turbulence. *Physics of Fluids A:Fluid Dynamics*, 4(5):1–11, 1991. doi:10.1063/1.858254.

[21] P. Santangelo, The generation of vortices in high-resolution, two-dimensional decaying turbulence and the influence of initial conditions on the breaking of self-similarity. *Physics of Fluids A:Fluid Dynamics*, 1(6):1027–1034, 1989. doi:10.1063/1.857393

[22] T. Moschandreou, A method of solving compressible Navier Stokes equations in cylindrical coordinates using geometric algebra. *Mathematics*, 7(2):126, 2019. doi:10.3390/math7020126

[23] C. Doran and A. Lasenby, *Geometric Algebra for Physicists*. Cambridge University Press, Cambridge, 2013. doi:10.1017/CBO9780511807497

[24] J.T. Beale, T. Kato and A. Majda, Remarks on the breakdown of smooth solutions for the 3-D Euler equations. *Communications in Mathematical Physics*, 94:61–66, 1984.

[25] K.C. Afas, Extending the Calculus of moving surfaces to higher orders. *arXiv preprint arXiv:1806.02335*, 2018.

[26] W. Ye and Z. Yin, On the Cauchy problem for the Hunter-Saxton equation on the line. *arXiv preprint arXiv:2012.15429v1*, 2020.

[27] T.E. Moschandreou, Navier Stokes equations for a 3D incompressible fluid on $\mathbb{T}^3$. *Submitted to Partial Differential equations in Applied Mathematics (June 10,2022)-currently under review*, 1–22, 2022.

[28] J. Leray and R. Terrell, On the motion of a viscous liquid filling space. *arXiv preprint arXiv:1604.02484v1*, 1–113, 6 April 2016.

[29] T.E. Moschandreou, On the 4th clay millennium problem for the periodic Navier Stokes equations millennium prize problems. *Recent Advances in Mathematical Research and Computer Science*, 4(12):79–92, 2021. doi:10.9734/bpi/ramrcs/v4/14379D

[30] T.E. Moschandreou and K.C. Afas, Existence of incompressible vortex-class phenomena and variational formulation of Rayleigh Plesset cavitation dynamics. *Applied Mechanics*, 2(3):613–629, 2021. doi:10.3390/applmech2030035

[31] A. Cheskidov, P. Constantin, S. Friedlander and R. Shvydko, Energy conservation and Onsager's conjecture for the Euler equations. *arXiv preprint arXiv:0704.0759v1*, 2007.

[32] P. Isett, A proof of Onsager's conjecture. *Annals of Mathematics*, 188(3):871–963, 2018. doi:10.4007/annals.2018.188.3.4

[33] D. DeTurck, H. Gluck, R. Komendarczyk, P. Melvin, C. Shonkwiler and D.S. Vela-Vick, Triple linking numbers, ambiguous Hopf invariants and integral formulas for three-component links. *Matematica Contemporânea*, 34:251–283, 2008.

[34] R.D. Benguria, C. Vallejos and H.V.D. Bosch, Gagliardo-Nirenberg-Sobolev inequalites for convex domains in $\mathbb{R}^d$. *arXiv preprint arXiv:1802.01740v2*, 1–16, 2018.

[35] R. Temam, Behaviour at time $t = 0$ of the solutions of semi-linear evolution equations. *Journal of Differential Equations*, 43:73–92, 1982.

[36] R. Temam, Suitable initial conditions. *Journal of Computational Physics*, 218:443–450, 2006.

[37] M.C. Cross and P.C. Hohenberg, Pattern formation outside of equilibrium. *Reviews of Modern Physics*, 65:851–1111, 1993.

[38] Y.C. Chen, C. Shi, J.M. Kosterlitz, X. Zhu and P. Ao, Global potential, topology, and pattern selection in a noisy stabilized Kuramoto–Sivashinsky equation. *Proceedings of the National Academy of Sciences of the United States of America*, 117(38):23227–23234, 2020.

[39] P. Brunet, Stabilized Kuramoto-Sivashinsky equation: A useful model for secondary instabilities and related dynamics of experimental one-dimensional cellular flows. *Physical Review E, Statistical, Nonlinear, and Soft Matter Physics*, 76:017204, 2007.

[40] J. Zhou, Instability analysis of saddle points by a local minimax method. *Mathematics of Computation(AMS)*, 74(251):1–21, 2005.

[41] K. He, Embedded moving saddle point and its relation to turbulence in fluids and plasmas. *International Journal of Modern Physics B*, 18(13):1805–1843, 2004.

[42] C.V. Tran, X. Yu and D.G. Dritschel, Velocity-pressure correlation in Navier-Stokes flows and the problem of global regularity. *Journal of Fluid Mechanics*, 911:A18, 2021. doi:10.1017/jfm.2020.1033

[43] Z.-M. Chen and W.G. Price, Time dependent periodic Navier-Stokes flows on a two-dimensional torus. *Communications in Mathematical Physics*, 179:577–597, 1996.

[44] C.L. Fefferman, Existence and smoothness of the Navier Stokes equation. In *Millennium Prize Problems*. Clay Mathematics Institute, Cambridge, MA, pp. 57–67, 2006.

[45] D. Nelson, T. Piran and S. Weinberg, *Statistical Mechanics of Membranes and Surfaces*. World Scientific, Singapore, 2004.

[46] R. Finn, Capillary surface interfaces. *Notices of the AMS*, 46(7):770–781, 1999.

[47] N.A. Kudryashov and D.I. Sinelshchikov, Analytical solutions of the Rayleigh equation for empty and gas-filled bubble. *Journal of Physics A: Mathematical and Theoretical*, 47(40):405202, 2014.

[48] T. Pedergnana, D. Oettinger, G.P. Langlois and G. Haller, Explicit unsteady Navier-Stokes solutions and their analysis via local vortex criteria. *Physics of Fluids*, 32:046603, 2020. doi:10.1063/5.0003245

[49] T.E. Moschandreou and K.C. Afas, Compressible Navier-Stokes equations in cylindrical passages and general dynamics of surfaces—(I)-Flow Structures and (II)-Analyzing Biomembranes under Static and Dynamic Conditions. *Mathematics*, 7(11):1060, 2019. doi:10.3390/math7111060

[50] H. Alzer and S. Ruscheweyh, The arithmetic mean - geometric mean inequality for complex numbers. *Analysis*, 22:277–283, 2002. doi:10.1524/anly.2002.22.3.277

[51] W. Kahan, Notes for Math. H110. Available online: https://people.eecs.berkeley.edu/ wkahan/MathH110/NORMlite.pdf (accessed on 26/6/2021).

[52] G.F.D. Duff, Navier Stokes derivative estimates in three dimensions with boundary values and body forces. *Canadian Journal of Mathematics*, 43(6):1161–1212, 1991.

[53] P. Grinfeld, *Introduction to Tensor Analysis and the Calculus of Moving Surfaces*. Springer, New York, 2013.

[54] A. Atland and B. Simons, *Condensed Matter Field Theory*. Cambridge University Press, Cambridge, 2010.

[55] M. Miklavčič and C.Y. Wang, Viscous flow due to a shrinking sheet. *Quarterly of Applied Mathematics*, 64(2):283–290, 2006.

[56] T. Tao, Quantitative bounds for critically bounded solutions to the Navier Stokes equations. *arXiv preprint arXiv:1908.04958v2*, 1–45, 2020.

[57] T. Moschandreou and K. Afas, Periodic Navier Stokes equations for a 3D incompressible fluid with Liutex vortex identification method. *IntechOpen*, 1–22, 2023. doi:10.5772/intechopen.110206

[58] E.J. Dubuc, A note on Cantor sets. *Real Analysis Exchange*, 23(2):767–771, 1997–1998.

[59] A. Weinberger, *Lambert W's Taylor Series*. Overleaf Creative Commons License, pp. 1–4, 2015.

[60] J.C. Robinson, J.L. Rodrigo and W. Sadowski, *The Three-Dimensional Navier–Stokes Equations*, pp. 1–471. Cambridge University Press, Cambridge, 2016.

[61] M.L. Kavvas and A. Ercan, Generalizations of incompressible and compressible Navier–Stokes equations to fractional time and multi-fractional space. *Scientific Reports*, 12:19337, 2022.

[62] L. Li and J.-G. Liu, Some compactness criteria for weak solutions of time fractional PDE's. *SIAM Journal on Mathematical Analysis*, 50(4), 3963–3995, 2018. doi:10.1137/17M1145549

[63] F.M. White, *Fluid Mechanics*, Fourth edition. WCB McGraw Hill, New York, 1999.

[64] M. Caputo, Linear models of dissipation whose Q is almost frequency independent II. *The Geophysical Journal of the Royal Astronomical Society*, 13:529–539, 1967.

[65] K. Diethelm, N. J. Ford, A. D. Freed and Y. Luchko, Algorithms for the fractional calculus: A selection of numerical methods. *Computer Methods in Applied Mechanics and Engineering*, 194(6–8):743–773, 2005.

[66] M. Greenberg, *Foundations of Applied Mathematics*. Prentice-Hall Inc., Englewood Cliffs, NJ, 1978.

[67] R. Gorenflo and F. Mainardi, Essentials of fractional calculus, Preprint submitted to MaPhySto Center, January 28, 2000.

[68] I. Podlubny, *Fractional Differential Equations*. Academic Press, San Diego, CA, 1999.

[69] I. Podlubny, Geometric and physical interpretation of fractional integration and fractional differentiation. *Fractional Calculus and Applied Analysis*, 5:367386, 2002.

Index

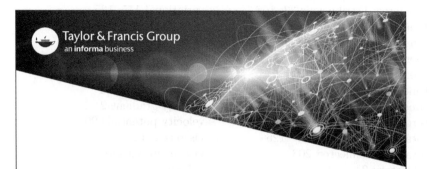

First published [...] Chapter [...] and Frontmatter [...] Copyright [...] [...]
[...] [...] © [...] [...] [...] [...] [...] Taylor & Francis
Verlag GmbH, Kaiserstrasse [...], [...] [...] Weinheim, Germany